# *The* SILVICULTURAL BASIS *for* AGROFORESTRY SYSTEMS

# *The* SILVICULTURAL BASIS *for* AGROFORESTRY SYSTEMS

Edited by
Mark S. Ashton, Ph.D., M.F.
Florencia Montagnini, Ph.D., M.S.

CRC Press
Boca Raton London New York Washington, D.C.

**Library of Congress Cataloging-in-Publication Data**

The silvicultural basis for agroforestry systems / edited by Mark S. Ashton, Florencia Montagnini,

p. cm.

Includes bibliographical references.

ISBN 0-8493-2206-5 (alk. paper)

1. Agroforestry. 2. Silvicultural systems. I. Ashton, Mark S. II. Montagnini, Florencia, 1950-

S494.5.A45 S525 1999

634.9′9—dc21 99-046106

CIP



International Standard Book Number 0-8493-2206-5

Library of Congress Card Number 99-046106

Printed in the United States of America 1 2 3 4 5 6 7 8 9 0

Printed on acid-free paper

# Preface

The silvicultural basis for agroforestry systems emphasizes a perspective on agroforestry that is based in large part on the knowledge of sivicultural practice in natural forests. We have obviously restricted the content and ideas of the book to the woody and perennial components of agroforestry systems. We also have a strong North American perspective built on our understanding of stand dynamics and forest development, but agroforestry examples are given that are worldwide. The stand dynamics perspective, as one can judge by the authorship, has mainly come from one institution, namely Yale School of Forestry and Environmental Studies. The stand dynamics perspective of the book also provides a set of ideas that we believe inject fresh thought into agroforestry practice; but our ideas are just that, they should be integrated into the much greater realm of agroforestry literature that has been published primarily over this last decade.

Our book can be considered an advanced text and reference for the silviculturist, forester and agroforestry specialist who is concerned with a deeper ecological basis for the management of agroforestry systems. We assume the reader has gained a sufficient knowledge of silviculture from other more introductory books to plunge straight into the ecological theory of tree mixtures. As such most of the book provides examples relevant to agroforestry on the competition for resources of plant-plant and plant-animal mixtures. Unlike many examples of modern agricultural practice we promote the idea of working with natural processes rather than continuously attempting to control or fight against them. We emphasize this primarily because we believe that most forests and agroforests will be restricted to lands that are considered too marginal for permanent or intensive agriculture. Marginal lands will therefore be too risky to invest in cost intensive practices. New ways of obtaining multiple integrative values, both goods and services, will need to be considered on the same site. Historically, we have often over-emphasized the use of one value on these more marginal lands, predisposing the use of the land to risk from both biological (disease, insect infestations, ice storms, fire) and social (market collapse, unemployment, illiteracy) catastrophe. We therefore hope that the framework and ideas provided in this book will be of help to the reader. We encourage feedback as we consider our attempt at this topic to be our first effort and for us this continues to be a work in progress.

MA and FM
August 28, 1999

# About the Editors

**Dr. Mark Ashton** received a B.S. degree in forestry from the University of Maine, and M.F. and Ph.D. degrees from Yale University School of Forestry and Environmental Studies. His area of expertise is in silviculture or applied forest ecology. He currently is the John Musser Director of Tropical Resources Institute.

Professor Ashton conducts research on the biological and physical processes governing the regeneration of natural forests and in the creation of their agroforestry analogs. His long-term work concentrates in the Asian tropical and American temperate forests. The results of his research have been applied to the development and testing of silvicultural techniques for restoration of degraded lands and for the management of natural forests for a variety of timber and non-timber products.

His work has been published in over 60 journals, books, and governmental publications. He is the author of several plant guides on the tropics, a coauthor of a silviculture textbook, and coeditor of several books covering resource issues of protected areas management.

**Dr. Florencia Montagnini** received a B.S. degree from the National University of Rosario, Argentina, an M.S. from the Venezuelan Institute for Scientific Research, and a Ph.D. from the Institute of Ecology, University of Georgia. Her area of expertise is silviculture and agroforestry systems with a particular emphasis on mixed-species plantations. She is currently head of the Department of Silviculture and Forest Conservation at CATIE, Costa Rica.

Dr. Montagnini's overall research goal is to integrate ecological principles, economic and social factors, and human health considerations into the design of sustainable land-use programs for rehabilitating degraded lands in humid tropical regions. She is currently working toward this goal with projects in three tropical regions of Latin America having similar environmental and socioeconomic problems; northeastern Costa Rica, northeastern Argentina, and the eastern Amazon of Brazil. Her research involves understanding the nutrient cycling properties of indigenous tree species and their potential for rehabilitation of degraded lands, while at the same time yielding economically valuable products.

She is the author of the Spanish agroforestry textbook used throughout South and Central America. She has published numerous articles in peer-reviewed journals and for government agencies.

# Contributors

**Mark S. Ashton, Ph.D., M.F.**
Yale University, New Haven, CT

**Peter S. Ashton, Ph.D.**
Harvard University, Cambridge, MA

**Andrew P. Beckerman, Ph.D., M.F.S.**
University of Sterling, Sterling, UK

**Graeme P. Berlyn, Ph.D., M.S.**
Yale University, New Haven, CT

**James A. Bryan, Ph.D., M.S.**
Yale University, New Haven, CT

**Jeannie Cho, M.S.**
Yale University, New Haven, CT

**David M. Smith, Ph.D., M.F.**
Yale University, New Haven, CT

**Mark J. Ducey, Ph.D., M.F.S.**
University of New Hampshire, Durham, NH

**Carl F. Jordan, Ph.D., M.S.**
University of Georgia, Athens, GA

**Matthew J. Kelty, Ph.D., M.S.**
University of Massachusetts, Amherst, MA

**Bruce C. Larson, Ph.D., M.F.S.**
Yale University, New Haven, CT

**Rodrigo M. Machada, Ph.D., M.S.**
University Federal de Minas Gerais, Brazil

**Florencia Montagnini, Ph.D., M.S.**
CATIE, Turrialba, Costa Rica

# Table of Contents

# 1 A Philosophical Approach to Silviculture in Agroforestry

*P. Mark S. Ashton[1] and Florencia Montagnini[2]*

## TABLE OF CONTENTS



## 1.1 INTRODUCTION

As the title implies, the chapters of this book concentrate on the silvicultural practices applicable to agroforestry systems. **Silviculture** itself has been variously defined as the scientific principles and techniques of controlling, protecting, and restoring the regeneration, composition, and growth of natural forest vegetation and its plantation analogs. The practice of silviculture can therefore be described as **applied forest ecology.** Its foundation requires a good knowledge of vegetation **autecology (silvics)** which concerns the influences of various environmental factors on the growth and development of plants as individuals.

Basic sources of information about autecology can be found in certain general text books. Examples of such books for North America are Spurr and Barnes (1980), Kimmins (1987), and Kozlowski, Kramer, and Pallardy (1991); and for moist tropical forests books include Richards (1954, 1996), Longman and Jenik (1974, 1987), Whitmore (1984, 1990), Jacobs (1987), and Mabberly (1992). Examples of more specific autecology for individual trees can be found in books by Burns and Honkala (1990) for North America; Joshi (1980) for India; and Corner (1988) for Malaya. For most forest regions of the world there is an autecological text to which readers can refer, although in many tropical regions, where the floristics are not well-known, this

[1] School of Forestry and Environmental Studies, Yale University, 360 Prospect Street, New Haven, CT 06511. Tel: (203) 432-9835; Fax: (203)432-3809.

[2] CentroAgronómico Tropical de Investigación y Enseñanza (CATIE), 7170 Turrialba, COSTA RICA. Tel.: (506)556-6440; Fax: (506)556-1533.

0-8493-2206-5/00/$0.00+$.50

might not be possible. Silviculture books are harder to find because, for many forest regions, the silviculture is even more poorly understood than the autecology. Certain regions have good silvicultural textbooks such as in North America (Daniels et al. 1979, Barrett 1995, Nyland 1996, Smith et al. 1997); Europe (Evans 1984, Burschel and Huss 1987, Matthews 1989, Dengler 1990, Savill 1991); and parts of tropical Asia (Wyatt-Smith 1963, Champion & Seth, 1968). There are also general texts on plantation systems for both tropical (Evans, 1992) and temperate systems (Shepherd, 1986). In summary, because silviculture can be regarded as that discipline concerned with the technology of managing vegetation, silviculture is to forestry what agronomy is to agriculture.

Since we have defined silviculture in the preceding paragraph we should also define what we mean by agroforestry. When woody plants (shrubs, trees) are organized in temporal and spatial mixtures with annual crop plants that may require tillage, with pasture for domestic animals, or with ponded water for aquaculture, then the term used to describe the management of the land is **agroforestry** (Smith et al. 1997). Nair (1993) provides a more classical description of agroforestry by defining it as a land use system in which woody species are grown intentionally in combination with agricultural crops or cattle on the same land, either simultaneously or in sequence. The objective is to increase total productivity of plants and/or animals in a sustainable manner, especially under levels of low technical inputs and in marginal lands.

Agroforestry was recognized as an important management system during the 1960s and 1970s by western aid organizations that were attempting to grapple with the development of marginal lands for production of food. At that time, during the Green Revolution in many developing countries, agroforestry systems tended to reflect modern agricultural systems by their intensive use of fertilizers, pesticides and site modifications to fit the desired crop. Since the 1980s agroforestry has learned from traditional indigenous systems to work more closely with the fertility of marginal lands through the use of less intensive cultivation and fallow periods. This has largely come about from foresters and agriculturists trained in the west, first recognizing, and then learning about indigenous systems. Knowledge about agroforestry has now been summarized in several comprehensive books such as those by Steppler and Nair (1987), Nair (1989), MacDicken and Vergara (1990), Montagnini, (1992) and Nair (1993). There are now journals (*Agroforestry Systems; International Tree Crops Journal; Agroforestry Abstracts*) devoted to the combinations of resources that agroforestry makes up.

The objective of this book is to provide a silvicultural framework for thinking about the design and practice of agroforestry systems. We use the textbook *The Practice of Silviculture: Applied Forest Ecology* by Smith et al. (1997) and its paradigm of thought about forest dynamics as a guide for this book. As such this book should be regarded as a collection of readings on autecology (silvics) and silviculture applicable to agroforestry systems for the advanced reader. The book does not provide an account of the techniques used (e.g., treatments for site preparation, release operations, thinning, regeneration methods, etc.) to manage vegetation. This can be obtained from any introductory text that describes silvicultural practices such as that by Matthews (1989), Evans (1992), Nyland (1996), Savill et al. (1997), or Smith et al. (1997).

Unlike many general agroforestry books, this book emphasizes research and thoughts from a forestry perspective rather than an agricultural one. Many of the examples are based on ecological theory of forests that concern the competition for resources of plant-plant and plant-animal mixtures. This book also uses the knowledge gained about the temporal and spatial dynamic and productivity of forests as the basis for silvicultural applications in agroforestry systems. By taking this perspective we seek to emphasize the dynamic nature of agroforestry systems. In so doing, this book promotes the ideas of: i) working with natural processes of succession rather than continuously trying to control or directly fight against them; and ii) selecting crop plants and cultivation systems that match the constraints of the site, rather than the intensive modification of the site to fit preselected crop plants and cultivation systems. This runs counter to modern agricultural perspectives that promote intensive cultivation which closely controls vegetation growth and arrests successional processes.

## 1.2 PART I

This book has three parts. The first part of the book concentrates on the dynamic autecological relationships between plant mixtures and the environment. In this part of the book authors of the various chapters emphasize the mechanistics of these relationships and, in so doing provide a solid theoretical basis for understanding competition among organisms relevant to silvicultural practice in agroforestry systems. Chapters sequentially build upon each other by first focusing on the physical environmental factors of light, soil moisture, and nutrients and their influences on plant mixtures (Chapter 2). None of these factors can be considered separately as far as productivity is concerned because of their strong interactions. Yet understanding the availability of these factors (light, soil moisture, and nutrients) is the basis for any agroforestry system. This is followed by a comprehensive discussion of nitrogen-fixation and the potential symbiotic relationships between plants and microbiota (Chapter 3). Evidence suggests that the role of plants in fixing nitrogen has been first utilized by people thousands of years ago for increasing agricultural production of crops, and it has been an area of special focus in agroforestry. The last two chapters of this part of the book concentrate on the relationships among plants, and plants and animals. The chapter on relationships among plants is focused on their speciation and co-existence as mixtures across forest landscapes (Chapter 4). This chapter illuminates what ecological theories underly the construction of plant mixtures for long-term resilience and productivity, the evidence to date from agroforestry that supports such theories, and also provides examples of how people might use this knowledge in their own agroforestry systems. The last chapter in this part of the book (Chapter 5) has a focus on the relationships among plants and animals and provides the theory for understanding the trophic interactions between plants, herbivores, and carnivores that are applicable to the design of agroforestry systems. In this chapter the focus is on two underpinnings of ecological theory that relate to agroforestry systems: livestock grazing in plantation forests and predator control of herbivore pests.

## 1.3 PART II

The second part of the book builds upon the mechanistic understanding of plant autecological relationships and describes their patterns and processes across resource gradients that vary with landscapes, and among different climatic regions. This part of the book focuses on providing the reader with an understanding of the dynamics of native forests and how they can be used as a template for design of silvicultural practices in agroforestry systems. As such the first chapter (Chapter 6) in this part provides an overall framework for understanding the pattern in relationships between climate and soil water availability within agroforestry systems. Chapter 7 continues with this theme but with a focus on forest productivity and soil nutrition and nutrient-use efficiencies of plant mixtures in time and space. These chapters relate differences in use efficiency of nutrients and water to different forest climates and provide examples of agroforestry systems where research has been reported on these phenomena. This pattern is elaborated upon in Chapter 8 but with special emphasis on agroforestry systems that utilize plants that fix nitrogen. Chapter 9 reviews the concepts and methods of agricultural intercropping using the stand development model of Oliver and Larson (1996). In particular the effect of species interactions on production in agroforestry plantations is considered. The last chapter in this part of the book (Chapter 10) concentrates on describing successional processes of native forests and suggests their reconstruction as agroforestry analogs in moist and dry forest regions. The paradigm of stand development described by Oliver and Larson (1996) is heavily relied upon in these last two chapters to suggest how silvicultural practices for agroforestry systems can work with natural processes of vegetation development rather than actively managing for their intensive suppression.

## 1.4 PART III

The last part of the book concerns the concept of the site and its classification. How do soil resources change across temporal and spatial scales? At what scales of area are appropriate design of agroforestry management units? Chapter 11 uses the silvicultural concept of a stand as the basic biological management unit and provides examples of its use in agroforestry systems. The chapter also discusses the arrangement of stands in time and space for agroforestry systems that make them compatible management units with resource gradients and disturbances that occur across landscapes.

The last chapter (Chapter 12) provides a protocol for thinking about appropriate silvicultural systems for the particular kind of agroforestry desired. The chapter gives a simple classification that summarizes the different ecological phenomena of forests that have been described in previous parts of the book and discusses their implications for different silvicultural manipulations in agroforestry.

Although we provide a simple classification with silvicultural examples it is not the intention of the authors to suggest cookbook designs for complicated sites. It is important to recognize the uniqueness of each site both from biological as well as social circumstances that demand custom designed agroforestry systems. However, these chapters provide a protocol for thinking about design compatibility based on knowledge of the stand dynamics of native forests.

## 1.5 REFERENCES


Barrett, J. W., ed., *Regional Silviculture of the United States,* 3rd ed. New York: John Wiley and Sons, 1994, 643.

Burns, R. M. and B. H. Honkala, eds., *Silvics of North America,* 2 vols. *USDA Agriculture Handbook No. 654* 1990, 1552.

Burschel, P. and J. Huss, *Grundriss des Waldbaus,* Hamburg, Germany: Verlag Paul Parey, 1987, 352.

Champion, H. G. and S. K. Seth, *General Silviculture for India,* Delhi: Government of India, 1968, 511.

Corner, E. J. H., *Wayside Trees of Malaya,* 2 vols. Kuala Lumpur, Malaysia, Malayan Nature Society, 1988, 774.

Daniel, T. W., J. A. Helms, and F. S. Baker, *Principles of Silviculture,* 2nd ed., New York: McGraw-Hill, 1979, 500.

Dengler, A., *Waldbau, auf okologischer Grunglage. II. Baumartenwahl, Bestandesgrundung, und Bestandspflege,* 6th ed., revised by E. Rohrig and H. A. Gussone. Hamburg, Germany: Verlag Paul Parey, 1990, 314.

Evans, J., *Silviculture of Broadleaved Woodland,* HMSO, *Forestry Commission Bulletin 62,* UK. 232.

Evans, J., *Plantation Forestry in the Tropics,* 2nd ed. Oxford: Oxford University Press, 1992, 403.

Jacobs, M., *The Tropical Rainforest: A First Encounter,* New York: Springer-Verlag, 1988, 295.

Joshi, H. B., *Troup's Silviculture of Indian Trees,* 3 vols. Delhi: Government of India, 1980, 1502.

Kimmins, J. P., *Forest Ecology,* New York: Macmillan, 1987, 531.

Kozlowski, T. T., P. J. Kramer, and S. G. Pallardy, *The Physiological Ecology of Wood Plants,* San Diego: Academic, 1991, 657.

Longman, K. A. and J. J. Jenik, *Tropical Forest and Its Environment,* 2nd ed. Colchester: Longman, 1987, 347.

Mabberly, D. J., *Tropical Rain Forest Ecology,* 2nd ed., New York: Chapman and Hall, 1992, 300.

MacDicken, K. G. and N. T. Vergara, eds., *Agroforestry: Classification and Management,* New York: John Wiley, 1990, 382.

Matthews, J. D., *Silvicultural Systems,* Oxford, U.S.: Oxford University Press, 1989, 284.

Montagnini, F., "Sistemas agroforestales." *Principios y aplicaciones en los trópicos,* 2da. ed. San Jose, Costa Rica: OTS/CATIE 1992, 622.

Nair, P. K. R., Ed., *Agroforestry Systems in the Tropics,* Dordrecht, The Netherlands: Kluwer Academic Publishers, 1989, 672.

Nair, P. K. R., *An Introduction to Agroforestry,* Dordrecht, The Netherlands: Kluwer Academic Press, 1993, 499.

Nyland, R. D., *Silviculture: Concepts and Applications,* New York: McGraw-Hill, 1996, 576.

Oliver, C. D. and B. C. Larson, *Forest Stand Dynamics.* New York: John Wiley and Sons, 1996.

Richards, P. W., *The Tropical Rain Forest. An Ecological Study,* Cambridge, U.K.: Cambridge University Press, 1954, 450.

Richards. P. W., *The Tropical Rain Forest,* 2nd ed. Cambridge, U.K.: Cambridge University Press, 1996, 575.

Savill, P. S., *The Silviculture of Trees Used in British Forestry,* Tucson, AZ: CAB International, 1991, 142.

Savill, P. S., J. Evans, D. Auclair, and J. Falck, *Plantations Silviculture in Europe,* Oxford, U.K.: Oxford University Press, 1997, 297.

Shepherd, K. R., *Plantation Silviculture,* Dordrecht, The Netherlands: Kluwer Academic Publishers, 1986, 322.

Smith, D. M., B. C. Larson, M. J. Kelt, and P. M. S. Ashton, *The Practice of Silviculture: Applied Forest Ecology,* New York: John Wiley, 1997, 537.

Spurr, S. H., and B. V. Barnes. *Forest Ecology,* 2nd ed. New York: John Wiley, 1980, 687.

Steppler, H. A. and P. K. R. Nair, Eds. *Agroforestry — A Decade of Development,* Nairobi, Kenya: ICRAF, 1987, 335.

Whitmore, T. C., *Tropical Rain Forests of the Far East,* Oxford, U.K.: Oxford University Press, 1984, 352.

Whitmore, T. C., *An Introduction to Tropical Rain Forests,* Oxford, U.K.: Oxford University Press, 1990, 226.

Wyatt-Smith, J., *Manual of Malayan Silviculture for Inland Forests,* 2 vols. Malay Forest Records 23, Forest Research Institute of Malaysia, 1963.

# Part I

## Understanding Plant Environment Interactions: A Basis for Silviculture in Agroforestry Systems

# Part 1 Introduction

Chapters in this first part of the book cover the autecological relationships between plant mixtures and the environment. The intention for this part of the book is to provide the ecological theory as a basis for understanding ecological pattern and process in tree-based agroforestry systems (Part II); and to provide an ecological framework for the practice of silviculture in agroforestry systems (Part III). Chapters in this part sequentially build upon each other. The first chapter has a focus on light, moisture, and nutrients (Chapter 2). It then considers these factors in agroforestry systems with an emphasis on woody plants. Chapter 3 concentrates on an important part of plant nutrition, namely nitrogen-fixation. This is an area that has played a fundamental role in the application of agroforestry as a tool to increase or restore soil productivity. This knowledge, based on plant responses to light, and acquisition of water and soil nutrients provides a platform to discuss the ecological theory of diversity relevant to understanding the role of plant mixtures in agroforestry (Chapter 4). Chapter 5 moves one step beyond by using ecological theory to explore two types of important interactions that are most relevant to agroforestry systems: livestock grazing on plantations and predator control of herbivore pests.

# 2 Light, Moisture, and Nutrient Use by Plants

*Graeme P. Berlyn and Jeannie Cho*[1]

**TABLE OF CONTENTS**



## 2.1 INTRODUCTION

Agroforestry is an application of agricultural practice and silviculture to grow trees and crops together in habitats that permit suitable yields of selected fiber and food crops. The intelligent application of agroforestry requires an understanding of the

[1] Yale University School of Forestry and Environmental Studies, New Haven, CT 06511. Tel.: (203)432-5142; Fax: (203)432-3809; e-mail:graeme.berlyn@yale.edu

0-8493-2206-5/00/$0.00+$.50

major physical factors determining plant distribution across the landscape, namely, light, moisture, temperature, nutrients, latitude, and altitude. These do not operate in a vacuum. There is also a spectrum of biological factors that interact with the physical factors and greatly affect plant growth, reproduction, and yield. These must be understood and their potentials utilized if agroforestry systems are to be economically and biologically sustainable. This chapter will focus on light, moisture, and nutrients as an underlying basis for agroforestry. None of these factors can be considered separately as far as productivity is considered because of their strong interactions in plant growth. We first examine the basic properties of light followed by water and nutrients and then discuss their importance in agroforestry.

## 2.2 THE LIGHT FACTOR

The primary source of energy for the biosphere is solar energy with relatively minor geothermal inputs (vents, volcanism, weathering). Sunlight is a flow of radiant energy that stems from rapidly varying (and mutually perpendicular) self-propagating electric and magnetic fields traveling in a vacuum at a velocity of $3 \times 10^8$ $ms^{-1}$. The transmittance of light through material substances such as glass or leaves is a function of the dielectric and permeability constants of the material. This is exemplified by the index of refraction of the transmitting material, N, which is equal to the quotient of the speed of light in a vacuum divided by the speed of light in the transmitting material ($n = c/v$). Since light generally travels more slowly in transmitting materials than it does in a vacuum the index of refraction will be greater than one ($n > 1$). The wave theory of light is useful for considerations of image formation, amplitude, frequency, phase, polarization, and interference of light. However, physicists have left us with particle-wave duality in the definition of light because the modern particulate theory of light is more useful for studying the interaction of light with matter such as in photosynthesis (see Asimov 1966, Clayton 1970, Sobel 1987). In this particulate theory of light, originally developed by Planck and Einstein shortly after the turn of the century, light is perceived of as a shower of photons, each containing a quantum of energy. A vestige of the wave theory still remains as each quanta (packet of energy) has an assigned wavelength. The energy per photon is $E = hC/\lambda$, where h is Planck's constant, C is the speed of light in a vacuum and $\lambda$ is the wavelength of the light. We normally think of the wavelength as the fundamental property of light that determines color, but actually the fundamental property is the frequency, $C/\lambda$, because it is invariant in transmitting materials of different indices of refraction while wavelength can change in passing through refractive materials. The equation shows that energy per photon increases as wavelength decreases and this is why ultraviolet light, UV, is so damaging to biological materials such as chlorophyll and other absorbing pigments in leaves. Light can be considered a shower of photons and each minuscule photon of UV light (200 nm) can have over threefold more the energy of a far red photon (770 nm).

According to the Stark-Einstein Law the absorption of one photon can cause a change only in one molecule in a strictly photochemical reaction, but in the multistage biological process of photosynthesis it takes a number of photons (8–20) of light for each molecule of carbon dioxide ($CO_2$) reduced. The quantum yield, $\phi$, is defined as the number of molecules affected divided by the number of photons absorbed and thus the quantum yield of photosynthesis is less than one. The number of incident photons of sunlight is generally expressed as an einstein or mole of photons ($6 \times 10^{23}$ photons per mole). In terms of photosynthesis, the number of moles of photons in the wavelength range of 400 to 700 nm is termed photosynthetically active radiation, PAR. Quantum units of intensity are generally used, i.e., the number of moles of PAR light per square meter of surface area per second. This is referred to as the photosynthetic photon flux density, PPFD. Full sunlight is about 1400–2000 $\mu Mm^{-2}sec^{-1}$ at the surface of the earth, varying with season of the year and location. This quantum terminology is now in wide use and has replaced the photometric units formerly used. Energy units such as watts $m^{-2}$ are also used for certain purposes, but generally not for photosynthesis. At the earth's surface the sunlight contains on the average about 2% UV (ultraviolet rays, 200–380 nm in wavelength), 45% visible light (380–780 nm), and the balance is infrared wavelengths greater than 780 nm.

Light reception by vegetation is influenced by altitude, canopy structure, clouds, latitude, time of year, time of day, and topography (Jones 1992). These factors affect the quality, quantity, and duration of light available to plant life. Canopy structure includes the number of canopy layers in the canopy and the distribution and orientation of leaves within each canopy layer. Topography primarily influences the quantity of light received by the vegetation because the Lambert Cosine Law indicates that the PPFD received by the canopy surface is equal to the incident PPFD multiplied by the cosine of the angle of incidence of the light. Light impinging on the tops of canopies may be direct or diffuse. Direct light is enriched in the green and red wavelengths compared to diffuse light, the latter having a peak wavelength in the blue range of the spectrum due to Rayleigh scattering by clouds. On clear days direct light may constitute more than 80% of the light received by canopies while on cloudy days diffuse light may be up to 100% of the light received. About 20% of the light impinging on the vegetation is reflected (albedo).

Morphological factors also influence light reception and reflection. These include leaf orientation, shape, pubescence, thickness, amount of palisade and spongy mesophyll, and thickness and properties (chemical, physical, optical) of the cuticle and epidermis. In addition to the external and internal leaf structure, physiological factors that affect light processing by vegetation primarily include the distribution and concentration of light absorbing pigments such as chlorophyll, carotenoids, and flavonoids. Leaf moisture content also influences reflection, transmittance, absorption, and utilization of light by leaves.

Light reflected from (and transmitted through) leaves is enriched in the green wavelengths as evidenced by their green color. Due to peak absorption in the blue and red by chlorophyll, light transmitted by leaves is depleted in these wavelengths and enriched in the yellow-green (500–600) and far red wavelengths (700–800). Light that impinges on the lower canopy strata and ultimately on the forest floor can be characterized by its red to far red ratio. Light arriving at the upper canopy surface at a particular location on the surface of the earth has a R/FR (red-far red) ratio of around 1.27 (accompanying a light intensity of 1400–1600 $\mu M\ m^{-2}\ s^{-1}$) while light lower in the upper canopy might have a R/FR ratio of 1.15 (800 $\mu$M m$^{-2}$ s$^{-1}$). The ratio declines as the volume of processing vegetation increases as one moves down the canopy. A typical forest floor R/FR ratio is ca. 0.20–0.46 (50 $\mu$M m$^{-2}$ s$^{-1}$ or less) (Ashton and Berlyn 1992, 1994). Solar radiation reaching the forest floor is also enriched in the near infrared wavelengths because very little light in these wavelengths is absorbed by plant leaves (Reifsnyder 1989). Sunflecks, as mentioned below, may modify the forest floor light environment in terms of both quality and quantity.

## 2.3 EFFECTS OF LIGHT ON PLANTS

Photosynthesis and its consequence, production of biomass, are not the only effects of light on plants. Light is also involved in orientation of plants in time and space (phototropisms, nastic movements), opening and closing of stomates and leaves (sleep movements), sun/shade leaf morphology and physiology, branching patterns, crown geometry and morphology, pigmentation types and quantities, photomorphogenesis/photoperiodism (flowering, seed germination, bud dormancy/activation, cambial activity), light activation of enzymes, high intensity light reactions (stem extension, anthocyanin formation), and cold hardiness (in combination with temperature).

In choosing plants for agroforestry it is important to consider the day-length requirements. Some plants are day-neutral and flower whether days are long or short. Other plants are short-day plants (SD) and require exposure to short days as in the fall to flower. Some plants require long days (LD) for floral induction while still others require sequences such as short days followed by long days (SDLD) while for others it is the reverse (LDSD). Crop breeding programs have selected many crop species to be day-neutral or as close to it as possible. Short-day plants arose or adapted in low latitudes where day length does not exceed 14 hours. LD plants are more typical of the temperate zone. Some seeds are inhibited by light and this ensures that they will not germinate unless they are buried. Ecotypes of a species may have different light requirements and some varieties of crop plants like maize, rice, and soybeans have been bred for different light requirements. All of this must be factored into species selection for specific areas and objectives.

The response to light is a function of the environmental signals and their receptors. In the case of photosynthesis, the light signal is received by pigments, namely, chlorophylls and carotenoids. The response is directly proportional to the quantity of the energy input up to the saturation point for a given leaf. In the case of cybernetic functions like photomorphogenesis the light signal is also perceived by pigments (phytochrome, cryptochrome), but the energy input is low while the output utilizes rather large amounts of stored energy. Generally this is a triggered or on-off response with amplification (see Berlyn and Battey 1985, Salisbury and Ross 1992, Mohr and Schopfer 1995). In some cases the light is not a trigger, but modulates the expression of the process being evoked. Phytochrome exists in two interchanging forms—the red absorbing and the far-red absorbing forms. The far-red absorbing form is the photomorphogenetically active form and it is converted back to the inactive red light absorbing form upon exposure to far red light. Although sunlight has about equal amounts of the red and far red wavelengths (actually there are slightly more photons in the red), the sun serves as a red source because the red absorbing form absorbs more effectively than the far red absorbing form ($\phi = P_{fr}/(P_{fr} + P_r) = 0.6$). This system provides a signal that effectively indicates whether a plant is being shaded by a canopy or not. When unshaded more of the phytochrome will be in the active form and photosensitive morphogenetic developmental processes can ensue.

## 2.4 PRIMARY PRODUCTION

From a physiological point of view crop production consists of: (1) optimizing the amount of net photosynthesis per unit of ground area per unit of time; (2) increasing the efficiency of the conversion of photosynthate into woody biomass or whatever output is desired (harvest index); and (3) increasing the water-use efficiency, if possible, of the harvest index conversion (in areas where water is limited). In some agroforestry systems nonwood products such as nuts, fruits, fodder, or thatch may constitute the harvest index and in these cases special spacing and pruning may be beneficial (Cannell 1983).

Photosynthesis is the process of transforming visible light energy into chemical energy. It also includes storing this chemical energy in the form of chemical bonds. Humans refer to certain useful aggregates of these bonds as wood and to others as food. The major factor in productivity is the plant leaf, and its morphology and physiology are greatly effected by light intensity. Sun leaves may or may not be less water efficient (Ashton and Berlyn 1992, 1994), but generally have much greater net photosynthesis. However, they do not function as well as shade leaves in the shade. The space-time continuum in light has resulted in the sun-shade and dry-wet continuum of leaf structure and function. Leaves evolved as an efficient mechanism to transform light energy into usable chemical bond energy in harmony with the space-time coordinates in light and the other site parameters: climatic, edaphic, physiographic, and biotic. Leaf structure encodes information on the health and vigor of the forest. While light is the dominant factor, nutrient and water status are also important and interactive with light. With appropriate sunlight and moisture that would normally result in a large or thick leaf, nutrient deficits may cause a much smaller and/or thinner leaf to form.

All of this is written into leaf structure and morphology. Leaf structure/function analysis can obviate the need for elaborate and continuous (data intensive) measurements of site conditions. Some general structural and functional properties of shade leaves over sun leaves include:

1. Higher photosynthetic efficiency in dim light
2. Inability to sustain high Ps in high light
3. Saturate (reached at the point of maximum net $CO_2$ uptake) at lower light intensities
4. Larger external surface in older trees and/or in seasonally arid areas

5. Thinner leaves, less deeply lobed
6. Larger specific leaf area, La/Lw (leaf area/leaf weight), at least in older trees or/and in seasonally dry areas
7. Lower ratio of internal to external surface, i.e., 10 vs 20–30 for sun leaves
8. Thinner cuticle and epidermis
   a. more cuticular transpiration
   b. greater sensible heat loss
9. Lower stomatal density, but larger stomates
10. Thinner palisade, lower palisade to spongy mesophyll ratio (subject to modifications in tropics)
11. Fewer but larger chloroplasts with larger grana
12. Lower accessory pigments and protective compounds
13. Lower chl a/chl b ratio, more PS I
14. Less conductive and supportive tissue
15. More cellulose and less lignin
16. Less starch, lipids, and organic acids
17. Higher water content of fresh tissue and higher cytoplasm to cell wall ratio (less stress resistance)

The leaves are aggregated to form tree crowns that are a mix of sun and shade leaves. This heterogeneity is further amplified by the aggregation of the tree crowns into canopies. Forest canopies are characterized by: (1) layering; (2) orientation; (3) density; (4) spatial continuity; (5) pigmentation; (6) degree and timing of deciduousness; and (7) reiteration. In agroforestry systems both the canopy structure of the trees and the planting geometry of all components determine the light environment of the crop species. As the tree component increases in height an increasing proportion of the light energy available to the crop species comes from sunflecks. Consequently, the nature of the associated trees and their planting patterns is important for crop productivity. Multilayered, heterogeneous canopies result in more smaller, more frequent sunflecks of shorter duration while a simplified canopy formed by fewer larger stems results in larger, less frequent, sunflecks of longer duration (Reifsnyder and Furnival 1970). In natural forests in the tropics sunflecks are extremely variable and generally of short duration (Pearcy 1988, 1990, Chazdon and Pearcy 1991). If both the trees and crops are planted in regular patterns, sunflecks become less variable and more controllable.

A plant in shade that is suddenly illuminated by a sunfleck takes time to respond with increased photosynthesis. This is called the induction effect (Salisbury and Ross 1992). The lag time is due to the time required for stomatal opening and for the accumulation of the biochemical intermediates required for photosynthesis. A number of enzymes, including the carbon dioxide binding enzyme, ribulose bis phosphate carboxylase/oxygenase (rubisco), require light activation before they become highly active. Once a leaf has been induced during the day the induction lagtime is reduced for succeeding sunflecks. Conversely, when the sunfleck is over, photosynthesis does not immediately drop back to the ambient light level. Rather, it continues at an accelerated rate until the pool of intermediates formed under the higher light environment are depleted.

Different crop species may vary in their ability to use sunfleck energy, i.e., some plants can achieve saturating photosynthetic rates more rapidly than others (Walker 1981). The slower species require sunflecks of longer duration and this must be achieved by adjusting the spacing (size of openings and spacing) and orientation of plants. It is possible that increased productivity of agroforestry systems may be achieved by taking these physiological features into account in species selection.

The effect of cardinal direction on the amount of energy received depends upon the angle of the solar beam at a given location on the earth. For areas near the equator row, orientation is still important. East to west rows will have the center of the row exposed to direct solar beam for most of the day while north to south orientation of the rows will result in the center of the opening receiving the direct solar beam radiation for only the middle portion of the day. The amount of time will depend upon the height of the surrounding trees (Reifsnyder 1989). For any opening to receive the full solar beam in the center of the opening the diameter of the opening approximates twice the height of the surrounding trees but varies with latitude and time of year (see Smith et al. 1997). The intensity and duration of sunflecks depend upon: (1) nature of rows of plants; (2) gaps in the surrounding tree canopy; (3) canopy length, width, layering, leaf orientation, density, and shape; (4) solar declination and time; (4) coincidence of rows and solar path (E-W, N-S, etc.); (5) clouds at any given space-time point; (6) Wind (leaf fluttering and cooling); and (7) altitude of site (light increases with altitude).

The choice of spacing may depend on many contingencies in addition to light and water. Wind shielding of the crops and firmness of the trees are factors that may be important under certain circumstances. Pollution and salt-spray shielding are additional factors that can influence choice of species as well as spacing. The harvest index of both the trees and herbaceous crop plants must also be considered.

Solar energy flows through the food chains and trophic levels of ecosystems. The producer, consumer, and decomposer organisms that comprise ecosystems are the conduits for the physicochemical cycles of those ecosystems. Any agroforestry systems will benefit from careful management of the light environment through self-correcting species selection, planting, thinning, and harvesting programs. Self-correcting means that light levels have to be periodically measured and steps taken to insure that the desired light levels are maintained for the species selected.

## 2.5 THE WATER FACTOR

Living organisms evolved in the presence of water and have utilized many of special physical and chemical properties of water in their physiological processes. Thus, it is not surprising that the spectrum of water properties appears to make it uniquely suited to support life as we know it. However, despite this circular effect water does indeed have a number of unusual properties that make it extremely suitable as the universal medium of life. Compared to many other common substances, water has a high boiling point, high heat of vaporization, high heat of fusion, high dielectric constant, high melting point, high specific heat, high surface tension, solid less dense than liquid (water freezes from top down), and low viscosity (see Salisbury and Ross 1992, Kramer and Boyer 1995).

All of these properties have been utilized to facilitate the establishment, growth and reproduction of living organisms. Water tends to stabilize plant temperature due to its high specific heat while it also cools organisms under heat stress because of the high heat of vaporization. Its low viscosity means that water with its dissolved solutes is easily transported, but it has sufficient internal cohesion and adhesion to plant cell wall surfaces to permit water to be transported to the tops of the tallest trees (and in fact the increasing difficulty of doing this as tree height increases may impose limits to tree height). The fact that water freezes from the top down and that ice floats permits survival of fish in bodies of water that would otherwise not support this aquatic life.

The integration of water into living organisms is manifested in the fact that it is the major constituent of protoplasm and is responsible for the turgidity of cells and organs as well as for the external morphology of the entire organism. An examination of mummies in museums gives insight into the appearance of the dehydrated animal organism while a bag of dried fruit gives the same perspective for plants. Water is also the major solvent for nutrient and solute uptake and transport. In plants growth in length or height requires a level of turgor sufficient to force the primary cell wall to expand. When the turgor drops below this yield point, cell enlargement cannot take place. Thus, relative volume growth of a plant cell

$$\left[\left(\frac{1}{V}\right)\frac{dV}{dt}\right]$$

is equal to the irreversible extensibility of the wall ($\phi$) times the difference between the turgor pressure of the cell (P) and the yield point of the primary cell wall (Y). As long as P exceeds Y, elongation can occur:

$$\left(\frac{1}{V}\right)\frac{dV}{dt} = \phi(\mathrm{P} - \mathrm{Y}) \text{ (after Lockhart, 1965)}$$

If the wall pressure, P, drops below about 0.2 MPa, growth ceases since this is the yield point (Y) for primary cell wall extension in most situations.

Water is a reactant in many metabolic processes and also supplies electrons for the reduction of $CO_2$ in photosynthesis, the process that "supplies the world with food and power." In general, plant biomass production is directly proportional to the supply, use, and distribution of water in the soil and plant. However, the major use of water by plants is in transpiration, the process whereby water moves from the soil through the plant, changes to the vapor state, and exits the plant through the stomata (stomatal transpiration) or directly through the leaf cuticle (cuticular transpiration). The latter usually accounts for only 1–10% of the water loss, but may be as high as 30% of the total loss. Cuticular transpiration proceeds day and night throughout the year while stomatal transpiration only occurs during daylight hours in most plants and only during the leaf-bearing periods of the year in deciduous trees (Berlyn et al. 1993). Transpiration is a two-edged sword; as the stomates open it increases greatly, but open stomates are necessary for $CO_2$ to enter the plant in order for photosynthesis to occur. In fact, photosynthesis and transpiration are generally highly correlated. Transpiration is often referred to as a necessary evil, but it also helps cool the plant and is one of the means for transport of solutes from the soil throughout the plant.

Water is often said to be transparent or colorless, but it has a slight absorption in the blue wavelengths. Thus large, deep bodies of water have a blue or blue-green appearance; generally, the deeper, the bluer. The hydrogen bond absorbs strongly in the infrared and liquid water appears jet black at a wavelength of 3,000 nm. This characteristic is used in photointerpretation to separate, for example, insect infested trees. The infested foliage has less water and is therefore lighter in color under infrared illumination.

The two most common ways of measuring the water status of plants are the relative water content and the water potential (see Boyer 1995). Relative water content (RWC) is the fresh weight ($W_f$) minus the dry weight ($W_d$) divided by the difference between the saturated weight ($W_s$) and the dry weight:

$$RWC = (W_f - W_d)/(W_s - W_d)$$

The relative water content is good for comparing the plant in time and in response to environmental conditions. It is not good if, for example, conditions change the water-holding capacity of the tissue (as under drought conditions). In this case the maximum relative water content is to be preferred (Li et al. 1996).

$$RWC_{max} = (W_s - W_d)/Ws = 1 - W_d/W_s$$

Water potential (Yw) is the chemical potential (change in free energy of water in the tissue) expressed in pressure instead of energy units by dividing the energy units by the partial molar volume of the water. Water potential is the difference between the chemical potential in the tissue ($\mu_w$) and the chemical potential of pure water without any solutes ($\mu_w{}^o$) divided by the molar volume ($V_w$):

$$\Psi_w = (\mu_w - \mu_w{}^\circ/V_w = RT/V_w \ln P_v/P_v{}^\circ = \Psi_\pi + \Psi_P + \Psi_H$$

where R = universal gas constant, T = temperature in degrees Kelvin, $P_v$ = the vapor pressure of the water in the tissue, $P_v{}^\circ$ = the vapor pressure of pure water, $\Psi_\pi$ = the osmotic potential (equal to the osmotic pressure, but negative in sign), $\Psi_P$ = the wall pressure, and $\Psi_H$ = the gravitational term used only for the rise of water in trees or the movement of water down into soils (Kramer and Boyer 1995, Boyer 1995). The water potential of pure water is zero and decreases with the addition of solutes to the solution. Thus, in general the water potential of the soil or plant cell is less than zero (negative values). As water content of tissue decreases, the water potential becomes increasingly negative, such that around −1.5 MPa, leaves of most angiosperms generally wilt; and below −2.0 to −3.0 MPa, they are unlikely to recover. Conifer leaves can withstand this level of water stress and leaves of some desert plants may be able to withstand water potentials of −6 MPa or more. Seeds are the most drought-resistant plant life forms; some can withstand water potentials as low as −10 MPa. Leaf water potential of well-watered plants may range from −0.1 to −1.0 MPa.

Although water potential is decreased by addition of solutes or concentration of the cell solution, it is increased by pressure and temperature. Different factors may dominate in a given set of environmental conditions, but for many plants a relative water content below 75% is lethal.

## 2.6 ECOPHYSIOLOGY OF WATER RELATIONS

Plants have two main adaptations with respect to water stress: drought avoidance and drought tolerance. Drought avoidance may be based on timing of growth (e.g., desert ephemerals), storage of water (cacti), structural/functional modifications that limit water loss. Structural/functional modifications in response to drought may be genetic or due to phenotypic plasticity and include: (1) stomatal opening only at night; (2) rapid stomatal closure; (3) reduced stomatal size; (4) modification of stomatal structure (pubescence, sunken stomates with wax filled antechambers); (5) thickened epicuticular wax and cuticle; (6) increased ability to remove soil water (compensation); (7) increased lateral or vertical root development; (8) changes in leaf angle or pigmentation to absorb less radiation; (9) increased thickness of plant cell walls that increase water storage capacity and modulus of elasticity that limits damage due to wilting; (10) increased synthesis of proline and other metabolites associated with drought resistance; and (11) increased vascularization per unit leaf area. Drought avoiders survive drought with relatively high tissue water potentials while true drought tolerators tend to endure drought with low (highly negative) tissue water potentials. Soil resources are necessary if plants are to utilize the available solar energy efficiently. On a well-watered and sufficiently fertilized site (see references in Marschner 1995), there will generally be a greater response to thinning of a stand in terms of both tree and interplanted crops because the below ground resources are not limiting. Therefore, spacing designs must consider the availability of water and nutrients as well as the increased light available as spacing is increased (Young 1989a).

Watanabe et al. (1988) describe four major tree-crop combinations in Thailand which in turn reflect the differences in the climates and soil fertility of various regions. For example: (1) teak (*Tectona grandis*) and upland rice, maize, or sorghum (Verinumbe and Okali 1985) are used in the northern region, (2) a fast-growing tree species (such as *Eucalyptus camaldulensis, Melia azederach,* or *Leucaena leucocephala*) and cassava (Ghosh et al. 1989) in the northeast and east regions, (3) a fast-growing species and maize or sorghum (Poschen 1986, Hocking and Gangadhar Rao 1990) in the west, and (4) para-rubber (*Hevea brasiliensis*) and cashew nuts or coffee in the south; all combinations which may be augmented with interplanting of other subcrops such as beans. This would correspond to the high rainfall levels and long rainy seasons in the south, with drought periods comparatively longer for the regions to the west, northeast, and north.

The ability of a plant to endure drought is a function of its previous water environment. Plants acclimate to droughty conditions by becoming increasingly drought resistant. Jarvis and Jarvis (1963) found that aspen held at a water stress of 0.1 bars for one week would endure a water stress of 48 bars before succumbing while aspen held at 4.0 bars could withstand 75 bars before dying. Intermediate stress training yielded intermediate $LD_{100}$ values. Water use conservation can be achieved through minimization of water application in irrigation systems. Planting stock survival can also be improved by drought conditioning prior to planting.

## 2.7 WATER AND LIGHT

Physiologists define water use efficiency (WUE) as the amount of dry matter produced divided by the amount of water used to produce it (kg water transpired per kg dry matter produced) (Kramer and Boyer 1995). Some authors use stomatal water vapor conductivity (g) instead of Kg water transpired so the values in the literature vary quite widely. Sometimes different units are used for numerator and denominator (mM/g or vice versa) so that values of WUE are difficult to compare. It is not really a measure of efficiency since its values are generally much greater than one. True efficiencies should vary between zero and one (i.e., 0–100%). Many woody plants adapted to somewhat droughty environments such as oaks and conifers, have high WUEs while most herbaceous crop plants, such as rice, oats, and beans, are much less water-efficient in terms of biomass conversion and therefore have much lower WUEs. Such knowledge of WUEs is advantageous in crop selection, particularly in regions where water is a limiting factor. Most annual crops display heightened sensitivity to moisture stress in the window of time between flower initiation and fruiting. For example: maize during tasseling, silking, and early grain formation; sorghum at the end of tillering; rice prior to flowering; and cotton before flowering and boll formation (Leyton 1983).

Most plants fix carbon through the binding of carbon dioxide from the air to a five-carbon sugar (ribulose bis phosphate). The enzyme that catalyzes this carbon fixation is known as rubisco (ribulose bis phosphate carboxylase/oxygenase). The first identifiable product is a three-carbon organic acid (phosphoglyceric acid) and thus the pathway is termed the C3 photosynthesis or the Calvin pathway (named for the leader of the group that mapped the path of carbon in this type of photosynthesis). However, there are two other types of photosynthesis, C4 and CAM (crassulacean acid metabolism). The C4 pathway was discovered much later and is also known as the Kortshak & Slack pathway (Hatch and Slack 1970, Kortshak et al. 1965). It initially binds $CO_2$ with a different enzyme, PEPcase (phosphoenolpyruvate carboxylase). The C4 pathway is an adaptation to the high temperature and arid environment and is common (ca 50%) in grasses, especially tropical grasses. In this pathway the fixed carbon accumulates in four-carbon organic acids (malate and aspartate). This function occurs only in the mesophyll cells of the leaf. The veins (vascular bundles) of the leaf are surrounded by bundle sheath cells. In C4 plants these cells (in the minor veins) are much larger than in other plants and these cells photosynthesize by the C3 pathway. The C4 (four carbon) acids release carbon dioxide as needed by the bundles of sheath cells, thereby functioning like a $CO_2$ pump. Because of this pump and storage function C4 photosynthesis is not as dependent on stomatal conductivity as C3 plants and C4 plants are enabled or programmed to be very water-efficient, with WUEs around twice that of most C3 plants. The C4 plants also have much higher net photosynthesis and much lower $CO_2$ compensation points because they do not have the seemingly wasteful process of photorespiration that may release up to 50% of the carbon fixed by C3 plants. Thus, in hot dry environments, C4 crop plants may be good choices for agroforestry. Some important crop plants with C4 metabolism are sorghum (*Sorghum bicolor*), millet (*Panicum miliaceum*), maize (*Zea mays*), grain amaranthus (*Amaranthus viridis*), and sugar cane (*Saccharum officinarum*).

Another way to look at water use efficiency over a whole growing season is by means of carbon isotope discrimination. The most abundant carbon isotope in atmospheric carbon dioxide is $^{12}C$ but there is also a small amount of the heavier $^{13}CO_2$ (1.1%). The carboxylating enzyme, rubisco, has a higher affinity for $^{12}O_2$ and, as a result, plant tissue has a lower proportion of $^{13}CO_2$ than that of the atmosphere by about 2%. The discrimination is highest when the stomates are fully open, i.e., at maximal stomatal conductance. As stomatal aperture decreases so does the discrimination, and at very low stomatal apertures nearly all the internal $CO_2$ is fixed with little or no discrimination between isotopes. Thus, the less discrimination for a given dry matter increase the greater the water use efficiency. These measurements require the use of a mass spectrograph and represent an average over the growth period of interest. Recently, some problems with this approach have been reported (Le Roux et al. 1996). Large variability in WUE was not explained by isotope discrimination values ($^{13}\delta$), indicating that $^{13}\delta$ may not be a reliable indicator of WUE. Crown and leaf morphology affect water use as does soil moisture availability and variability. Some plants have high WUE, but grow very slowly (CAM). CAM crops, like pineapple for example, avoid stress by closing their stomata during the day and opening them at night (Leyton 1983).

Net photosynthesis is generally directly proportional to transpiration because $CO_2$ is limiting and as the stomates open wider more $CO_2$ may be taken up; however, the wider the stomatal aperture, more water vapor leaves the plant in transpiration. Early successional species may be profligate in their use of water and have high respiration rates, but they still have high productivity because of their high rate of net photosynthesis (Ashton and Berlyn 1992, 1994, Strauss and Berlyn 1994). Sun leaves, however, generally have higher WUE than shade leaves of a given tree. Analogies of Ohm's law of electrical resistance are used extensively in evaluating transpiration (Jones 1992); stomatal resistance to transpirational water loss is a chief controller of water use, especially in moving air. In still air at larger stomatal aperture openings ($\geq$25% of maximum), the degree of air friction induced by boundary layer resistance may be the main control of water loss. Stomatal resistance decreases as stomatal aperture opening and frequency increase. The cuticular resistance can also dominate in certain situations as in winter drying where the soil is frozen. This may be important in high altitude agroforestry systems. Thus, a variety of interacting environmental and biotic factors affect water use. Among the physical factors are light intensity (stomates open in the light and with a spectrum similar to that of photosynthesis), vapor pressure difference between leaf and air, temperature (also influenced by light), wind velocity, and plant and soil moisture supply. The sensitivity of stomates to wind and water potential as well as their size, density, and structure is also important. Certain trees like some of the eucalypts have the reputation of being high water users and caution should be exercised in selecting such trees where water is limited, as streams may actually be pumped dry by fast-growing plantations of trees that are profligate water users.

## 2.8 NUTRIENTS

### 2.8.1 MINERAL NUTRIENTS

Plants are remarkably selective but similar in their mineral-nutrient requirements. There are 92 naturally occurring mineral elements on earth, but only 16 are universally required by all plants (Table 2.1). Some nutrients are required in larger quantities and these are termed macronutrients (carbon, oxygen, hydrogen, nitrogen, potassium, calcium, magnesium, phosphorus, and sulfur), the most common of these used as fertilizer being nitrogen, phosphorus, and potassium (NPK). Those nutrients required in lesser quantities are termed micronutrients or trace elements (chlorine, iron, boron, manganese, zinc, copper, and molybdenum). There are several more mineral elements that are required by some plants (sodium, vanadium, selenium, nickel, and cobalt) or by their endophytes such as *Rhizobium* and *Frankia* that require cobalt for nitrogenase activity in nitrogen-fixing plants. These minor micronutrients will not be discussed here. Certain crops accumulate more of specific minerals than others, e.g., dicots accumulate more boron than monocots, legumes contain more calcium than grasses, and some varieties of soybeans require more iron than others. Sodium is required by many plants, especially those that have become adapted to high sodium environments like alkali flats. Turnips, beets, celery, and sugar beets do better in the presence of sodium and it may actually be essential for them. Some blue-green algae also require sodium and it is required by all C4 plants where it induces the uptake of pyruvate in the mesophyll. The pyruvate is then converted into phosphoenolpyruvate, the compound that binds with $CO_2$ from the atmosphere in C4 plants in the presence of the enzyme phosphoenolpyruvate carboxylase (PEPcase) (C4 pathway). Rice can grow in the absence of silicon, but it has greater stalk stiffness in its presence. Erect stems are promoted by stalk stiffness and this results in increased yields especially in densely planted rice fields. Diatoms on the other hand use silicon as an essential part to their cell wall structure and cannot survive well without it.

The limited selection of mineral nutrients by plants is quite efficient because most of the essential nutrients participate in a number of different processes and often in several places within major metabolic pathways like photosynthesis and respiration. Unfortunately, this makes the diagnosis of mineral deficiencies by humans very difficult. Mineral nutrients participate in plant metabolism in three general ways: electrochemical, structural, and catalytic (cofactors for enzymes). Macronutrients tend to participate in all three types, but trace elements tend to be limited to catalytic activities.

The use of fertilizers, especially inorganic ones ("chemical fertilizers") is complicated by the fact that nutrients may interact with each other in various ways. Thus the agroforester must be aware of what is in the ground and how it will interact with any added fertilizer. If excess concentration of nutrient $X_1$ reduces the absorption of

**TABLE 2.1**
**Mineral Elements Universally Required by Plants (Adapted from Larcher, 1995; Latshaw and Miller, 1924)**

| Mineral | Form taken up from soil | Function in plant | Plant requirement ($g\ kg^{-1}$ dry matter) | Deficiency symptoms |
|---|---|---|---|---|
| C | Organic, $CO_2$ | | 450 | nonspecific |
| O | $H_2O$, organic | Structural, metabolic | 450 | nonspecific |
| H | $H_2O$, organic | Structural, metabolic | 60 | nonspecific |
| N | $NO_3^-$, $NH_4^+$ | • Protoplasm and enzyme component | 15–25 | • Chlorosis<br>• Stunted growth |
| P | $HPO_4^{2-}$, $H_2PO_4^-$ | • Basal metabolism and synthesis | 1.5–3 | • Necrosis without prior chlorosis<br>• Disturbs reproductive process<br>• Stunting |
| K | K+ | • Maintenance of membrane potential and stomatal<br>• Protein synthesis<br>• Carbohydrate metabolism<br>• Enzyme activation role<br>• Regulation of hydration | 5–20 | • Disturbed water balance<br>• Manifests in mature leaves first<br>• Dark necrotic spots |
| Ca | $Ca^{2+}$ | • Regulation of hydration<br>• Enzyme activator<br>• Growth regulator | 3–15 | • Chlorosis<br>• Drying of leaf tips<br>• Multinucleate cells |
| Mg | $Mg^{2+}$ | • Basal metabolism<br>• Regulation of hydration<br>• Chlorophyll component | 1–3 | • Stunted growth<br>• Chlorosis of older leaves |
| S | $SO_4^{2-}$ | • Protoplasm, enzyme, and amino acid component | 2–3 | • Chlorosis of young leaves |
| Cl | $Cl^-$ | • Enzyme activation role<br>• Increases hydration | >0.1 | • Bronzing of leaves<br>• Necrotic and chlorotic spots<br>• Wilting<br>• Reduced growth |
| Fe | $Fe^{2+}$, Fe(III)-chelate | • Basal metabolism<br>• N metabolism<br>• Chlorophyll synthesis | 0.1 | • Chlorosis of young leaves<br>• Apical bud formation suppressed |
| B | $HBO_3^{2-}$, $H_2BO_3^-$ | • Growth regulator activator<br>• Carbohydrate transport and metabolism<br>• Phenol metabolism | 0.01–0.04 | • Impaired growth<br>• Reduced branching |
| Mn | $Mn^{2+}$, $Mn^-$, chelate | • Basal metabolism<br>• Chloroplast stabilizer<br>• Nucleic-acid synthesis | 0.03–0.05 | • Inhibited growth<br>• Chlorosis and necrosis of young leaves<br>• Numerous abiotic diseases |

| | | | | |
|---|---|---|---|---|
| Zn | $Zn^{2+}$, $Zn^{-}$, chelates | • Chlorophyll formation<br>• Enzyme activator<br>• Basal metabolism<br>• Protein breakdown<br>• Growth regulator synthesis | 0.01–0.05 | • Stunted growth<br>• Chlorosis then necrosis of young leaves |
| Cu | $Cu^{2+}$, $Cu^{-}$ | • Basal metabolism<br>• N metabolism<br>• Necessary for SOD activity<br>• Necessary for lignification | 0.005–0.01 | • Chlorosis of young leaves<br>• Twisted leaf<br>• Necrotic spots<br>• Citrus die back |
| Mo | $MoO_4^{2-}$ | • N fixation<br>• P metabolism<br>• Fe absorption and translocation | <0.0002 | • Less stress resistant<br>• Interveinal chlorosis possible |

nutrient $X_2$ then $X_1$ is said to be exerting *ion antagonism* to $X_2$. If the antagonism is sufficiently strong and the level of $X_2$ is sufficiently low, the plant may become deficient in $X_2$ and yield may be reduced despite the fact that its level in the soil does not suggest it is limiting. Conversely, high concentrations of calcium or lithium may accelerate the absorption of other ions (Viets effect). There is also an interaction based on soil pH. For example, raising the pH can induce an iron deficiency even if iron is abundant in the soil, while lowering the pH below 5.5 increases the solubility of aluminum. For example, the increased acidity of lakes due to acid rain resulted in increased aluminum concentration that directly caused fish kills in Scandinavia and the northern Adirondacks of the U.S. as fish are extremely sensitive to aluminum.

As previously mentioned, the many uses for each nutrient and the participation of many nutrients in each pathway makes nutrient deficiencies difficult to diagnose and treat. For example, chlorosis, a common deficiency symptom characterized by yellowing of the leaves, may be caused by several different nutrient deficiencies (e.g., N, Mg, Fe, Mn). If the cause is iron, adding more nitrogen will not alleviate the chlorosis. Micronutrients are responsible for a number of putative abiotic "diseases." Calcium deficiency can cause blossom or bottom rot of fruits; zinc deficiency is responsible for little leaf or "rosette" disease in fruit trees; manganese deficiency results in marsh spot of peas; boron deficit causes celery crack, heart and dry rots of sugar beets, and browning of cauliflower; and copper deficiency can be manifested as loss of stiffness (lazy plant disease or pendula form) resulting from depression of lignin content. Pollutants may induce symptoms of mineral deficiency by: (1) competitive or noncompetitive inhibition of essential nutrients; (2) destruction of carrier molecules or whole membranes; (3) interference with phloem transport (phloem necrosis, immobilize transport forms); (4) alteration of nutrient forms and interference with nutrient function; and (5) leaching nutrients out of plants, especially from the leaves. Many nutrient deficiencies cause detrimental effects, including increased susceptibility to insects, disease, and environmental stressors, long before actual deficiency symptoms appear. The challenge is to provide good cultural practice that avoids deficiencies before they manifest themselves in loss of stress resistance and ultimately in loss of harvest index.

In agroforestry systems geared toward soil conservation, erosion control, and fertility maintenance, species selection according to this soil nutrient status may be important. Young (1989b) notes that in such systems, included among the species with the highest potential for soil improvement are: *Acacia albida, Acacia tortilis, Calliandra calothyrsus, Casuarina equisetifolia, Erythrina poeppigiana, Gliricidia sepium, Inga jinicuil, Leucaena leucocephala, Prosopis cineraria, Sesbania sesban.* It is important, however, to note that not all multipurpose tree (MPT) species contribute equally to agroforestry systems since these woody species may benefit the microclimate to varying extents. For example, *Inga* and *Calliandra* genera may perform similarly in Central America what is observed of *Acacia albida* in Africa (Huxley 1985). Likewise, some may more optimally be used as shade species; for example, *Erythrina poeppigiana, Gliricidia sepium,* and *Leucaena leucocephala* in the humid tropics (MacDicken 1990).

#### 2.8.1.1 Nitrogen (N)

This is the mineral nutrient required in the largest amount and tends to be the second most limiting factor (after water) for plant growth although phosphorus tends to be more limiting in some tropical forest areas. Nitrogen comprises about 1–5% of plant tissue by weight. The predominant form taken up in agricultural soils is $NO_3^{1-}$, but $NH_4^{1+}$ is the main uptake form in many forest situations (Melillo et al. 1982). Nitrogen has a valence of +5 in $NO_3$, but in the amino form found in proteins it has a valence of −3, resulting in a very stable peptide bond because of the strong covalent linkage between carbon and nitrogen.

A common recommendation for woody plants is 3–4 lb. of nitrogen per 1000 $ft^2$ = 178 lb. $Acre^{-2}$ = 20 $gm^{-2}$ = 200 kg $ha^{-1}$. Unfortunately as much as 50% of added nitrogen may be lost to leaching unless slow release fertilizers or specific soil amendments are added to reduce this source of ground water contamination that emanates not only from agriculture fields, but home lawns and gardens, golf courses, parks, and cemeteries.

The optimum nitrogen to add varies with different harvest indices. In the case of sugar beets, the per acre yield of sucrose may decrease with increased nitrogen fertilization even though the yield of sugar beets in tons of beets per acre may increase with increasing nitrogen (Walker et al. 1950). Thus, it is important to determine the harvest index (total mass, protein content, mineral content, sugar content, cellulose, other carbohydrate, tons of beef, etc.) and the effect of the fertilizer on it. The source of the nitrogen may also affect the response of the plants. Nitrogen in green manures may differ from inorganic sources and different green manures may also have different effects (Black 1968, Kershnar and Montagnini 1998).

High nitrogen promotes shoot growth at the expense of root growth. The extra nitrogen is packaged into leaf protein and higher pigment content. Theoretically, when conditions are favorable for growth, high nitrogen promotes protoplasm formation over that of cell walls. Thus the cells produced tend to be large, thin-walled and high in water content. Such cells have lower stress resistance to drought and low temperatures. Their storage capacity for water and extra cellular ice is compromised.

Mechanical resistance is also reduced. Whether this scenario materializes depends on the ambient level of nitrogen as well as the species-specific response. In some experiments red spruce that was fertilized with nitrogen was more stress resistant than unfertilized seedlings (Fowler et al. 1989).

#### 2.8.1.2 Phosphorus (P)

Next to nitrogen, phosphorus is the most limiting mineral nutrient element and in many situations, especially in tropical areas, may be the limiting substance for plant growth. Most P is taken up actively in anionic form as either the monobasic (monovalent) or dibasic (bivalent) phosphate compounds, $H_2PO_4^{1-}$ or $HPO_4^{2-}$. It can be transported in the xylem to the tops of plants and then converted to phloem mobile organic forms or converted to organic compounds in the roots and utilized in situ or transported in the phloem. Turnover is also rapid, especially in adenosine triphosphate (ATP). P is essential for ATP and other sugar nucleotides (GTP, UTP, UDP, etc.), nucleic acids, phospholipids, B-vitamins, and other coenzymes. There is a strong correlation between phosphorus and nitrogen in that many of the phosphate compounds also require nitrogen. The major metabolic pathways of photosynthesis and respiration require relatively large amounts of both substances. Membranes are critically important structures and require nitrogen for the protein components and also have phospholipids in the lipid phase.

Phosphorus deficient plants, unlike nitrogen deficient plants, may appear dark green. Stunting is another common symptom of phosphorus deficiency. Phosphorus is easily distributed within the plant, but younger leaves are stronger sinks than mature leaves so symptoms first appear in the older leaves. Phosphorus is often present in sufficient amounts in the soil, but its availability varies with pH and soil hydration. Somewhat acidic pHs leads to increased availability as P does not tend to be as readily bound to insoluble iron and aluminum phosphates. Wet and inundated soils often have decreased redox potentials and P is more available due to release from iron and aluminum phosphates.

#### 2.8.1.3 Potassium (K)

This is the third most important macronutrient. Like N and P, it is involved in many major activities of the plant. It is required for all protein synthesis although its mechanism of action is unknown. It is necessary for pyruvate kinase activity, an enzyme essential for carbohydrate metabolism. It has important osmotic functions in the opening and closing of stomates and in the sleep movements of leaves.

Potassium is the most mobile element in the plant and phloem mobility provides increased flexibility. Like N and P, potassium deficiencies tend to be manifested in the mature leaves before younger ones. In dicots potassium deficiency is often expressed as dark necrotic spots, enriched in the polyamines putrescine and cadaverine ($NH_2(CH_2)_4NH_2$ and $NH_2(CH_2)_5NH_2$). In monocots, K deficiency is often indicated by a basipetally spreading tip necrosis, weak stalk, and accelerated root rot.

#### 2.8.1.4 Calcium (Ca)

This is another essential mineral requirement of plants, but it is abundant in most soils and therefore is not often deficient unless the soils are heavily disturbed, acid-leached, etc. Calcium is necessary for detoxification (especially for Mg, Cu, and Fe) and for membrane integrity. Membranes become leaky with a paucity of calcium because the calcium is necessary to bind the hydrophilic sections of the membrane phospholipids to one another. This binding to the negatively charged groups on membranes maintains the membrane conformation necessary for ion selectivity. Roots will grow in moist air but they will not grow into calcium-free soil. Without the detoxification and structural functions of calcium, roots cannot function in their normal manner. Calcium is required for the middle lamella, pectin, microtubules, spindle fibers, alpha amylase activity, and many other essential structures and functions.

The concentration of calcium of ground plant tissue ranges from ca. 0.2% to 3% or more. Maximum yield of grasses may occur when the above ground biomass contains as little as 0.1 to 1%, while for legumes the dry weight percent is about 1.5%. In cereals, the calcium concentration is ca 0.6%. For many crop plants like tomato the calcium at optimum yield is around 2.5%. It is absorbed as the divalent cation $Ca^{2+}$, but once inside cells it must be maintained at very low concentrations (ca. 1 μM) to prevent it from tying up ATP and other organic phosphates. Much of the cytoplasmic calcium is bound to the protein calmodulin, but free calcium precipitates both as organic and inorganic phosphorus compounds.

Structures with high concentrations of membranes like chloroplasts and mitochondria are enriched in calcium and over half of the calcium of plants is located in these regions. Calcium is only xylem mobile and young leaves are affected first when calcium is deficient (unlike N, P, and K). Because of its requirement for spindle fibers and therefore cell division, a common symptom of calcium deficiency is multinucleate cells.

#### 2.8.1.5 Sulfur (S)

Most soils contain sufficient sulfur for optimal plant growth. Its primary importance is its incorporation into essential amino acids (methionine, cysteine, cystine, S-adenosyl methionine [methyl donor]). Sulfur is actively taken up as the sulfate anion, $SO_4^{2-}$. It must undergo an eight-electron change before being incorporated into proteins. As with calcium, its mobility is restricted to the xylem and deficiencies manifest in young leaves first. It is a structural component of several coenzymes (coenzyme-A-SH, thiamin), the cytochrome system, nitrogen fixation (nitrogenase reductase), and ferredoxin, an essential constituent for photoelectron flow in photosynthesis. Sulfur dioxide can enter plants through the stomates and react with water to form the toxic bisulfite ion ($HSO_3^{1-}$) or sulfuric acid ($H_2SO_4$). The concentrated use of high sulfur fuels can cause significant damage through this process.

#### 2.8.1.6 Magnesium (Mg)

This element is also required in large amounts, but is much less abundant in soils than other macronutrients, thereby defining many areas of natural deficiency. For example, many sandy and acid soils around the world are low or deficient in magnesium. Crushed limestone is often used as a source of magnesium to alleviate deficiency. Magnesium is a structural component of chlorophyll where it is coordinately bonded to the central nitrogens of the four pyrrole residues that constitute the main ring of the porphyrin. Thus, magnesium deficiency will obviously result in chlorosis, generally first observed on older leaves due to its mobility within the plant. Nonchlorophyllous tissue and organisms also require much magnesium because it is essential for ATP function and for ribosome integrity (i.e., in the binding of the subunits that comprise the ribosome). It serves as an activator for one or more phosphate transfer reactions (except phosphorylases) and for nucleic acid synthesis. It is also involved in $CO_2$ transfers and is localized in cell walls as a part of the Ca-Mg pectate complex.

#### 2.8.1.7 Iron (Fe)

In a typical crop iron comprises about 5–10 Kg per ha as compared to about 200 kg per ha for nitrogen. However, iron is a critical component of all the major metabolic pathways of plants: photosynthesis (ferredoxin, cytochromes, chlorophyll synthesis), respiration (cytochromes, Krebs cycle enzymes), nitrogen fixation (nitrogenase), lignification (peroxidase), cell wall formation and growth (hydroxyproline formation), nitrogen assimilation (nitrate reductase), and peroxisomes (catalase). It is usually abundant in soils, but in situations of limited availability chelation may be necessary in some cases. The bulk of the iron is located in the chloroplast, but its relative immobility tends to first affect new leaves when in deficient quantities. Some plants produce acid exudates under deficiency of nitrogen and this increases the solubility and availability of iron in the soil. Some roots also exude compounds like caffeic acid that reduce $Fe^{3+}$ to $Fe^{2+}$ and most if not all iron is absorbed by plants in the ferrous ($Fe^{2+}$) state.

The transport form of iron in many plants is ferric hydroxide, $Fe(OH)_3$ and it is not very soluble. Plants that use ferric citrate are less prone to iron deficiency. Soybean varieties that are resistant to iron deficiency have been shown to use the citrate transport form (Ambler and Brown 1972). A negative consequence of liming soils can be iron immobilization due to complexing with carbonate or bicarbonate. Again some plants like grasses (e.g., barley, oats, and wheat) can produce inducible phytosiderophores that facilitate iron uptake when iron is in low supply. Mugineic acid was among the first phytosiderophores to be isolated. They have been shown to be 1,000-fold more effective than synthetic iron chelators like Fe-EDDHA in some cases. Other known phytosiderophores are avenic acid and the hydroxamates. Another class of compounds, the phytochelatins, also complex with heavy metals. They are generally cysteine-rich peptides (review in plant growth regulators). These compounds may serve to detoxify iron when the concentration is too high and there is a danger of iron toxicity. Rice plants in flooded soils may oxidize iron at their root surfaces and the oxidized iron is not taken up. This can be visually apparent as rust-colored coating on the root surface. Rosaceous fruit trees, maize, and sorghum are very sensitive to iron deficiency.

#### 2.8.1.8 Manganese (Mn)

Deficiencies are not common. When deficient, it causes a number of troubling abiotic diseases such as marsh spot of pea, speckled yellows of sugar beets, and gray spot of oats. Several forms (2+, 3+, and 3+) exist in the soil, but it is largely taken up as the manganous divalent action, $Mn^{2+}$. It is generally taken up from natural chelators or through reduction at the root surface. In addition to being a metal activator for a number of enzymes, manganese, along with chloride, is required for the splitting of water in photosynthesis that takes place in photo-system II. The production of a number of amino acids is also manganese-dependent.

#### 2.8.1.9 Zinc (Zn)

According to Vallee and Auld (1990) over 80 enzymes require the presence of zinc for activity. Like manganese, zinc is taken up as the divalent cation and probably from natural chelators. Zinc is required for membrane and ribosome stabilization, protein synthesis, the formation of auxin, superoxide dismutase (SOD) activity, and for the function of mycorrhizae. Deficiency can be induced by high phosphorus fertilization and it may form in rice paddies under alkaline conditions where the $HCO_3{}^{1-}$ concentration builds up.

#### 2.8.1.10 Copper (Cu)

This element is required in very small amounts and, therefore, is generally not limiting in most parts of the world. Certain soils in Australia, however, require it along with zinc and molybdenum. It is generally taken up in the cupric form ($Cu^{2+}$) but may be absorbed in the reduced state ($Cu^{1+}$) in wet soils. It is required by crucial enzymes of respiration, viz., cytochrome oxidase, and for photosynthetic electron flow, viz., in plastocyanin. Copper uptake is strongly correlated with ascorbate oxidase activity and this activity can be used as a measure of uptake. It is necessary for the process of lignification. There are both blue and nonblue copper-containing proteins and even multicopper proteins. It is necessary for SOD activity in detoxification of the superoxide radical, especially important in the chloroplast. The lower bound for Cu in leaves is 3–5 $\mu$mg per g dry weight, but at 20–30 $\mu$mg per g dry weight, it becomes toxic, a four- to tenfold range. Some plants are more resistant to copper toxicity than others, probably due to exclusion and immobilization in cell walls and soluble complexes (Marschner 1995).

#### 2.8.1.11 Molybdenum (Mo)

This element is required in extremely small amounts (only cobalt may be lower) and quickly becomes toxic at higher levels. Nevertheless, there are known molybdenum deficiencies in New Zealand, Australia, and in 16 states in the U.S. (Salisbury and Ross 1992, Marschner 1995). Molybdenum deficiencies tend to occur in sandy soils, serpentines, strongly acid soils, sphagnum peats, and laterites. The usual occurrence of molybdenum is in $MoO_4^{2-}$ where it has a valence of $Mo^{6+}$. Molybdenum is a component of a number of enzymes such as those required in nitrogen fixation and nitrate reduction. Plants suffering Mo deficiency generally have lowered levels of ascorbate, various amino acids, and sugars, making them more susceptible to insects, disease, and environmental stressors (Berlyn and Sivaramakrishnan 1996). Whiptail disease of cauliflower and broccoli is a symptom of Mo deficiency which can often be eliminated by liming. The older leaves are affected first and become twisted and eventually die. There may be some interveinal chlorosis present. Molybdenum deficiency can also cause yellow spot in citrus.

In the case of nitrogen-fixing trees and crops, Mo deficiency results in smaller and less effective nodules. Nitrogen-fixing trees require more Mo than other plants and can tie up the Mo of the soil, making it deficient for other plants in agroforestry mixtures. Thus, planting nitrogen-fixing trees in soils that are low in Mo may cause a problem that needs to be considered and alleviated before deficiency causes damage. Excess liming can free up Mo, but can also make it more susceptible to leaching.

#### 2.8.1.12 Boron (B)

This nutrient is deficient in many areas of the world (Gupta 1979) and where it is deficient considerable crop damage is incurred. Cell-wall quality is affected (less lignin and more carbohydrates) leading to stem cracks (celery), "heart rots" (beets), drought spots, hollow stem of broccoli, and other wall disruptions. Pollen tube and root tip growth are also arrested. Despite these catastrophic effects the exact mechanisms of boron function in plants is not known—only the results of its deficiency are clear (Marschner 1995). It is a major source of lowered productivity in *Pinus radiata* stands in Australia (Crane and Borough 1987). Dicots appear to accumulate considerable boron, more than any other micronutrient, but monocots do not accumulate nearly as much. There is some implication that boron is necessary for nucleic acid synthesis, membrane function, cellulose formation, lignification, and cell elongation, and this would explain why it is able to affect many aspects of the plant's structure and function. It is absorbed from the soil as boric acid $B(OH_3)$, also presented as $H_3BO_3$. It forms complexes with diols and polyols and this may be the way it links to cell walls. Teasdale and Richards (1989) have shown that in vitro there is an uptake carrier for boron that can only function when calcium and boron are both bound, but at separate sites. $Mg^{2+}$ competitively displaces $Ca^{2+}$ and inactivates the carrier with respect to boron. The level of Mg in added fertilizers could induce boron deficiency if its ambient concentration in the soil is too low.

## 2.9 ORGANIC BIOSTIMULANTS

In addition to the mineral elements, carbon, oxygen, and hydrogen are commonly considered plant nutrients. It is well known, however, that plants can benefit from organic supplements in vitro. Thus, advances in plant tissue culture, embryo culture, protoplast culture, micropropagation, somatic embryogenesis, and genetic engineering require ever more organic supplements to be added to the regenerating media. The senior author first realized that such compounds could benefit plants in soil ca. 1980 after working in biotechnology for 20 years. Organic biostimulants are often defined as non-nutritive supplements that enhance plant growth and stress resistance. Under optimal conditions green plants are thought to be the ideal autotrophs in the sense that they bind the energy of the sun into the chemical energy required to build their primary and secondary bodies from the elemental substances of the biosphere (earth, atmospheric gases, and water). Despite their potential autotrophic capacity, plants are seldom situated in optimal conditions where such capacity can be fully utilized to provide optimal plant and growth development. Cultured roots in vitro must be supplied with B vitamins because they do not synthesize them (at least in sufficient quantities). Instead, in the intact plants they obtain these substances from the leaves. However, under sufficient stress the leaves may not be able to produce and transport optimal amounts of B vitamins to the roots.

ROOTS™ is a vitamin-marine algal-humic acid product. There are other products on the market that only contain humic acids and marine algal extracts and these are not as effective. The vitamin components are in a sense fertilizers since they are substances that are essential to plant growth and development. While plants normally manufacture these substances, exogenous application is beneficial in the field when plants are under stress and are not able to produce the optimal amounts of these essential substances. This is not all that dissimilar to adding NPK-type fertilizers to soils that already contain these materials at some suboptimum level. It has been shown that these substances can enter the plant (Russo and Berlyn 1990, 1992, Berlyn and Sivaramakrishnan 1996). It has also been shown that these substances can provide optimal yields with up to a 50% reduction in fertilizer application, increased drought and low temperature resistance, and increased resistance to insect attack through increased leaf toughness and production of protective compounds (Berlyn and Sivaramakrishnan 1996). ROOTS™ was shown to increase the uptake of mineral nutrients, especially phosphorus and nitrogen in coffee plants (Russo 1990). In most cases the plants also grow faster, and in woody plants xylem production is accelerated. ROOTS™ also promotes turnover of thatch and associated mineral cycling. For these reasons ROOTS™ and those biostimulants that utilize similar organic supplementation as developed in biotechnology have great potential in agroforestry.

## 2.10 PLANT PROPERTIES AND AGROFORESTRY

Species selection and use in agroforestry systems is a process that involves many parameters, but the underlying thread is the imperative nature with which one must understand the plant's physiological and ecological requirements in order to connect the thousands of relationships existing between ecosystem components. Any mixture of agroforestry components selected modifies the microenvironment in which they grow as a result of their morphological and physiological differences (Connor 1983, Hadgar and Beer 1993). This growth is in turn limited by the amount of light harvested in canopy interactions, and the levels of available water and nutrient storages. To exercise control over any given component of a given ecosystem requires that each decision involved in model design be functional and deliberate, thus ensuring an ideal match of technology with system dynamics (Sanchez 1995, Gillespie et al. 1993). For example, Grewal (1995) found that manipulating plantations of eucalypt (*Eucalyptus tereticornis*) and bhobar grass (*Eulaliopsis binata*) improved the infiltrability of the soil, which resulted in less runoff and increased efficiency of rainwater for biomass production.

Use of multipurpose trees such as a eucalypt is a common "tool" of agroforestry systems (Table 2.2). They are maintained in intercropping systems for a number of reasons, including conservation, wood products, and fuelwood. Regardless of the ultimate intended use, however, different tree/crop combinations have distinct interactions that affect the growth rate of the establishing trees (Williams and Gordon 1995). *Acacia albida* is one of the most widely known soil-improving trees, commonly used in the semiarid regions of West Africa and in the subhumid zone (as in Senegal, Malawi and Ethiopia), mainly for its characteristic 50–100% increase in soil organic matter and nitrogen below the trees, coupled with higher water-holding capacities. One of their most important characteristics is that they allow sun rays to pass through to undercrops. During the dry off-season, these trees provide foliage when other trees are bare, and in rainy seasons, crop growth is uninhibited since the trees have shed their leaves (Gupta 1993). The latter is ideal and timely for the planting of food crops, supplying soil nutrients quickly, in view of the hastened decomposition rate accompanied by the heat and humidity of the rainy season.

*Calliandra calothyrsus* is another multipurpose tree, used primarily for improvement of degraded soils, due to its nitrogen-fixing capabilities, production of litter upon rapid decay, and efficient nutrient uptake resulting from a deep root system. *Erythrina poeppigiana* and *Inga jinicuil* are other nitrogen-fixing species valued for their use in combination with coffee (Fassbender 1987) and cacao (Rosand et al. 1987), providing both shade and nutrients in their litterfall to compensate for what is exported in harvest (Glover and Beer 1986, Herrera et al. 1987, Beer 1988, Imbach et al. 1989). *Prosopis cineraria* is a tree used typically in semiarid and subhumid areas, offering improvement on soil organic matter, the soil's water-holding capacity, and soil physical conditions. Nutrient content is purported to be higher beneath these trees when compared with that of adjacent open land, a possible explanation for the enhanced growth of pastures and crops under *Prosopis* than on adjoining uncovered land (Yadav and Khanna 1992, Young 1989b).

Previously confined to the humid tropics, the range of *Leucaena leucocephala* use has broadened considerably since it has been found to equal and often even to excel the performance of other species in moist and dry subhumid climates (Young 1989b). This flexibility of home environments is largely due to its accommodating nature under differing conditions, namely its capacity to withstand large variations in rainfall, temperature, wind, and drought (Gupta 1993). This species is distinguished by a deep taproot system that exploits subsurface nutrients by aggressively penetrating difficult and impervious soil types, ideal for the purposes of increasing soil moisture penetration and decreasing surface run-off problems (Lal 1989a, Lal 1989b).

**TABLE 2.2**
**Summary of Some Multi-Purpose Tree Characteristics**

| Tree Species | Climate | Common Associated Characteristics | Reference |
|---|---|---|---|
| *Acacia albida* | semiarid/ subhumid | • Supplies soil nutrients quickly<br>• Increases soil water-holding capacities<br>• Provides shade to undercrops when other trees shed their leaves | Gupta 1993, Young 1989 |
| *Calliandra calothyrsus* | humid tropics | • N-fixing<br>• Improves degraded soils<br>• Produces litter for soil nutrients<br>• Efficient nutrient uptake with deep root system | Young 1989, Huxley 1985 |
| *Prosopis cineraria* | semiarid/ subhumid | • Increases soil water-holding capacities<br>• Improves soil organic matter | Yadav and Khanna 1992, Young 1989 |
| *Erythrina poeppigiana* | humid tropics | • N-fixing<br>• Widely coupled with coffee and cacao<br>• Commonly used as a shade tree | Fassbender 1987, Rosand et al. 1987, Glover and Beer 1986, Herrera et al. 1987, Beer 1988, Imbach et al. 1989 |
| *Leucaena leucocephala* | moist and dry subhumid, tropics | • Used widely in hedgerow intercropping<br>• Capacity to withstand large variations in rainfall, temperature, wind, drought<br>• Deep taproot system to penetrate difficult soil types<br>• Helps maintain soil fertility | Lal 1989a, Lal 1989b, Gupta 1993, Budelman 1989 |
| *Gliricidia sepium* | semiarid/ humid tropics | • Used widely in hedgerow intercropping<br>• Associated with both annual or perennial crops<br>• Commonly used as a shade tree (i.e., with coffee, tea, and cacao) | Moreno 1987 |

There are some limiting factors to this species' growth, most importantly pH. Species like *L. diversifolia* and *L. shannoni* are more acid tolerant, where *L. leucocephala* is more constrained to soils having pH levels above 5.0 or low alumina content (Young 1989b). In arid regions, *Leucaena* fare comparably worse than other species like *Brasilettia mollis, Hardwickia binata, Acacia aneura, Sesbania grandiflora* in meeting fodder, fuel, and community requirements (Gupta 1993). However, in semiarid zones, this species seems to exceed other top suppliers such as *Dichrostachys cinerea, Acacia, Eucalyptus* (Malik and Sharma 1990), *Morus,* and *Cordia. Leucaena leucocephala* is used most commonly, but not exclusively, in hedgerow intercropping (alley cropping), a system which seems to improve both the physical and chemical properties of soil. Additionally, it possesses several characteristics that contribute to the maintenance of soil fertility: high biomass production, high nitrogen fixation, high nitrogen levels in the leaves, and high biomass in the root system (Young 1989b, Budelman 1989). Much of the species' utility may be attributed to the spacial arrangement associated with hedgerow intercropping; the increased productivity is possibly resultant of the increased soil fertility from the additive effect of nitrogen near the root systems of paired rows of *Leucaena.*

While *Leucaena* appears to have many possibilities in agroforestry systems, it is not always appropriate in all systems. Recent studies have shown that psyllid (*Heteropsylla cubana*), a noxious insect of the tropical and subtropical regions, greatly reduces its forage yield (Austin et al. 1995). Furthermore, the insect pests have also been associated with some fuelwood and multi-purpose trees including some *Acacia* and *Melia* species (Neupane 1992). Some varieties of *Leucaena,* however, are resistant, and show growth that greatly correlates to environmental factors (Hardi and Santosa 1992, Patil et al. 1994).

*Gliricidia sepium* is one of the few species thus far identified to have potential equaling *Leucaena* in hedgerow intercropping (Young 1989b). It is a good component for an agroforestry system in the humid and semiarid tropics and may be associated with either annual or perennial crops. For example, it may be used as a shade tree in coffee (*Coffea arabica*) cultivation, with tea (*Camellia sinensis*), and cocoa (*Theobroma cacao*) (Moreno 1987).

Thus, degree of competition over light, moisture, and nutrients enters as another consideration in species selection. Marshall et al. (1992) reported an example of species selection by examination of the effects of two different spatial arrangements on tree and crop water use and competition, from the International Center for Crop Research in the semiarid tropics at Hyderabad, India. In a comparison between hedgerow intercropping and a traditional agroforestry system in the semiarid tropics, competition for moisture between the trees and crops was observed to significantly reduce crop yield, a factor which could be minimized by reducing the population of the tree species as in the case of a traditional agroforestry system (Singh et al. 1989).

Marshall et al. (1992) exemplified this finding in their experiment between groundnut (*Arachis hypogaea*) and pigeonpea (*Cajanus cajan*), a species that contrasts distinctly with *Leucaena leucocephala* since its rapid growth does not begin until a month after the onset of rain, thereby avoiding strong competition with peak crop growth. In an effort to ascertain whether the close planting of the hedgerow intercropping method was simply unsuited to arid zones or whether fast growers like *L. leucocephala* exhausted available water, they tested an arrangement of scattered pigeonpea (traditional system) with a hedgerow-intercropping arrangement. It was observed that during the dry season, the scattered system was able to extrapolate 45% more water and produce twice as much fodder per hectare than the line system, confirming the hypothesis that a dispersed tree system extends fodder production more efficiently into the dry season, under limited resources, than a hedgerow intercropping arrangement. Likewise, the groundnut crop exhibited a difference in transpiration of 20% between the two respective arrangements. However, upon further analysis, pigeonpea's ability to maximize limiting resources was in fact, not sufficient to compensate for any competition it would impose on the crops. While no appreciable difference was observed in the first year of growth between the two arrangement patterns, by the second year, there was a 40% reduction in crop yield in the traditional arrangement, while the hedgerow intercropping design resulted in only a 20% reduction (Marshall et al. 1992). Therefore, pigeonpea's ability to maximally exploit water is not ideal, as the extra biomass produced under the scattered arrangement is poor compensation for the loss in groundnut yield when market prices of the fodder and groundnut are compared. Rather, it has been suggested that a more beneficial plan would involve a 50% reduction in the pigeonpea population and a 50% reduction in leaf area by pruning in the first 60 days of the second crop year (Marshall et al. 1992).

Similar conclusions have been suggested against the row planting of *Eucalyptus tereticornis* adjacent to field crops in deep water table conditions such as in arid or semiarid zones (Malik and Sharma 1990), and that competition for nutrients and water may likely exist between maize and tree species like *Cassia siamea, Eucalyptus tereticornis, Leucaena leucocephala,* and *Prosopis chilensis* (Jonsson et al. 1988, Singh et al. 1989). Such constraints between the roots of a tree or shrub and its associated food crop is a common concern in alley cropping. However, it may be possible to minimize some of the competition, for example, through knowledge of rooting behavior in species selection (Ruhigwa et al. 1992), or by harvesting the trees at appropriate times (Dhyani et al. 1990). Deep rooting is an advantage to those plants during dry periods, and is a function of soil and climatic conditions. The relative root profiles and degree of exploitation of component species of an agroforestry system are important to consider, but either alone is insufficient as the sole parameter by which to measure root activity. In relative terms, however, rooting depths observed in well-drained soils are very deep for fruit and forest trees; deep for cotton, sorghum, sugar cane, and citrus; and medium for rice, millet, maize, groundnut, tobacco, cereals, and banana (Leyton 1983).

We have seen that modifications to existing agricultural practices is facilitated by comparison of the structure and function of natural plant systems with present crop systems. For example, it is not always effective to simply pair knowledge of the multipurposes of a tree species in an agroforestry context with the desires and objectives of a managed system. Maximizing a species' inherent potential in one parameter is not always desirable, profitable, or even coincident with the initial aims of the system. It is for this reason that light, moisture, and nutrients are necessary components of the foundation underlying agroforestry systems. These factors govern plant distribution and interaction, and may best be utilized when system components are managed to maximize their availability.

## 2.11 REFERENCES

Ambler, J. E. and J. C. Brown. Iron-stress response in mixed and monocultures of soybean cultivars, *Plant Physiol.* 50 (1972):575–578.

Ashton, P. M. S. and G. P. Berlyn. A comparison of leaf physiology and anatomy of *Quercus* (section *Erythroblanus-Fagaceae*) species in different light environments. *Am. J. of Bot.* 81 (1994):589–597.

Ashton, P. M. S. and G. P. Berlyn. Leaf adaptations of some *Shorea* species to sun and shade. *N. Phytologist* 121 (1992):587–596.

Asimov, I. *Understanding Physics: Light, Magnetism, and Electricity.* New York: Mentor, 1966.

Austin, M. T., C. T. Serensson, J. L Brewbaker, W. Sun, and H. M. Shelton. Forage dry matter yields and psyllid resistance of thirty-one *Leucaena* selections in Hawaii, *Agroforestry Syst.* 31, 3 (1995):211–222.

Beer, J. Litter production and nutrient cycling in coffee (*Coffea arabica*) or cacao (*Theobroma cacao*) plantations with shade trees. *Agroforestry Syst.* 7 (1988):103–114.

Berlyn, G. P. and Y. C. Battey. Metabolism and synthetic function of cambial tissue, in: *Biosynthesis and Biodegradation of Wood Components,* T. Higuchi, ed. (New York, Academic Press, 1985), 63–85.

Berlyn, G. P. and Sivaramakrishnan. The use of organic biostimulants to reduce fertilizer use, increase stress resistance, and promote growth, in: *National Proceedings, Forest and Conservation Nursery Associations, Gen. Tech. Rep. PNW-GTR-389,* T. O. Landis and D. B. South, eds. (Portland, OR: Department of Agriculture, Forest Service, Pacific Northwest Research Station, 1996), 106–112.

Berlyn, G. P., A. O. Anoruo, A. H. Johnson, D. R. Vann, R. Strimbeck, R. L. Boyce, and W. L. Silver. Effects of filtered air and misting treatments on cuticles of red spruce needles on Whiteface Mountain, NY. *J. Sustainable For.* 1 (1993):25–47.

Black, C. A. *Soil-Plant Relationships.* 2nd ed. New York: Wiley, 1968.

Boyer, J. S. *Measuring the Water Status of Plants and Soils.* N. Y. New York: Academic Press, 1995.

Budelman, A. The performance of selected leaf mulches in temperature reduction and moisture conservation in the upper soil strata. *Agroforestry Syst.* 8 (1989):53–66.

Cannell, M. G. R. Plant management in agroforestry: manipulation of trees, population densities, and mixtures of trees and herbaceous crops, in: *Plant Research and Agroforestry,* P. A. Huxley, ed. Nairobi, Kenya: International Council for Research in Agroforestry (1983):455–487.

Chazdon, R. L. and R. W. Pearcy. The importance of sunflecks for forest understory plants. *BioScience.* 41 (1991):760–766.

Clayton, R. K. Light and living matter, Vol. 1: *The Physical Part.* New York: McGraw-Hill, 1970.

Connor, D. J. Plant stress factors and their influence on production of agroforestry plant associations, in: *Plant Research and Agroforestry,* P. A. Huxley, ed. (Nairobi, Kenya: International Council for Research in Agroforestry, 1983), 401–426.

Crane, W. and C. Borough. Boron—A micronutrient of importance to forest growers. *Australian For. Grower.* 10 (1987):24–26.

Dhyani, S. K., P. Narain, and R. K. Singh. Studies on root distribution of five multipurpose tree species in Doon Valley, India. *Agroforestry Syst.* 12 (1990):149–161.

Fassbender, H. W. Nutrient cycling in agroforestry systems of coffee (*Coffea Arabica*) with shade trees in the central experiment of CATIE, in: *Advances in Agroforestry Research: Proceedings of a Seminar,* J. W. Beer, H. W. Fassbender, and J. Heuveldop, eds. (Turrialba, Costa Rica: CATIE, 1987), 189–196.

Fowler, D., J. N. Cape, J. D. Deans, I. D. Leith, M. B. Murray, R. I. Smith, L. J. Sheppard, and M. H. Unsworth. Effects of acid mist on frost hardiness of red spruce seedlings. *New Phytologist* 113 (1989):321–335.

Ghosh, S. P., B. Mohan Kumar, S. Kabeerathumma, and G. M. Nair. Productivity, soil fertility, and soil erosion under cassava-based agroforestry systems. *Agroforestry Syst.* 8 (1989):67–82.

Gillespie, A. R., D. M. Knudson, and F. Geilfus. The structure of four home gardens in the Peten, Guatemala. *Agroforestry Syst.* 24, 2 (1993):157–170.

Glover, N., and J. Beer. Nutrient cycling in two traditional Central American agroforestry systems. *Agroforestry Syst.* 4 (1986):77–87.

Grewal, S. A eucalypt and bhobar grass plantation system to optimize resource use for biomass production in the foothills of subtropical North India. *J. of Trop. For. Sci.* 8, 1 (1995):87–106.

Gupta, R. K. *Multipurpose trees for agroforestry and wasteland utilisation.* New York: International Science Publisher, 1993.

Gupta, U. C. Boron nutrition of crops. *Adv. in Agron.* 31 (1979):273–307.

Hadgar, J. P. and J. W. Beer. Effect on maize growth of the interaction between increased nitrogen availability and competition with trees in alley cropping. *Agroforestry Syst.* 21, 3 (1993):239–249.

Hardi, T. and S. Santosa. The growth and the attack of psyllid on some *Leucaena* varieties. *Buletin Penelitian Hutan* 0, 549 (1992):19–26.

Hatch, M. D. and C. R. Slack. Photosynthetic $CO_2$ fixation pathways. *Annu. Rev. of Plant Physiol.* 21 (1970):141–162.

Herrera, R., J. Aranguren, G. Escalante, G. Cuenca, A. Accardi, E. Navidad, and M. Toro. Coffee and cacao plantations under shade trees in Venezuela, in: *Advances in Agroforestry Research: Proceedings of a Seminar,* J. W. Beer, H. W. Fassbender, and J. Heuveldop. eds., Turrialba, Costa Rica: CATIE, 1987, 189–196.

Hocking, D. and D. Gangadhar Rao. Canopy management possibilities for arboreal *Leucaena* in mixed sorghum and livestock small farm production systems in semiarid India. *Agroforestry Syst.* 10 (1990):135–152.

Huxley, P. A. The basis of selection, management, and evaluation of multipurpose trees—an overview, in: *Attributes of Trees as Crop Plants,* M. G. R. Cannell and J. E. Jackson, eds. (U.K.: Institute of Terrestrial Ecology, 1985), 13–35.

Imbach, A. C., H. W. Fassbender, R. Borel, J. Beer, and A. Bonnemann. Modelling agroforestry systems of cacao (*Theobroma Cacao*) with laurel (*Cordia Alliodora*) and cacao with poro (*Erythrina Poeppigiana*) in Costa Rica. *Agroforestry Syst.* 8 (1989): 267–287.

Jarvis, P. G. and M. S. Jarvis. The water relations of tree seedlings. IV. Some aspects of the tissue water relations and drought resistance. *Physiologia Plant.* 16 (1963):501–516.

Jones, H. G. *Plants and Microclimate.* 2nd ed. Cambridge: Cambridge University Press, 1992.

Jonsson, K., L. Fidjeland, J. A. Maghembe, and P. Höberg. The vertical distribution of fine roots of five tree species and maize in Morogoro, Tanzania. *Agroforestry Syst.* 6 (1988):63–69.

Kershnar, R. and F. Montagnini. Leaf litter decomposition, litterfall, and mulch performance in mixed and monospecific plantations in Costa Rica. *J. of Sustainable For.* 7 (1998):55–78.

Kortshak, H. P., C. E. Hartt, and G. O. Burr. Carbon dioxide fixation in sugar cane leaves. *Plant Physiol.* 40 (1965):209–213.

Kramer, P. J. and J. S. Boyer. *Water Relations of Plants and Soil.* (New York: Academic Press, 1995), 495.

Lal, R. Agroforestry systems and soil surface management of a tropical alfisol: IV. Effects on soil physical and mechanical properties. *Agroforestry Syst.* 8 (1989a):197–215.

Lal, R. Agroforestry systems and soil surface management of a tropical alfisol: V. Water infiltrability, transmissivity, and soil water sorptivity. *Agroforestry Syst.* 8 (1989b):217–238.

Larcher, W. *Physiological Plant Ecology,* 3rd ed. New York: Springer-Verlag, 1995.

Latshaw, W. L. and E. C. Miller. Elemental composition of the corn plant. *J. of Agric. Res.* 27 (1924):845–861.

Le Roux, D., W. D. Stock, W. J. Bond, and D. Maphanga. Dry mass allocation, water use efficiency, and $^{13}$ d C in clones of *Eucalyptus grandis, E. grandis X camaldulensis,* and *E. grandis X nitens* grown under two irrigation regimes. *Tree Physiol.* 16 (1996): 497–502.

Leyton, L. Crop water use: Principles and some considerations for agroforestry, in: *Plant Res. and Agroforestry.* P. A. Huxley, ed. (Nairobi, Kenya: International Council for Research in Agroforestry, 1983), 377–400.

Li, W-I., G. P. Berlyn, and P. M. S. Ashton. Polyploids and their structural and physiological characteristics relative to water deficit in *Betula papyrifera* (Betulaceae). *Amer. J. of Bot.* 83 (1996):15–20.

Lockhart, J. An analysis of irreversible plant cell elongation. *J. of Theor. Biol.* 8 (1965):264–275.

MacDicken, K. G. Agroforestry management in the humid tropics, in: *Agroforestry: Classification & Management.* K. G. MacDicken and N. T. Vergara, eds. New York: John Wiley & Sons, 1990.

Malik, R. S. and S. K. Sharma. Moisture extraction and crop yield as a function of distance from a row of *Eucalyptus tereticornis. Agroforestry Syst.* 12 (1990):187–195.

Marschner, H. *Mineral Nutrition of Higher Plants.* 2nd ed., (New York: Academic Press, 1995), 889.

Marshall, F., C. Ong, and C. Black. Tree arrangements affect competition with crops. *Agroforestry Today.* 4, 4 (1992):9–10.

Melillo, J. M., J. D. Aber, and J. F. Muratore. Nitrogen and lignin control of hardwood leaf litter decomposition dynamics. *Ecology* 63 (1982):621–626.

Mohr, H. and P. Schopfer. *Plant Physiology.* Berlin: Springer-Verlag, 1995.

Moreno, A. H. Agroforestry systems with *Gliricidia sepium,* in: *Advances in Agroforestry Research: Proceedings of a Seminar.* J. W. Beer, H. W. Fassbender, and J. Heuveldop, eds. (Turrialba, Costa Rica: CATIE, 1987), 189–196.

Neupane, F. P. Insect pests associated with some fuelwood and multipurpose tree species in Nepal. *J. of Trop. Forest Sci.* 5, 1 (1992):1–7.

Patil, N. G., P. S. Baker and G. V. Pollard. Life history parameters of the leucaena psyllid *Heteropsylla cubana* (Crawford) (Homoptera: Psyllidae) under various temperature and relative humidity regions. *Insect Sci. and its Appl.* 15 (1994):293–299.

Pearcy, R. W. Photosynthetic utilization of lightflecks by understory plants. *Australian J. of Plant Physiol.* 15 (1988):223–238.

Pearcy, R. W. Sunflecks and photosynthesis in plant canopies. *Annu. Rev. of Plant Physiol. and Mol. Biol.* 41 (1990):421–453.

Poschen, P. An evaluation of the *Acacia albida*-based agroforestry practices in the Hararghe Highlands of Eastern Ethiopia. *Agroforestry Syst.* 4 (1986):129–143.

Reifsnyder, W. E. Control of solar radiation in agroforestry practice, in: *Meteorology and Agroforestry,* W. E. Reifsnyder, and T. O. Darnhofer, eds. (Nairobi, Kenya: ICRAF, 1989), 141–156.

Reifsnyder, W. E. and G. M. Furnival. 1970. Power-spectrum analysis of the energy contained in sunflecks, in: Proc. Third Forest Microclimate Symposium. Kananaskis Forest Experiment Station, Seebe, Alberta, Canadian Forestry Service. 32–26 September 1969.

Rosand, P. C., M. Santana, and A. C. Zevallos. Associations between cacao (*Theobroma cacao*) and shade trees in Southern Bahia, Brazil, in *Advances in Agroforestry Research: Proceedings of a Seminar,* J. W. Beer, H. W. Fassbender, and J. Heuveldop, eds. (Turrialba, Costa Rica: CATIE, 1987), 189–196.

Ruhigwa, B. A., M. P. Gichuru, B. Mambani, and N. M. Tariah. Root distribution of *Acica barteri, Alchornea cordilfolla, Cassia siamea,* and *Gmelina arborea* in an acid ultisol. *Agroforestry Syst.* 19, 1 (1992):67–78.

Russo, R. O. Effects of a new humic-algal-vitamin biostimulant (Roots™) on vegetative growth of coffee seedlings. Unpublished doctoral dissertation, Library, Yale University School of Forestry and Environmental Studies, 1990.

Russo, R. O. and G. P. Berlyn. The use of organic biostimulants to help low input sustainable agriculture. *J. of Sustainable Agri.* 1 (1990):19–42.

Russo, R. O. and G. P. Berlyn. Vitamin-humic acid-algal biostimulant increases yield of green beans. *HortScience.* 27 (1992):847.

Salisbury, F. and C. Ross. *Plant Physiol.* Belmont, CA: Wadsworth, 1992.

Sanchez, P. A. Science in agroforestry. *Agroforestry Syst.* 30, 1–2 (1995):5–55.

Singh, R. P., C. K. Ong, and N. Saharan. Above and below ground interactions in alley-cropping in semiarid India. *Agroforestry Syst.* 9 (1989):259–274.

Smith, D. M., B. C. Larson, M. J. Kelty, and P. M. S. Ashton. *The Practice of Silviculture.* 9th ed., New York: Wiley, 1997.

Sobel, M. I. *Light.* Chicago, IL: University of Chicago Press, 1987.

Strauss-Debenedetti, S. and G. P. Berlyn. Leaf anatomical responses to light in five trophical moraceae of different successional status *Amer. J. of Bot.* 81 (1994):1582–1591.

Teasdale, R. D. and D. K. Richards. Boron deficiency in cultured pine cells. *Plant Physiol.* 93 (1989):1071–1077.

Vallee, B. L. and D. S. Auld. Zinc coordination, function, and structure of zinc enzymes and other proteins. *Biochemistry* 29 (1990):5647–5659.

Verinumbe, I. and D. U. U. Okali. The influence of coppiced teak (*Tectonia grandis* L.F.) regrowth and roots on intercropped maize (*Zea mays L.*). *Agroforestry Syst.* 3 (1985):381–386.

Walker, A. C., L. R. Hoc, A. Ulrich, and F. J. Hills. Nitrogen fertilization of sugar beets in the woodland area of California. Proceeding of the American Society for Sugar Beet Technology 6 (1950):362–371.

Walker, D. A. Photosynthetic induction. Proceedings of the International Congress for Photosynthesis Research 4 (1981):89–202.

Watanabe, H., P. Sahunalu, and S. Khemnark. Combinations of trees and crops in the taungya method as applied in Thailand. *Agroforestry Syst.* 6 (1988):169–177.

Williams, P. A. and A. M. Gordon. Microclimate and soil moisture effects of three intercrops on the rows of a newly-planted intercropped plantation. *Agroforestry Syst.* 29, 3 (1995):265–302.

Yadav, J. P. and P. Khanna. Root pruning enhances crop production under *Prosopis cineraria. Agroforestry Today* 4, 4 (1992):13–14.

Young, A. The environmental basis of agroforestry, in *Meteorology and agroforestry,* W. E. Reifsnyder and T. O. Darnhofer, eds. (Nairobi, Kenya: ICRAF, 1989), 29–48.

Young, A. *Agroforestry for Soil Conservation.* U.K.: CAB International, ICRAF, 1989b.

# 3 Nitrogen-Fixing Leguminous Trees and Shrubs: A Basic Resource of Agroforestry

*James A. Bryan*[1]

## TABLE OF CONTENTS



## 3.1 INTRODUCTION

Trees of the Leguminosae have been grown together with grains and other crops since the early prehistory of agriculture: Seeds of leguminous trees have been found together with grains, whether from wild or cultivated plants, in archeological sites in the Mediterranean region, Iran, southern Syria, Palestine, southern Jordan, and west-central Anatolia. Included with other cereal and legume seeds are the seeds of *Prosopis* sp. from a site in southwest Iran, accompanied by apparently cultivated grains and lentils and a variety of wild legumes, grasses, and other plants, during the period between about 9500 b.p. and 8000 b.p. (Cohen 1977). Seeds of various *Leucaena* and *Prosopis* species were found with the wild progenitors of maize and

[1] Program Director, Tropical Resources Institute, Yale University, School of Forestry and Environmental Studies, New Haven CT 06511. Tel.: (203)432–3660; Fax: (203)435–5942; e-mail: james.bryan@yale.edu.

0-8493-2206-5/00/$0.00+$.50

cultivated *Phaseolus* species in the Tehuacan Valley of southern Mexico (Kaplan 1967, Mangelsdorf et al. 1967, Smith 1967); *Prosopis* seeds were found at the earliest levels of the excavated sites; *Leucaena* and *Prosopis* pods, whether or not they were cultivated, furnished regular components of human diets (Smith 1967). *Prosopis* species also provided part of the diets in the dry regions of South America throughout the archeological record (Cohen 1977). Seeds of *Hymenaea* species provided a food source to humans since the earliest known stages of Amazonian prehistory (Roosevelt et al. 1996). At various times in agricultural history, the growth of leguminous trees with other crops became intentional in many regions of the world, as farmers recognized the trees beneficial effects on plant growth.

Now, with human populations and demands for agricultural products increasing, greater demands are placed on farm land; the fallows needed to restore soil fertility are being shortened or eliminated in many regions cover crops and hedgerows are no longer cost-efficient in many modern farming systems and unprecedented amounts of marginal land are being brought into cultivation. A variety of traditional and modern agroforestry systems are now being examined for their potential to intensify agriculture by permitting continued cropping with good yields, as an alternative to the periods of forest regrowth that have been used worldwide since pre-history to restore fertility to land exhausted by clearing and crop production (Nye and Greenland 1960, Sanchez and Hailu 1996). As a decline in soil nitrogen status is generally a factor in the loss of soil fertility under cultivation, $N_2$-fixing trees are widely regarded as an important resource in efforts to maintain or restore soil fertility; in fact, the young science of agroforestry often seems like the science of growing leguminous trees in conjunction with annual crops (Mulongoy et al. 1992).

In this chapter, the reasons for the emphasis on $N_2$-fixation and leguminous trees in agroforestry will be presented, some assumptions will be examined, and some of the benefits and costs of mixing $N_2$-fixing trees with production of annual crops will be considered.

## 3.2 IMPORTANCE OF NITROGEN IN AGRICULTURE AND NUTRITION

Nitrogen is required for every amino acid of every protein, for genetic material (DNA and RNA), for chlorophyll and hemoglobin, and for cellular transfer molecules such as adenosine triphosphate (ATP), nicotinamide adenine dinucleotide phosphate (NADP), flavin adenine dinucleotide (FAD), and the cytochromes (Goodwin and Mercer 1983). Biologically usable nitrogen is therefore required, in abundance, by every living cell. Nitrogen is often the limiting factor for growth of humans, other animals, plants, and bacteria. Nitrogen is needed in agriculture both as a raw material for plant growth and consequently for animal growth, and as a major agricultural product, the protein in both plant and animal foods. Without adequate nitrogen, growth stops and living tissues are damaged (reviewed by Waterlow 1995); nitrogen depletion leads to chlorosis in plants, to protein deficiency and arrested growth in animals, and to loss of vigor and ultimately death in both.

The supply of biologically available nitrogen declines in the soils of most fields during the course of agricultural use (Nye and Greenland 1960; reviewed by Stevenson 1982). The increased exposure of soil to rain, wind, and sun caused by land clearing greatly increases erosion and aeration and exposure of organic material, resulting in higher microbial activity, leading to accelerated loss of nitrogen by volatilization and leaching as well as erosion. Crop removal also removes nitrogen. The decrease in organic matter in most cleared soils must necessarily be accompanied by loss of nitrogen, since more than 90% of surface nitrogen is organically combined (Stevenson 1982a); in some soils nitrogen is lost to deeper soil horizons; some of this may be recovered by deep-rooted plants (Bormann and Likens 1979). Crop yields decline with the decrease of nitrogen; and the available nitrogen must be replenished if agricultural production is to continue. Much of the effort in agriculture, both traditional and modern, is devoted to maintaining a sufficient supply of nitrogen for crop production needs.

## 3.3 SOURCES OF NITROGEN

Enormous reserves of nitrogen reside in the earth's minerals and atmosphere; most, however, are not available to plant or animal growth. While nitrogen is considered an atmophile, a gaseous element that does not readily form compounds and therefore occurs mainly in the atmosphere (Berry et al. 1983), about 98% of the earth's nitrogen remains in the lithosphere, most in the igneous rocks of the mantle (Stevenson 1982b). Some nitrogen also resides in sediments, coal, and rocks of the earth's crust (Stevenson 1982b), but it is assumed that little nitrogen is added to the biosphere by weathering of rocks (Bormann and Likens 1979).

The second largest and biologically most important nitrogen reserve is the atmosphere. Nitrogen, in the form of triple-bonded di-nitrogen ($N{\equiv}N$ or $N_2$), comprises about 78% of the atmosphere. Atmospheric di-nitrogen, however, is stable, inert and, therefore, also unavailable for plant use: the triple-bond of atmospheric $N_2$ must be broken by a high-energy input, such as lightning, the combination of high heat and high pressure supplied industrially in the Haber-Bosch process, or the energy collected from the sun by photosynthesis, concentrated and stored in photosynthetic organisms, and used in conjunction with the nitrogenase enzyme catalysts that are components of several groups of $N_2$-fixing bacteria and cyanobacteria (blue-green algae). After various transformations, nitrogen can then be supplied to plants in mineral forms such as nitrate or ammonium and subsequently to animals in organic forms such as proteins.

Soils contain variable amounts of nitrogen. Nitrogen normally comprises from 0.02 to 0.5% of surface mineral soils, and about 5% of soil organic matter (Brady and Weil 1996). Most (>90%) of the soil nitrogen resides in soil organic matter (Stevenson 1982b), a varying and generally unknown amount of which is biologically available. Nitrogen is removed from the soil by crop removal, leaching, volitilization, and erosion and is resupplied by rain, dust, manure, leaf- and seed-fall, return from the subsoil, the continuing degradation of rocks and the nitrogenase enzyme, in what became known a century ago as the nitrogen cycle (reviewed by Sanchez 1976; Stevenson 1982; Brady 1984).

Biologically available nitrogen in the soil, both now and in the agricultural and geological past, has generally been in short enough supply to limit plant growth. Symbiotic $N_2$-fixation, at least until recently, has been the chief replenisher of biologically available nitrogen, and is therefore, like photosynthesis, one of the fundamental life processes. Like photosynthesis, biological $N_2$-fixation is also one of the fundamental processes of agroforestry, and much of the modern practice of agroforestry appears to be an effort to increase the capacity of rhizobial bacteria, with energy supplied by their tree symbionts, to provide needed nitrogen for associated crops.

## 3.4 SYMBIOTIC NITROGEN FIXATION

In symbiotic $N_2$-fixation, solar energy captured in photosynthesis is used as a source of electrons to break the triple-bonds of atmospheric $N_2$. The pathway by which solar energy is used to power bacterial nitrogenase enzymes for the reduction of atmospheric nitrogen is represented in Figure 3.1. Energy from the sun is harvested by chlorophyll in the plant leaves, transformed to chemical energy with the addition of water-derived hydrogen, and stored as sugars. The sugars are translocated in the phloem stream to the rest of the plant, including the roots. On the roots of $N_2$-fixing legumes, specialized organelles, the root nodules, develop by the organized division of root cortical cells in response to entering rhizobial bacteria. Plant sugars transported to the root nodules move through unique and specialized nodule vascular systems, and into the central tissues of the nodules, where the bacteria, in somewhat amorphous (bacteroidal) shapes, reside in specialized peribacteroid membranes of plant origin, which are formed under the influence of the bacteria (reviewed by Werner 1992). Using intricately controlled levels of oxygen within the peribacteroid membranes (reviewed by Evans and Burris 1992), the imported sugars are broken down to provide chemical and electron energy, in the form of ATP and NADH, to bacteroid nitrogenase enzymes, which catalyse the reduction of $N_2$ to ammonia ($NH_3$). The $NH_3$ is exported from the peribacteroid membrane into nonbacteroid cell regions in the nodule, where it is assimilated into the carrier amino acids glutamine and asparagine and, in some cases, ureides (reviewed by Werner 1992), and translocated throughout the plant for synthesis of proteins, genetic material, energy transfer molecules, chlorphyll, and other requisites of life processes. It should be noted that the two major inputs into the nitrogen reduction process, water and light-energy, bring the $N_2$-fixing plant-bacteria combination into direct competition with other plants for both water and photosynthetic area or growing space.

## 3.5 NITROGEN REQUIREMENTS IN AGROECOSYSTEMS

The amounts of nitrogen needed vary greatly from ecosystem to ecosystem and from agroecosystem to agroecosystem. Crop nitrogen requirements depend on environmental factors such as rainfall, temperature, soil texture, soil depth, soil structure, and soil nutrient status. The total nitrogen taken up from the soil by plants is always far

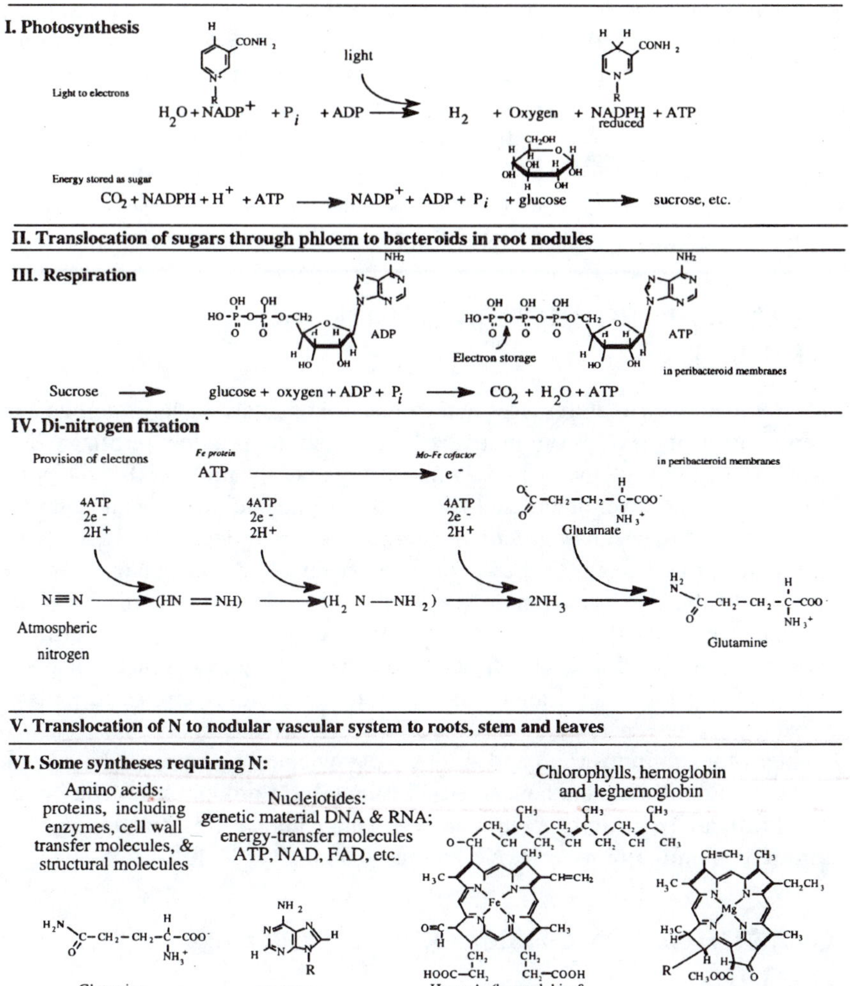


**FIGURE 3.1** Symbiotic nitrogen fixation reactions: inputs and some outputs.

less than the potentially available nitrogen in their soil. Soil nitrogen is removed from cleared and tilled fields and generally but not always from grasslands (Nye and Greenland 1960, Sanchez 1976, Bartholomew 1977), through volitilization, leaching, erosion, crop removal, and other known and unknown factors. Soil nitrogen usually increases under tree cover due to symbiotic $N_2$-fixation, root uptake from the soil strata, in many cases an increase in animal-provided nitrogen, and unknown factors (Bartholomew 1977, Bormann and Likens 1979).

Nitrogen needs of crops also vary by the crop species, cultivars, management practices, and yields. Sanchez (1976) presents ranges of nitrogen requirements of various crops in relation to yield expectations. Maize, rice, wheat, and sorghum, when grown at the average low-input production level in the tropics of 1 ton of grain per hectare, remove about 30 kg N per hectare; when yielding 4 to 5 tons grain per hectare in intensive input agriculture, these cereals' higher-yielding varieties, with fertilization, take up 100 to 150 kg N per hectare; and the same species take up 200 kg N per hectare or more at very high yield levels of 8 to 10 tons per hectare.

As in the case of immediate production needs, the amounts of nitrogen needed for the long-term productivity of soils depend on many factors such soil cover, soil structure and chemistry, rainfall, temperature, and desired production rates.

## 3.6 INDUSTRIAL PRODUCTION OF NITROGEN FERTILIZER

Using high temperatures and pressures in the presence of catalysts, large amounts of atmospheric nitrogen have been industrially reduced to provide fertilizer. The amounts of nitrogen fertilizer were rapidly increasing until the mid-1980s, when production leveled off and even decreased (FAO 1992). The addition of concentrated, industrially supplied nitrogen facilitated great increases in agricultural productivity, enabling rapidly growing numbers of people to obtain food, and making possible great urban population growth. In many cases, the increases in agricultural productivity have now leveled off, however, or even been reversed, as synthetic fertilizer and other aspects of high production agriculture and changing farm practices appear to have reached or passed their limits in many farming situations and the benefits of organic matter, organic matter of woody origin, and particularly the nitrogen-rich woody matter of leguminous trees and shrubs are now being given renewed consideration by agricultural planners concerned with long-term productivity. The varied effects of $N_2$-fixation in agroforestry are presented here as $N_2$-fixation cycles, derived from the familiar nitrogen cycle, in Figure 3.2.

## 3.7 QUANTIFICATION OF BIOLOGICAL NITROGEN FIXATION

Nitrogen is a limiting factor in plant growth in many environments, including most lands where agroforestry is used or considered, and the $N_2$ fixed by woody legumes may determine the functioning of many such ecosystems. However, no experimental method to accurately quantify nitrogen fixed by woody perennials has yet been developed; serious methodological problems continue to seriously limit the accuracy, both theoretically and practically, of determination of amounts nitrogen fixed in forested ecosystems, whether by the acetylene reduction method, by isotopic nitrogen ($^{15}N$) through natural abundance, depletion or fertilizer enrichment, by ureides, or by total N (reviewed by Danso et al. 1992; Sanginga et al. 1995; see also Sanginga et al. 1996).

Estimates of amounts of nitrogen fixed in various circumstances, difficult to quantify accurately, are reviewed by Dommergues (1987), Giller and Wilson (1991), Ledgard and Giller (1995); and Peoples et al. (1995). For tree legumes, depending on

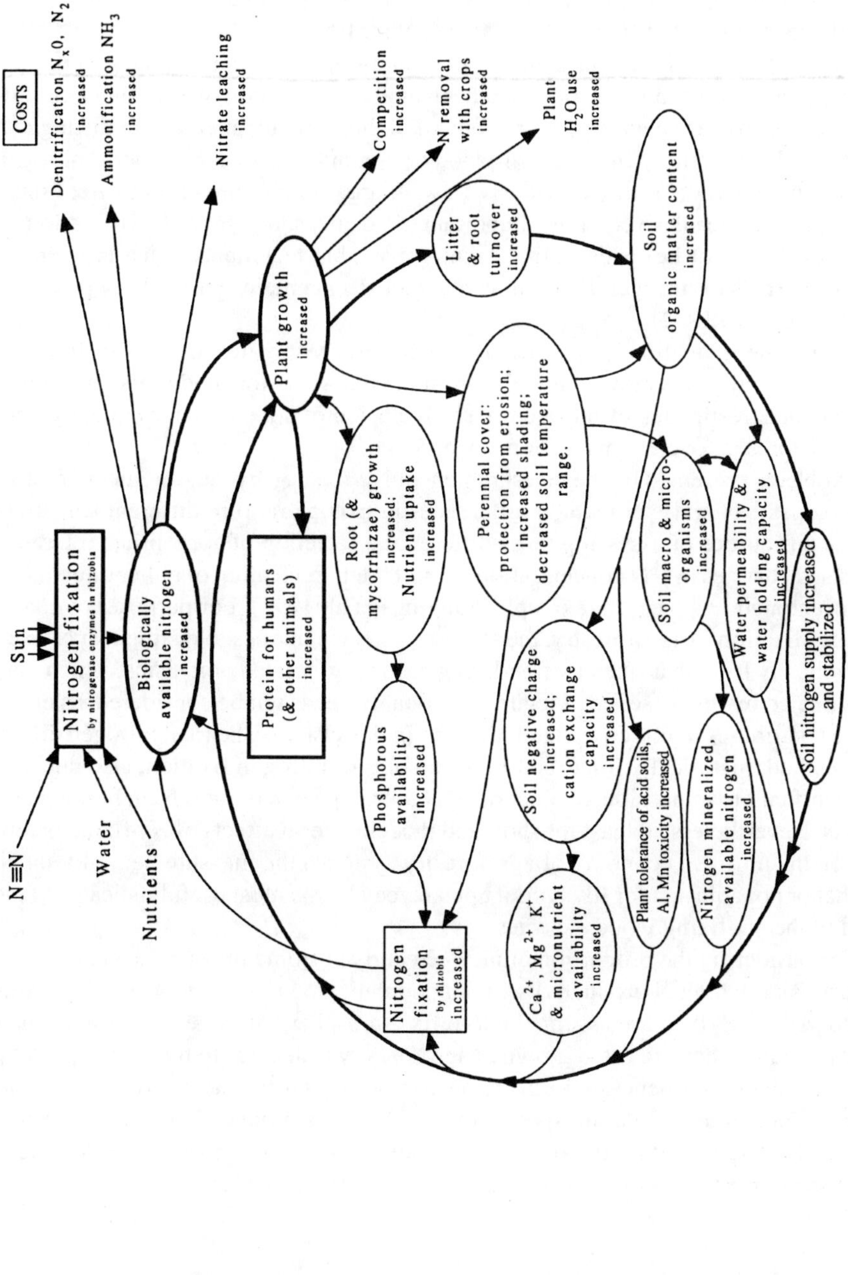


**FIGURE 3.2** Nitrogen fixation cycles in agroforestry systems: effects of the nitrogen-fixing tree component.

species, climate, and growth conditions, nitrogen fixation may reach a maximum of 300, 400, 500 kg N $ha^{-1}$ $yr^{-1}$, or even more for a fast-growing, high $N_2$-fixing tree like *Leucaena leucocephala* in favorable conditions, though the often-reported 600 kg N $ha^{-1}$ $yr^{-1}$ may include nitrogen taken up from the soil and is contested (Giller and Wilson 1991). While the amount of nitrogen fixed by leguminous trees is difficult to quantify, it is clearly on a continuum from high $N_2$-fixers to nodulating legumes that fix little nitrogen (Dommergues 1987) to nonnodulating legumes, which either fix no nitrogen, as is generally assumed, or small amounts of nitrogen, as has recently been suggested (Bryan 1995, Bryan et al. 1996).

While it is known that leguminous trees and shrubs contain abundant nitrogen, the proportion of this nitrogen obtained from biological fixation of atmospheric nitrogen continues to defy clear determination: all of the available methods to distinguish soil-derived nitrogen in field-grown woody perennials from atmospheric nitrogen reduced in their roots remain seriously flawed, and the interpretation of field measures appears speculative. Many important field measurements of $N_2$-fixation in agroforestry cannot be made with currently available techniques: measurement of changes in rates of nitrogen fixation with age of trees remains particularly problematic (Danso et al. 1992).

Acetylene reduction trials, from which early estimates of $N_2$-fixation are derived, generally separate the nodules from the tree roots for field trials, presenting probable underestimates of nitrogen fixed. Inclusion of soils in the sample, by contrast, may present overestimates of nitrogen fixed.

Problems presented by the measurement of $N_2$-fixation by measurement of nitrogen isotopes $^{15}N$ and $^{14}N$ include differences in isotope ratio in different soil strata and the difficulties of assessing these differences, selection of appropriate reference plants and changes in $^{15}N$ concentrations in soil after application of either diluted $^{15}N$ or concentrated $^{15}N$ (see, for example, Sanginga et al. 1996). For purposes of analysis of $N_2$-fixation in agroforestry, the failure of $^{15}N$ techniques to distinguish N fixed symbiotically from that fixed by free-living or root-associated organisms, which may be facilitated by the presence of legumes, probably needs not be considered, as either source contributes to effective fixed nitrogen in the system. The total nitrogen difference method, historically important in pot studies of nitrogen fixation, also depends on the difficult to impossible distinction of N taken up from the soil from biologically fixed N. Nevertheless, at least for short and moderate term effects of $N_2$-fixing perennials on the nitrogen economy of agroforestry systems, the measurement of total N, whether or not adjusted for likely root uptake, may be the most useful indicator of the effect of the $N_2$-fixing woody species.

In considering the proportion of nitrogen derived from atmospheric $N_2$-fixation, the surprisingly high N productivity of nonnodulating *Senna* species cannot be overlooked. Although they are assumed not to fix atmospheric nitrogen because of their lack of nodules, they are often grown in agroforestry trials, due to their rapid growth, high N content, and beneficial effects on companion crops. *Senna* trees often outcompete compared nodulating species in both biomass produced and above-ground biomass N. The use of *Senna* trees as non-nitrogen-fixing controls for $^{15}N$ experiments sometimes leads to negative results in nodulating species.

If not the most productive system of modern agroforestry, or the most accepted by farmers, the use of $N_2$-fixing legumes in hedgerows is probably the most widely used agroforestry system of agroforestry researchers in the past two decades. Whatever its benefits as an agroforestry system, the use of legumes in hedgerows has provided abundant data that can now be extrapolated to other agroforestry and forest systems, to provide badly needed rough estimates of the potential or probable amounts of nitrogen fixed.

## 3.8 NITROGEN FIXATION CYCLES IN AGROFORESTRY

Nitrogen fixation cycles in agroforestry may be said to begin when symbiotic $N_2$-fixation increases the amount of nitrogen available for the $N_2$-fixing host plant (Figure 3.2). The exudation of nitrogen from host plant roots, decay of roots, and fall and decay of nitrogen-rich leaves, twigs, and other plant parts increase the amount of biologically available nitrogen; since nitrogen limits plant growth in most agricultural systems, crop growth generally increases with the increase in available nitrogen. The amount of organic matter in the soil increases, with accompanying improvements in water permeability, water-holding capacity, aeration, friability, soil negative charges, and cation exchange capacity, which may also increase the availability of other limiting nutrients (Swift and Woomer 1993).

Since nitrogen is held in the soil predominantly in association with organic matter, an increase in soil organic matter increases the soil capacity to store nitrogen. Nitrogen available for plant growth is supplied in stable and gradually available organic forms, remaining in the soil much longer than the more labile industrially supplied nitrogen fertilizers. In addition, the organic matter provided by woody plant parts provides a stable lignin- and polyphenol-based organic fraction that releases nutrients more gradually and is degraded more slowly than the organic matter of herbaceous plants or animal manures (Mulongoy et al. 1992, Mulongoy and Merckx 1993), thus providing longer-lasting soil amendment. The turnover of plant minerals directly in root exudates and root, fine root, and nodule decay provide important sources of N and other minerals for those agroforestry systems in which the tree component is grown near the companion crops. In addition to their extensive spread, tree roots' mycorrhizae may provide mechanisms unavailable to roots themselves in phosphorus uptake, particularly on badly leached soils. Study of the changing mycorrhizal demography of older trees has apparently not been applied to agroforestry but may be implicated in the still unidentified benefits of traditional shifting cultivation systems on the maintenance of soil fertility. Feed-back loops should be noted here: the increased root and mycorrhizal growth afforded by $N_2$-fixation (Palm 1995) appears to lead to increased uptake of other minerals such as phosphorus by mycorrhizae; calcium by increasing the root zone; and other essential nutrients through these and other mechanisms, thereby increasing the productivity of the system as a whole. This also leads to the increased long-term presence and gradual availability of these nutrients in the soil organic matter, and enhanced physiological ability of plant roots to take up minerals; nitrogen-induced increase in root growth (Olson and Kurtz 1982) also facilitates the release of mucilage and sluffing of root parts into the soil, providing protection from acidity. The acidifying effects that follow application of industrially supplied nitrogen fertilizers (Kang 1993) may also accompany biological $N_2$-fixation. The acid-accompanying Al and Mn toxicity, however, appear to be at least partially neutralized by the provision of organic matter (Wong and Swift 1995). It would be worthwhile to more thoroughly investigate whether tree-derived organic matter is more effective in neutralizing Al and Mn toxicity than herbaceous or animal supplied organic matter (reviewed by Kang 1993).

Much of the world's agricultural expansion and intensification of recent decades has occurred on acid soils of the tropics, which are generally low in cation exchange capacity, deficient in base cations, and traditionally dependent, for maintenance of their fertility, on regularly renewed application of ash derived from woody fallows. Soil organic matter provides cation exchange sites to such soils and protects plant roots from the effects of low pH and Al toxicity. The extent to which the fast-growing legumes of agroforestry can provide sufficient organic matter to mitigate the phytotoxicity of low pH and accompanying high levels of soil solution Al and Mn still remains to be determined for most acidic soils. The use of fast-growing $N_2$-fixing leguminous trees to provide substitute organic matter is complicated by the decreased viability of *Rhizobium* in acid conditions and the diminished growth of many leguminous trees, as has been discovered most dramatically in failed *Leucaena leucocephala* plantations.

Earthworms and their population changes provide an illustration of the soil fertility cycles induced by $N_2$-fixation. The environment, especially the shade provided by $N_2$-fixing legumes such as *L. leucaena* in hedgerows, can increase the population of earthworms and production of earthworm casts by a factor of 5 over inter-row spaces and over fields without trees (Hauser 1993), leading in turn to improvements in soil physical qualities such as greater permeability to air and water, maintenance of soil organic material and friability, improvement of chemical qualities such as content and availability of nutrients, increase in microbial biomass, and positive response in plant growth (reviewed by Hauser and Asawalam 1998). As $N_2$-fixing trees enhance earthworm populations, cast production, and effects on soil fertility, plant growth, including that of $N_2$-fixing species, is augmented (Figure 3.2). It has been suggested that both earthworm activity and the importance of earthworm effects may be greater in the tropics than in temperate conditions where earthworm effects have been more studied (Ramakrishnan 1992). Research to evaluate the effects of the $N_2$-fixation itself on earthworm production, in distinction from other environmental benefits of woody perennials, such as provision of shade, appears to be required. Following increases in the nitrogen available for plants, the amount of protein produced by plants and, therefore, the protein available for animals, generally increases. Thus, the increased availability of soil nitrogen results in additional protein for humans in three ways: directly in vegetable crops; indirectly in meat; and, more indirectly but still of vital importance, in subsequent increases in mulch- and manure-derived nitrogen fertilizer for future production.

As $N_2$-fixing trees provide nitrogen for annual crop plants, annual herbaceous legume cover plants may be used to fix nitrogen and fertilize trees such as rubber; annual legumes increase root growth of the rubber trees, which enhances yields (reviewed by Humphreys 1994). The use of herbaceous legumes in conjunction with non-$N_2$-fixing trees in silvopastoral systems is widespread where non-$N_2$-fixing windbreaks are included in pastures; the agronomic combination of herbaceous legume-derived N with ligneous organic matter from non-$N_2$-fixing trees appears to have been little-studied, however, as a potentially attractive strategy for combining $N_2$-fixation with both the protective and economic benefits of trees.

In opposition to the benefits of $N_2$-fixation in agroforestry, losses in production also occur, and may cause either failure or rejection of a proposed agroforestry system. Potential losses, also indicated in Figure 3.2, include increased leaching of $NO_3$- and accompanying counter-ions, as well as loss in crop plant growth to competition for water, soil nutrients, irradiation and, in general, growing space. The source of energy for the function of the nitrogenase enzymes is sunlight, and to acquire sunlight, $N_2$-fixing plants are in competition with other plants for growing space. Since many of the leguminous trees recommended for agroforestry systems are vigorous pioneer species, their strong competitive abilities may make them undesirable to farmers and therefore unsuitable for agroforestry applications, in spite of their $N_2$-fixation.

Water, limiting agricultural production in many environments, is also needed by $N_2$-fixing plants, and the use of $N_2$-fixing trees and shrubs in these agricultural systems comes into direct competition with other crop plants for the water. In these regions, although the nitrogen fertility may be necessary for crop yields, agroforestry systems must be selected to avoid or minimize loss of crop yield by avoiding competition for water.

Minerals other than N may limit plant growth, and deficits in other minerals may be either alleviated or worsened by the presence of $N_2$-fixing trees. Minerals are generally recycled in the litter and root-turnover of $N_2$-fixing trees, which therefore do not reduce the long-term availability of P, Ca, etc., for crop plants. While the uptake of nutrients by the roots of $N_2$-fixing trees can lead to a temporary shortage of other nutrients, the long-term turnover of such nutrients from soil minerals to the stable and plant-available form of wood-based organic material may be the greatest benefit of growing $N_2$-fixing trees in conjunction with annual crops in poor soils.

## 3.9 EFFECTS OF $N_2$-FIXING TREES ON DURATION OF FERTILITY AND PRODUCTION OF CROPS

In well-managed traditional slash-and-burn or swidden fallow agricultural systems, forest regrowth during the fallow period causes restoration of the soil fertility depleted during cropping. Modern agroforestry is an array of attempts to capture the benefits of fallows without the lengthy removal of the land from production. The question is therefore raised, how can the $N_2$-fixing trees and shrubs of agroforestry provide for the restoration or augmentation of available N without excessive loss of crop space, crop growth or labor, and, specifically under what ecological circumstances the incorporation of $N_2$-fixing trees into cropping systems can increase the desired crop yields and not only the total biomass production of cropland. Belowground nitrogen production by legumes and its subsequent transfer both by root-part turnover and mycorrhizal transfer has increasingly been demonstrated to benefit non-legume companion plants. In respect to $N_2$-fixing trees, a positive change in $yield_{crop}$ would be a function of such benefits as improved N status from $N_2$-fixation + other secondary benefits such as improved status of other nutrients + improved water infiltration and water holding capacity + benefits of shade to some plants − loss of water and/or nutrients to competition − loss of photosynthesis to competition − loss of crop space.

It may be that the best evidence for potential benefits of $N_2$-fixing trees and shrubs on associated crops comes from the mixed results of the many experiments on hedgerows and alley cropping conducted during the past two decades, and from the widespread increases in livestock production, in animal weight gains, in milk production, and in increased pasture carrying capacity, resulting from inclusion of a variety of woody legumes in grass pastures.

The provision of nitrogen-rich organic matter generally benefits soil fertility and may increase crop production in alley farming, as demonstrated in numerous experiments, especially but not only on fairly rich soils such as the moist tropical lowlands of Ibadan, Nigeria, and the volcanic uplands of eastern Costa Rica. However, it has become increasingly apparent that the inclusion of $N_2$-fixing hedgerow trees in agricultural programs can also reduce crop growth, as also has been found in numerous alley cropping experiments on dry sites, on nutrient-poor, acid soils and in a wide variety of other conditions. Thus, in alley cropping under hedgerows, the theoretical benefits of growing $N_2$-fixing leguminous trees and shrubs in close spacial and temporal proximity with crops are only sometimes realized in actual farming situations, which remain difficult to identify with adequate certainty.

## 3.10 LEGUMINOUS TREES USED IN AGROFORESTRY

While the leguminous trees and other $N_2$-fixing trees used in agroforestry systems do not generally provide main staple crops, their varied products may be very important to rural people. In addition to the widely recognized fodder, fuelwood, green manure, gums, bee-pasture, building materials, and shade (reviewed in Felker and Bandurski 1979, NRC 1979, Allen and Allen 1981), $N_2$-fixing trees provide vitamins, medicinals, and foods. Some $N_2$-fixing trees provide foods only eaten in famine, but of historic life-saving value. Recently, over 275 species of leguminous trees with edible pods or seeds have been identified, most of which are used in some form of agroforestry (Bryan 1995). These and other "minor crops" have been systematically underestimated in their importance to some rural people, especially those for whom resources are scarcest. While often having little or no monetary value, even in local markets, the widespread use of $N_2$-fixing trees and shrubs in traditional agroforestry often provides valuable options for use, such as using the prunings for mulch or for fodder, as needed most. In contrast to alley cropping and modern pasture systems, however, very little experimental attention has been given to quantifying the agronomic benefits to staple crops of the $N_2$-fixing trees and shrubs used in the rich diversity of traditional agroforestry systems. In fact, most of the many leguminous trees used in long traditions of agroforestry have received no modern scientific attention beyond botanical description; many remain a vast resource for the future.

In contrast to the diverse minor crop leguminous trees and shrubs used locally in traditional agroforestry systems worldwide, a group of fast-growing multipurpose $N_2$-fixing tree legumes has now been made available internationally. Since the 1970s, useful $N_2$-fixing tree species have been identified and reviewed as particularly valuable resources for agroforestry (Felker and Bandurski 1979, NRC 1979, Le Houérou 1980, NRC 1980, Duke 1981, NRC 1983, von Maydell 1983, Burley 1987, Dommergues 1987, Giller and Wilson 1991, von Carlowitz 1991, Nair 1993). A number of these tree species have since been included in genetic selection, seed production, and extension programs worldwide. A series of growth trials for several species have been conducted under varying field conditions, throughout the tropics (Balasundaram 1987, Mathew et al. 1992, Abebe 1994, Duguma and Tonye 1994, Kamara and Maghembe 1994, Maghembe and Prins 1994, Okorio et al. 1994, Bertalot et al. 1998). Nodulation status, taken as an indication of $N_2$-fixation capacity, has been determined for all of the most important multipurpose trees (Allen and Allen 1981, Halliday 1984, Faria et al. 1989), and nodulation is sometimes a prerequisite for being considered a useful multipurpose tree for agroforestry. Some species or genera of tree legumes have now been described in some detail for their value, generally including their contribution to the nitrogen economy, as resources for agroforestry.

The most widely introduced multipurpose tree species is *Leucaena leucocephala*. Long an important tree in Mexico and Central America, it was transported around the tropics by the Spanish since the early 1600s, and has now been introduced throughout the tropics for its use for fodder, fuelwood, food, mulch, and erosion control (NRC 1984, Brewbaker 1987, Shelton and Brewbaker 1994). The complex and disputed taxonomy of *Leucaena* species has recently been been clarified in a revision by Hughes (1998b), who used both morphological and chloroplast DNA data to show the genus to comprise 22 species. The species' size, form, growth, natural distribution, ecology, and uses have been presented with admirable clarity by Hughes (1998a).

Several other important genera of N-fixing leguminous trees have also received botanical and practical attention, often in the form of workshop presentations. Examples include *Calliandra* (Evans 1996), *Erythrina* (Westley and Powell 1993), *Albizia, Paraserianthes,* and *Samanea* (Zabala 1997).

As a group, the international species have several benefits: They are generally easy to produce, fast-growing, and adaptable. Most have now been well-studied: their growth requirements, root patterns, and productivity have been reported widely. Seeds of most of these species are available commercially or through international organizations (von Carlowitz 1991), though many do not appear readily available within their native ranges.

The international attention to selected species appears to have accompanied a long-term loss of social status for local species in many areas, with the result that some species are officially accepted in foreign areas and ignored in their countries of origin. *Calliandra callothyrus, Gliricidium sepium, Leucaena leucocephala,* and *Pithecellobium dulce,* for example, natives of Central America, form the core of agroforestry extension efforts in many parts of the world, while the native legumes of these areas are often overlooked. These same species, however, are often ignored or weeded out in their own native regions in Central America, sometimes to make way for plantations of preferred species of exotic $N_2$-fixing trees. As part of the renewed interest in woody plant resources, however, attention has also been directed to the resources available in favorite trees of farmers (Le Houérou 1980, Morrison et al. 1996).

Several legumes presumed not to fix atmospheric nitrogen, such as *Tamarindus indica* (tamarind), *Ceratonia siliqua* (carob), various *Senna* species, *Delonix regia* (flamboyant), *Bauhinia purpurea,* and other nonnodulating legumes have been favorite and important trees in many traditional and modern agroforestry systems, appreciated by farmers worldwide and at least in some cases contributing to soil nitrogen status like nodulating species (see, for example Jambulingam and Fernandes 1986). A widely planted nonnodulating legume, *Senna siamea,* has been found to produce as much or more nitrogen as many nodulating species evaluated in agroforestry growth trials worldwide (Garrity and Mercado 1994, Bryan 1995). Many nonnodulating legumes grow successfully in infertile sites, some, such as *Mora excelsa* Benth., *Gilbertiodendron deweri, Cynometra* sp., *Prioria copaifera* Griesb., and *Burkea africana* Hook., as dominant species or in nearly monospecific stands. A $N_2$-fixing symbiosis within the roots of nonnodulating legume species has been proposed (Bryan, Berlyn, and Gordon 1996), which would explain the taxonomic anomalies of nodulation (Allen and Allen 1981, Corby 1981, Faria et al. 1989); the resulting phylogenetic puzzles of the origins of $N_2$-fixation in the Leguminosae (Young and Johnston 1989, Sprent 1994); the environment-based evolutionary successes of nonnodulating taxa of legumes (McKey 1994); and the stimulating effect of nonnodulating legume roots such as those of *Peltophorum* spp. on the root growth of other plants (Van Noordwijk et al. 1990, Van Noordwijk et al. 1992). A $N_2$-fixing symbiosis in nonnodulating legumes would also explain the surprising production of nonnodulating legumes such as *Senna siamea,* and *S. spectabilis* in numerous agroforestry trials (Garrity and Mercado, 1994; reviewed in Bryan, 1995). To the extent that nonnodulating legume species fix atmospheric nitrogen, the perceived value of $N_2$-fixed by nodulating legumes should be enhanced rather than reduced by comparison with *Senna,* where *Senna* spp. are used as non-$N_2$-fixing references for nodulating species (Ladha et al. 1993).

## 3.11 BASIC RESEARCH NEEDED IN $N_2$-FIXATION IN AGROFORESTRY

In the past 50 years, as agricultural land has become scarcer for populations in many parts of the world, many agricultural research and extension programs have examined the issue of $N_2$-fixation, $N_2$-fixing trees and their potential to make farming systems sustainable, particularly on the fragile soils of farmers whose economies require minimizing chemical, labor, and/or mechanical inputs. It has been clear that $N_2$-fixing trees may help with some of the chemical requirements and may greatly help with physical protection of the soil resource. Some fast-growing legumes, however, become serious weeds (Hughes and Styles 1989); and many mixed cultivation systems require additional labor.

One of the questions about the potential of $N_2$-fixation in agroforestry requiring increased attention is the extent to which the use of $N_2$-fixing trees in agriculture can be part of and contribute to the high-production agriculture that is expected, desired, and needed by many farmers, and which depends on inputs of fertilizer and other agro-chemicals. While nitrogen fertilizer can decrease nodulation and $N_2$-fixation, amelioration of other soil nutrient deficiencies has been shown to enhance nodulation and increase N-fixed (Dakora and Keya 1997). The need for combining increased food production with protection of the land base requires consideration of the potential that combining $N_2$-fixing trees with application of industrial fertilizer may result in greater crop growth than either fertilizer source alone. In some cases, synthetic fertilizers are more effective when used in conjunction with organic fertilizers (Van der Meersch et al. 1993, Siem and Phien 1995, Rosemeyer et al. 1999). Manures, crop residues, or mulch can enhance crop uptake of applied fertilizer if, for example, they increase soil water storage and decrease soil temperature flux, leading to a more favorable environment for root functioning (Doran and Smith 1987, Agbenin and Goladi 1997). With opposite effect, however, soil organic matter can tie up applied nitrogen fertilizer and thereby decrease immediate plant growth. $N_2$-fixing fallows in eastern Zambia were found to increase maize grain yields with low fertilizer input, but decrease the effectiveness of fertilizer at higher fertilization rates (Kwesige and Coe 1994, Torquebiau and Kwesiga 1996). Within the context of erosion-prevention, fast-growing $N_2$-fixing trees and shrubs have clear and unquestionable benefits to the sustainability of soil fertility. By slowing or preventing the loss of soil organic matter, soil structure, and the chemicals of surface soils, the woody component both protects native soil fertility and enhances the benefits of other inputs such as chemical fertilizers (Young 1989). The extent to which the benefits of trees and shrubs on the sustainance of soil productivity derives directly from $N_2$-fixation remains an important but challenging subject of research. The relatively controlled research conditions of agroforestry, such as alley cropping trials, may provide the opportunity for greater insights into the difficult to evaluate basic roles of $N_2$-fixation in trees.

Two factors of the contribution of $N_2$-fixing symbioses to agroforestry still requiring additional study are the cumulative long-term accumulation and release of residual nitrogen in tree-derived soil biomass (Yamoah et al. 1986), as opposed to soil organic matter of herbaceous origins; and the duration and changes in rates of $N_2$-fixation with age of plant and size of root system (Kadiata et al. 1996), especially in the context of the widely held assumption that most $N_2$-fixation occurs in young plants. Since the various systems of agroforestry using $N_2$-fixing trees use trees of various sizes and ages, rates of $N_2$-fixation may not be comparable across systems.

Finally, recent studies indicate that, while recognizing that $N_2$-fixation may not be the greatest need in many agroforestry systems, the combined and separate secondary benefits of rapid growth of woody $N_2$-fixing species may be more important than previously recognized. The recently discovered enhancement of worm growth by $N_2$-fixing legumes may represent numerous other benefits in need of evaluation.

## 3.12 REFERENCES

Abebe, T. Growth performance of some multipurpose trees and shrubs in the semiarid areas of Southern Ethiopia. *Agroforestry Syst.* 26 (1994):237–248.

Agbenin, J. O. and J. T. Goladi. Carbon, nitrogen, and phosphorus dynamics under continuous cultivation as influenced by farmyard manure and inorganic fertilizers in the Savanna of Northern Nigeria. *Agriculture, Ecosystems, & Environ.* 63 (1997):17–24.

Allen, O. N. and E. K. Allen. *The Leguminosae: A Source Book of Characteristics, Uses, and Nodulation.* Madison, WI: University of Wisconsin Press, 1981.

Balasundaram, V. R. Studies on the native nodulation and biomass production of some tree legumes. *Indian J. of For.* 10 (1987):94–96.

Bartholomew, W. V. Soil nitrogen changes in farming systems in the humid tropics, in: *Biological Nitrogen Fixation in Farming Systems of the Tropics,* A. Ayanaba and P. J. Dart. eds. (Wiley-Interscience: Chichester, 1977), 101–123.

Berry, L. G., B. H. Mason, and R. V. Dietrich. *Mineralogy.* San Francisco: W. H. Freeman and Company, 1983.

Bertalot, M. J. A., I. A. Guerrini, and E. Mendoza. Performance of four nitrogen-fixing species in cerrado soil conditions. *For., Farm, and Community Tree Res. Rep.* 3 (1998):1–3.

Bormann, F. H. and G. E. Likens. *Pattern and Process in a Forested Ecosystem.* New York: Springer-Verlag, 1979.

Brady, N. C. and R. R. Weil. *The Nature and Properties of Soils.* Upper Saddle River, NJ: Prentice-Hall, 1996.

Brewbaker, J. L. Leucaena: a multipurpose tree genus for tropical agroforestry, in: *Agroforestry: A Decade of Development.* H. A. Steppler, and P. K. R. Nair. Eds. (Nairobi, Kenya: ICRAF, 1987), 289–323.

Bryan, J. A. *Leguminous Trees with Edible Beans, with Indications of a Rhizobial Symbiosis in Nonnodulating Legumes.* New Haven, CT, Yale University, 1995.

Bryan, J. A., G. P. Berlyn, and J. C. Gordon. Toward a new concept of the evolution of nitrogen fixation in the leguminosae. *Plant and Soil* 186 (1996):151–159.

Burley, J. Exploitation of the potential of multipurpose trees and shrubs in agroforestry, pp. 273–287. in: *Agroforestry: A Decade of Development.* H. A. Steppler and P. K. R. Nair. Eds. (Nairobi, Kenya: ICRAF, 1987), 273–287.

Cohen, M. N. *The Food Crisis in Prehistory: overpopulation and the Origins of Agriculture.* New Haven, CT: Yale University Press, 1977.

Corby, H. D. L. The systematic value of leguminous root nodules, Part 2, in: *Advances in Legume Systematics.* R. M. Polhill, and P. H. Raven. Eds. (Kew, England: Royal Botanic Gardens, 1981), 657–670.

Dakora, F. D. and S. O. Keya. Contribution of legume nitrogen fixation to sustainable agriculture in Sub-Saharan Africa. *Soil Biol. and Biochem.* 29 (1997):809–817.

Danso, S. K. A., G. D. Bowen, and N. Sanginga. Biological nitrogen fixation in trees in agro-ecosystems. *Plant and Soil* 141 (1992):177–196.

Dommergues, Y. R. The role of biological fixation in agroforestry, in: *Agroforestry: A Decade of Development,* H. A. Steppler and P. K. R. Nair. Eds. (Nairobi, Kenya: ICRAF, 1987), 245–271.

Doran, J. W. and M. S. Smith. Organic matter management and utilization of soil and fertilizer nutrients, 53–72. in: *Soil Fertility and Organic Matter as Critical Components of Production Systems.* R. F. Follett, J. W. B. Stewart, and C. V. Cole. Eds. (Madison, WI: Soil Science Society of America, 1987), 53–72.

Duguma, B. and J. Tonye. Screening of multipurpose tree and shrub species for agroforestry in the humid lowlands of Cameroon. *For. Ecol. and Manage.* 64 (1994):135–143.

Duke, J. A. *Handbook of Legumes of World Economic Importance.* New York: Plenum Press, 1981.

Evans, D. O., Ed. *International Workshop on the Genus* Calliandra. Morrilton, AR: Winrock International and the Taiwan Forestry Research Institute, 1996.

Evans, H. J. and R. H. Burris. Highlights in biological nitrogen fixation during the last 50 years, in: *Biological Nitrogen Fixation.* G. Stacey, R. H. Burris, and H. J. Evans. Eds. (New York & London: Chapman & Hall, 1992), 1–42.

FAO. *Fertilizer Yearbook.* Rome, Italy: FAO, 1992.

Faria, S. M. de, G. P. Lewis, J. I. Sprent, and J. M. Sutherland. Occurrence of nodulation in the leguminosae. *New Phytologist.* 111 (1989):607–619.

Felker, P. and R. S. Bandurski. Uses and potential uses of leguminous trees for minimal energy input agriculture. *Econ. Bot.* 33 (1979):172–184.

Garrity, D. P. and A. R. Mercado. Nitrogen fixation capacity in the component species of contour hedgerows: how important? *Agroforestry Syst.* 27 (1994):241–258.

Giller, K. E. and K. J. Wilson. *Nitrogen Fixation in Tropical Cropping Systems.* Wallingford, U.K.: CAB International, 1991.

Goodwin, T. W. and E. I. Mercer. *Introduction to Plant Biochemistry.* Oxford: Pergamon Press, 1983.

Halliday, J. Register of nodulation reports for leguminous trees and other arboreal genera with nitrogen fixing members. *Nitrogen Fixing Tree Res. Rep.* 2 (1984):38–45.

Hauser, S. Distribution and activity of earthworms and contribution to nutrient recycling in alley cropping. *Biol. and Fertil. of Soils* 15 (1993):16–20.

Hauser, S. and D. O. Asawalam. Effects of fallow system and cropping frequency upon quantity and composition of earthworm casts. *Z. für Pflanzenernährung und Bodenkunde* 161 (1998):23–30.

Hughes, C. Leucaena: *A Genetic Resources Handbook.* Oxford, U.K.: Oxford Forestry Institute, Department of Plant Sciences, University of Oxford, 1998a.

Hughes, C. *Monograph of* Leucaena *(Leguminosae-Mimosoideae).* American Society of Plant Taxonomists, 1998b.

Hughes, C. E. and B. T. Styles. The benefits and risks of woody legume introduction, *Advances. in Legume Biology.,* Vol. 29 in: *Monographs in Systemic Botany from the Missouri Botanical Garden.* C. H. Stirton and J. L. Zarucchi. eds. n.p. 1989, 505–531.

Humphreys, L. R. *Tropical Forages: Their Role in Sustainable Agriculture.* Essex, England: Longman Scientific & Technical, 1994.

Jambulingam, R. and E. C. M. Fernandes. Multipurpose trees and shrubs on farmlands in Tamil Nadu State (India). *Agroforestry Syst.* 4 (1986):17–32.

Kadiata, B. D., K. Mulongoy, and N. O. Isirimah. Time course of biological nitrogen fixation, nitrogen absorption, and biomass accumulation in three woody legumes. *Biol. Agric. and Hortic.* 13 (1996):253–266.

Kamara, C. S. and J. A. Maghembe. Performance of multipurpose tree and shrub species 28 months after planting at Chalimbana, Zambia. *For. Ecol. and Manage.* 64 (1994):145–151.

Kang, B. T. Changes in soil chemical properties and crop performance with continuous cropping on an entisol in the humid tropics, in: *Soil Organic Matter Dynamics and Sustainability of Tropical Agriculture,* K. Mulongoy, and R. Merckx. eds. (Chichester: John Wiley and Sons, 1993), 297–305.

Kaplan, L. Archaeological *Phaseolus* from Tehuacan. in: D. S. Byers. ed. *Environment and Subsistence.* Vol. 1. (Austin, TX: University of Texas Press, 1967), 201–211.

Kwesige, F. and R. Coe. The effect of short rotation *Sesbania sesban* planted fallows on maize yields. *For. Ecol. and Manage.* 64 (1994):199–208.

Ladha, J. K., M. B. Peoples, D. P. Garrity, V. T. Capuno, and P. J. Dart. Estimating dinitrogen fixation of hedgerow vegetation using the nitrogen-15 natural abundance method. *Soil Sci. Soc. of America J.* 57 (1993):732–737.

Le Houérou, H. N., ed. *Browse in Africa.* Addis Ababa: ILCA, 1980.

Maghembe, J. A. and H. Prins. Performance of multipurpose trees for agroforestry two years after planting at Makoka, Malawi. *For. Ecol. and Manage.* 64 (1994):171–182.

Mangelsdorf, P. C., R. S. MacNeish, and W. C. Galinat. Prehistoric wild and cultivated maize, in: *Environ. and Subsistence.* Vol. 1, D. S. Byers. ed. (Austin, TX: University of Texas Press, 1967), 178–200.

Mathew, T., B. M. Kumar, K. V. S. Babu, and K. Umamaheswaran. Comparative performance of four multipurpose trees associated with four grass species in the humid regions of Southern India. *Agroforestry Syst.* 17 (1992):205–218.

McKey, D. Legumes and nitrogen: the evolutionary ecology of a nitrogen-demanding lifestyle, in: *Advances in Legume Systematics: The Nitrogen Factor.* Vol. 5. J. I. Sprent and D. McKey, eds, (Kew, England: Royal Botanic Gardens, 1994), 211–228.

Morrison, B. J., M. A. Gold, and D. O. Lantagne. Incorporating indigenous knowledge of fodder trees into small-scale silvopastoral systems in Jamaica. *Agroforestry Syst.* 34 (1996):101–117.

Mulongoy, K., M. Gueye, and D. S. C. Spencer. *Biological Nitrogen Fixation and Sustainability of Tropical Agriculture.* London: John Wiley and Sons, 1992.

Mulongoy, K. and R. Merckx, eds. *Soil Organic Matter Dynamics and Sustainability of Tropical Agriculture.* Chichester: John Wiley and Sons, 1993.

Nair, P. K. R. *An Introduction to Agroforestry.* Dordrecht, The Netherlands: Kluwer Academic Publishers, 1993.

NRC. *Tropical Legumes: Resources for the Future.* Washington, DC: National Academy Press, 1979.

NRC. *Firewood Crops: Shrub and Tree Species for Energy Production.* Washington, DC: National Academy Press, 1980.

NRC. *Firewood Crops: Shrub and Tree Species for Energy Production,* Vol. 2. Washington, DC: National Academy Press, 1983.

NRC. Leucaena: *Promising Forage and Tree Crop for the Tropics.* Washington, DC: National Academy Press, 1984.

Nye, P. H. and Greenland, D. J. *The Soil Under Shifting Cultivation.* Bucks, England: Commonwealth Agricultural Bureaux, Farnham Royal, 1960.

Okorio, J., S. Byenkya, N. Wajja, and D. Peden. Comparative performance of 17 upperstory tree species associated with crops in the highlands of Uganda. *Agroforestry Syst.* 26 (1994):185–203.

Olson, R. A. and Kurtz, L. T. Crop nitrogen requirements, utilization, and fertilization, in: *Nitrogen in Agricultural Soils.* F. J. Stevenson. ed. Madison, WI (American Society of Agronomy, 1982), 567–604.

Palm, C. A. Contribution of agroforestry trees to nutrient requirements of intercropped plants. *Agroforestry Syst.* 30 (1994):104–124.

Ramakrishnan, P. S. *Shifting Agriculture and Sustainable Development.* Paris: UNESCO, 1992.

Roosevelt, A. C., M. L. d. Costa, C. L. Machado, M. Michab, N. Mercier, H. Valladas, J. Feathers, W. Barnett, M. I. d. Silveira, A. Henderson, J. Sliva, B. Chernoff, D. S. Reese, J. A. Holman, N. Toth, and K. Schick. Paleoindian cave dwellers in the Amazon: the peopling of the Americas. *Science* 272 (1996):373–384.

Rosemeyer, M., K. Schlather, and J. Kettler. The *frijol tapado* agroecosystem: the survival and contribution of a managed fallow system to modern Costa Rican agriculture, in: *Managed Ecosystems: The Mesoamerican Experience,* L. U. Hatch and M. E. Swisher. Eds. (New York: Oxford University Press, 1999), 138–144.

Sanchez, P. A. *Properties and Management of Soils in the Tropics.* New York: John Wiley and Sons, Inc., 1976.

Sanchez, P. A. and M. Hailu, Eds. Alternatives to slash-and-burn agriculture. *Agriculture, Ecosystems, and Environment.* Amsterdam: Elsevier, 1996.

Sanginga, N., S. K. A. Danso, and F. Zapata. Field measurements of nitrogen fixation in leguminous trees used in agroforestry systems: influence of $^{15}N$-labeling approaches and reference trees. *Biol. and Fertility of Soils* 23 (1996):26–32.

Shelton, H. M. and J. L. Brewbaker. *Leucaena leucocephala*—the most widely used forage tree legume, in: *Forage Tree Legumes in Tropical Agriculture,* R. C. Gutteridge and H. M. Shelton. Eds. (Wallingford, England: CAB International, 1994), 15–29.

Siem, N. T. and Phien, T. Acid upland soil in Vietnam and their management for agriculture, in: *Plant-Soil Interactions at Low pH: Principles and Management,* R. A. Date, N. J. Grundon, G. E. Rayment, and M. W. Probert. Eds. (Dordrecht, The Netherlands: Kluwer Academic, 1995), 771–774.

Smith, C. E. Plant remains, in: D. S. Byers. ed. *Environment and Subsistence.* (Austin, TX: University of Texas Press, 1967), 220–225.

Sprent, J. I. Evolution and diversity in the legume-rhizobium symbiosis: chaos theory? *Plant and Soil* 161 (1994):1–10.

Stevenson, F. J. Organic forms of soil nitrogen, in: *Nitrogen in Agricultural Soils,* F. J. Stevenson. ed. (Madison, WI: American Society of Agronomy, 1982a), 67–122.

Stevenson, F. J. Origin and distribution of nitrogen in soil, in: *Nitrogen in Agricultural Soils,* Ed. F. J. Stevenson. (Madison, WI: American Society of Agronomy, 1982b), 1–42.

Swift, M. J. and Woomer, P. Organic matter and the sustainability of agricultural systems: Definition and measurement, in: *Soil Organic Matter Dynamics and Sustainability of Tropical Agriculture,* K. Mulongoy, and R. Merckx. Eds. (Chichester: John Wiley and Sons, 1993), 3–18.

Torquebiau, E. F. and Kwesiga, F. Root development in a *Sesbania sesban* fallow-maize system in Eastern Zambia. *Agroforestry Syst.* 34 (1996):191–211.

Van der Meersch, M. K., R. Merck, R., and K. Mulongoy, Evolution of plant biomass and nutrient content in relation to soil fertility changes in two alley cropping systems, In: *Soil Organic Matter Dynamics and Sustainability of Tropical Agriculture,* K. Mulongoy, and R. Merck eds. (Chichester: John Wiley and Sons, 1993), 143–154.

Van Noordwijk, M., Y. R. Dommergues, and M. Van Noordwijk. Agroforestry and soil fertility. Root nodulation: the twelfth hypothesis. *Agroforestry Today* 2 (1990):9–10.

Van Noordwijk, M., K. Hairiah, S. M. Sitompul and M. S. Syekhfani. Rotational hedgerow intercropping + *peltophorum pterocarpum* = new hope for weed-infested soils. *Agroforestry Today* 4 (1992):4–6.

von Carlowitz, P. G. *Multipurpose Trees and Shrubs: Sources of Seeds and Inoculants.* Nairobi, Kenya: ICRAF, 1991.

von Maydell, H.-J. *Arbres et Arbustes du Sahel: Leurs Charactéristiques et leurs Utilisations.* Eschborn: Deutsche Gessellschaft für Technische Zusammenarbeit, 1983.

Waterlow, J. C. Whole-body protein turnover in humans—past, present, and future. *Annu. Rev. of Nutr.* 15 (1995):57–92.

Werner, D. Physiology of nitrogen-fixing legume nodules: compartments and functions, in: *Biological Nitrogen Fixation,* G. Stacey, R. H. Burris, and H. J. Evans. Eds. (New York: Chapman & Hall, 1992), 399–431.

Westley, S. B. and Powell, M. H., Eds. *Erythrina* in the new and old worlds. *Nitrogen Fixing Tree Association Special Report.* Paia, Hawaii: Nitrogen Fixing Tree Association, 1993.

Wong, M. T. F. and R. S. Swift. Amelioration of aluminium phytotoxicity with organic matter, in: *Plant-Soil Interactions at Low pH: Principles and Management,* R. A. Date, N. J. Grundon, G. E. Rayment, and M. W. Probert. eds. (Dordrecht, The Netherlands: Kluwer Academic Press, 1995), 41–45.

Yamoah, C. F., A. A. Agboola, and K. Mulongoy. Decomposition, nitrogen release and weed control by prunings of selected alley cropping shrubs. *Agroforestry Syst.* 4 (1986): 239–246.

Young, A. *Agroforestry for Soil Conservation.* Wallingford, U.K.: CAB International, 1989.

Young, J. P. W. and A. W. B. Johnston. The evolution of specificity in the legume-*Rhizobium* symbiosis. *Trends in Ecol. and Evol.* 4 (1989):341–349.

Zabala, N., Ed. *International Workshop on Albizia and Paraserianthes Species: Forest, Farm, and Community Tree Research Reports.* Morrilton, AR: Winrock International and the Taiwan Forestry Research Institute, 1997.

# 4 Ecological Theory of Diversity and its Application to Mixed Species Plantation Systems

*P. S. Ashton*[1]

**TABLE OF CONTENTS**



## 4.1 INTRODUCTION

It is an axiom that modern agriculture has increasingly involved amendment and homogenization of site conditions to suit one or a few genotypes. This has been made possible through the enormous advances in crop genetics. In traditional agriculture, crops were selected to suit the site, which generally experienced only the amendment that one farmer could achieve with his own hands. The early invention of irrigation technology heralded the first step towards prevailing current practice. Although rotations of several species remain an important part of modern commercial farming, interplanting of more than one species has until recently been practiced on a more limited scale. Increasing realization of the hazardous side effects of chemical

[1] Department of Organismic and Evolutionary Biology, 22 Divinity Avenue, Harvard University, Cambridge, MA 02138. Tel.: (617) 495-2363

0-8493-2206-5/00/$0.00+$.50

pesticides, and the eutrophication of water bodies by excessive runoff of fertilizers, has led to a shift in objectives of plant breeding and the design of cropping systems with the aim of reducing chemical inputs yet without loss of yield. To what extent is this possible on theoretical grounds, and what evidence is there that traditional systems, in this case mixed-species agroforestry systems, have succeeded in sustaining relatively high levels of yield in relation to inputs, both as chemicals and labor? Mixed-species woody plantings have also been advocated as the preferred means of both ecologically restoring degraded land, and bringing it back into productive use (see, e.g., Hart 1980, Ewel 1986, Jacob and Alles 1987 Ashton et al. 1997a, 1997b). To what extent do species-mixtures achieve this more successfully than appropriately selected monocultures.

This chapter examines the extent to which the practice, traditional and contemporary, of a mixed-species agroforestry system is supported by ecological theory and empirical research. There are three ways in which theory might underpin practice. First, it is believed that the number of species populations that can coexist is determined by resource availability and the frequency and intensity of catastrophic disturbance. Second, this is thought to be because plant mixtures, particularly of different species as opposed to genotypes, can utilize above- and below-ground resources, both in space and over time, more efficiently than single species crops. And third, plants in mixture, be they of species or solely genotypes, may experience lower infestation rates from pathogens, herbivores, and seed predators.

It has to be said from the outset that experimental tests of these hypotheses in both ecology and agroforestry are extraordinarily difficult, and the literature is replete with inconclusive results. One reason is the agriculturist's tendency to quote results by value rather than yield. Values are somewhat easier to obtain indirectly, farmers worldwide being notoriously secretive about their yields, but they confine the usefulness of the result, even as a comparison, to the market conditions when the study was made. The other, and yet more intractable reason is the large number of variables involved, both biotic and abiotic, the long period from planting to full harvest of the majority of woody mixtures, and the difficulty of establishing controls and a satisfactory basis for comparisons.

Ecologists remain divided concerning the extent to which the floristic structure of plant communities is deterministic, which is to say that they follow a predictable sequence of species succession following disturbance, and are in an equilibrium of composition and rank order of abundance of species at the scale of the whole community. On one hand, Tilman (1982) has proposed predictable patterns of species richness in relation to soil resource availability, and Connell (1974, 1978) in relation to intensity and frequency of canopy disturbance, while Huston (1980) has developed a model which incorporates both. Janzen (1970) and Connell (1972) proposed that post-dispersal host-specific seed and seedling predators may reduce survivorship where plant density is high and close to parent trees, thereby lowering competition with other species and enhancing diversity. S. P. Hubbell (1979, 1997; Condit et al. 1996) instead proposed that the exigencies of fruiting periodicity and dispersal limitation, combined with stochastic processes of extinction and immigration from local to widespread scales, overwhelms deterministic processes to the extent that community structure is essentially unpredictable.

There seems to be little evidence that farmers themselves choose to plant in mixture or to selectively thin or lop their plantations for any of these theoretical ecological reasons. Only quite recently have pure strains of cereals and annual vegetable crops become available, although a number of perennial crops, notably root crops but also, for instance, bananas, have habitually been vegetatively propagated. Traditionally therefore, farmers have planted mixed-genotype annual crops, and mixed-genotype annuals with variably mixed perennial crops. To the extent that the farmer was aware that a mixture is being planted, it seems to have been preferred because only small amounts of the products of the species in mixture were needed for home use rather than trade, and because the products can be harvested year-round or over a long season, including slack periods between the planting, weeding, and harvesting times of annual crops.

## 4.2 RESOURCE AVAILABILITY AND THE COMPLEXITY OF MIXED-SPECIES PLANTATIONS

Variation in climate is regional, but in soil is principally at landscape scale and related to geology and topography. Resource availability therefore affects structure and function between as much as within plantations.

Tilman (1982) hypothesized that species mixtures, in the absence of disturbance, can be sustained at greatest diversity where essential soil resources are moderately limiting, and that diversity declines both where resources are severely limited, and where they are abundant. In the case of plant communities, these resources are soil nutrients and water. When all resources are severely limited, opportunity for competitive differentiation for their exploitation is also limited and the number of species that can coexist is small. In Tilman's view, each surviving species has the capacity to reduce one or more resources below the tolerance threshold of others (Figure 4.1). Coexistence occurs where resource requirements differ among the species, and resources are replenished to a level which permits competitors to survive; independent small scale spatial variation in resource availability will further enhance species diversity (Figure 4.2). As resources increase in availability, particularly if this increase varies between resources and also varies on a small spatial scale within the community, such opportunities will also increase, leading to stable coexistence of an increasing number of species populations. Eventually, Tilman argued, that resources cease to be limiting, with the exception of light. Then, those species which can competitively exclude others from light will dominate, and the total number which can coexist will again decline (Figure 4.2). This theory was first tested among freshwater algae (Tilman 1977). In terrestrial plant communities, with limits to seed dispersal, there are strict limits to the level of species diversity in sessile, poorly dispersed organisms that can be explained by this theory. Nevertheless, it is well-known that limestone grasslands, in which available nutrients are highly variable in space and water more or less limiting, are exceptionally diverse, even when not grazed, in comparison to acid grasslands or mesic meadows. Even a tropical rain forest varies according to the same pattern of species diversity, and apparently for similar reasons. Davis and Richards (1933–34) first recorded that tree species' richness peaked in mixed forest on moderately low nutrient yellow sand humult ultisols, being somewhat lower on alluvial loams and much lower on podsols, in wet seasonal lowland Guyana. A very similar pattern has been observed in northwest Borneo (Ashton 1992a, b, 1998, Ashton and Hall 1992).

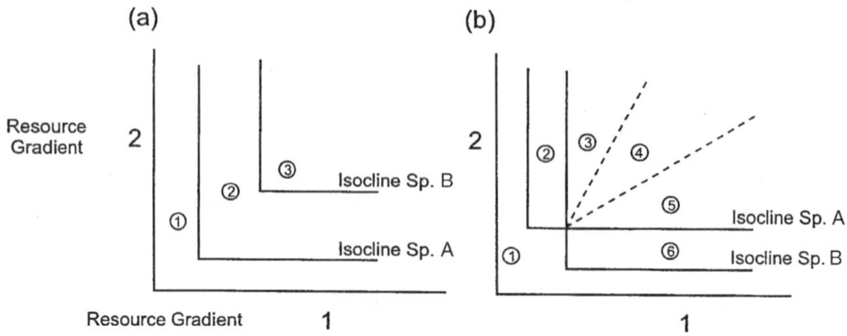


**FIGURE 4.1** a) Tilman's hypothesis of competitive exclusion: The isocline (limiting resource level) of Species A lies closer to the resource axis than the isocline of Species B. If the initial resource levels are in Region 1, then neither species can exist. But if the resource levels are in Regions 2 or 3, then Species A reduces the resource levels to a point on its own isocline (where Species B cannot survive and reproduce) so Species A excludes Species B. b) Potential coexistence of two competitors limited by two resources. The isoclines of Species A and B overlap. Neither species can survive when resource levels are in Region 1; Species A excludes Species B at Levels 2 and 3, while Species B excludes Species A in Regions 5 and 6. But both species can coexist in Region 4. (modified after Begon, et al., 1996).

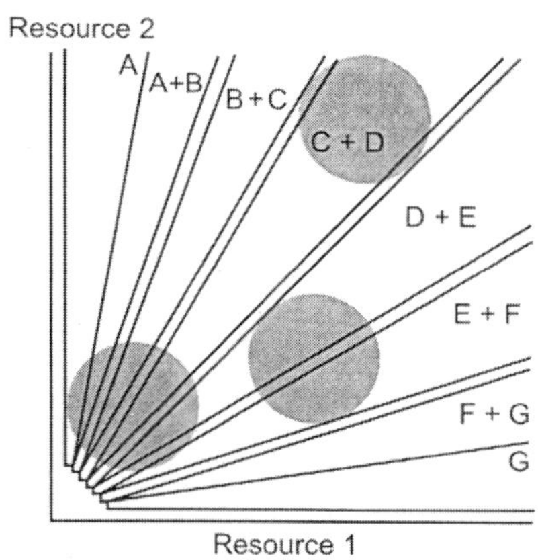


**FIGURE 4.2** Seven-species competition for two resources according to Tilman (1982). The regions of coexistence never have more than two species. When local environmental variability is considered (within the shaded disks), habitats with lower overall resource levels are likely to support more species over their range of variability than habitats with the same range of variability but higher resource levels.

Traditional mixed-species plantations vary enormously in complexity but, broadly, species diversity parallels that of natural vegetation in the same area. The most complex mixtures have always been confined to the tropics, and also the milder moist temperate regions of the world notably East Asia, Eastern North America, and moister parts of southwest Europe. All regions noted for the species diversity of their indigenous plant communities. In some cases, plantations originate through modification of the original vegetation, in which useful species are encouraged through natural regeneration or planting, others culled, and frequently additional species added. Examples include the *Pet Kot* of the Maya of southern Mexico (Gomez-Pompa 1987) and the *Shorea javanica* gardens of southern Sumatra (Torquebiau 1984). Such methods ensure, in a low-input system, that the species are pre-adapted to the prevailing site conditions.

Exotic species, however, have long been included in traditional mixed-species cultures. Although the spread of exotics was apparently on a relatively limited geographical scale in pre-Colombian America (albeit staples such as maize spread far beyond their original range), the bas-reliefs at Borobudur, Java, confirm that *Terminalia arjuna* had been brought from India in classical times and it appears likely that fruit such as jambu (*Syzygium* spp.), mango (*Mangifera indica*) and jack (*Artocarpus heterophyllus*), had been similarly imported. South Asia apparently gained the durian (*Durio zibethinus*), rambutan (*Nephelium lappaceum*), and banana (*Musa sapientum*) from the east in early times. These exotics are predominantly species of better soils, and little conscious selection is necessary in more fertile areas. On the sandy, less fertile yet extremely species diverse, landscapes of regions such as northwest Borneo the home gardens include many exotics but their enormous diversity is due to an abundance of indigenous species and land races. In the same region, and also on the sandstone of the Guyana Shield, gardens on the poorest, podsol, and peat soils can support few robust species but still include exotics such as, in Asia, binjai (*Mangifera foetida*). In Asia therefore, it would appear that at least the tree component of mixed-species plantations broadly follows a trend of maximum species richness according to Tilman's predictions where soil resources are limiting. But this is not so on high nutrient soils, where the farmer periodically harvests canopy individuals in order to maintain a relatively open canopy (but see the case of *Shorea javanica* gardens above). In fact, the documented cases of outstandingly species rich home gardens seem to originate from regions of generally fertile soils. Examples include the Hanúnoo old fields of Mindoro, Philippines (Conklin 1957), the home gardens of Kandy, Sri Lanka (Jacob and Alles 1987) and Java (Michon et al. 1983, 1985), and the chagga systems of East Africa (Fernandes et al. 1986). It is interesting that the humid sites among these are not known for their high natural floristic diversity, an issue to be addressed in the next section. However, it would seem certain that the correlation between soil droughtiness, fertility, and the number of species in plantations is because the number of available species for the site conditions follows the predicted trend, rather than that it is a consequence of competitive mortality in relation to soil resources following planting: Farmers know their land and their species, and plant accordingly.

## 4.3 MAXIMIZING USE OF THE SITE

Species-mixtures can in theory enhance the productivity of a site relative to monocultures in several direct ways, with or without addition of fertilizers and pesticides.

### 4.3.1 Sequential Mixing: Availing of Manmade Catastrophic Intervention

Mixtures may more efficiently exploit the resources of a site over time, by simulating succession in a manner which maximizes the yield of a sequence of harvested products. Because of the difficulty of harvesting crops separately in mixed, as opposed to relayed, annual cropping systems, these forms of mixed plantations more generally involve perennial crops. They have traditionally been associated with some form of swidden, that is, periodic annual cropping followed by supra-annual woody fallow. The now celebrated *taungya* method of the Karens of Burma was early recognized by Brandis (1876) and his staff as potentially compatible with timber production, and therefore a basis for involving rural communities in forest management. Similar examples are known elsewhere where swidden agriculture is practiced. (Posey 1984, Boonkird et al. 1984, Oduol et al. 1986). First, a cereal is generally planted, often corn or millet that will grow tall and compete successfully with other tall-growing crops in mixture. The cereal may be interplanted with crops which may depend on the cereal for support such as beans, or vegetables such as squash or melons. From the beginning, or in the years following according to the primary objective, trees and perhaps other woody crops are interplanted, and eventually overtop and may exclude the annual crops. Exclusion depends on planting distance, and soil water economy. In arid and semiarid regions, seasonality of rainfall and tree deciduousness may permit annual crops to continue to be grown between lady's finger (*Moringa oleifera*) or taller woody plants. Coconut (*cocos nucifera*), date, and other tall palms lend themselves to permanent herbaceous-woody mixtures.

Connell (1978) on the basis of coral reef ecology, suggested in his intermediate disturbance hypothesis that frequent or intense disturbance favors pioneer and successional against climax species, and rare or only mild disturbance favors the reverse. Greatest species diversity will occur at intermediate levels of disturbance where both climax and pioneer species are well represented (Figure 4.3). Huston (1979) incorporated Tilman's and Connell's species diversity gradients, of growth rate and "frequency of reduction," into a single model (Figure 4.4).

Subsistence smallholders may know these relationships from empirical experience, and use them to maximize diversity of species, thereby increasing ecological efficiency of their gardens, spreading risk, and perhaps increasing productivity. Modern attempts to increase the sustainable productivity of traditional systems have concentrated on replacing natural selection in the woody fallow with a judicious choice of successional species, planted to yield valued products including commodities (Raintree and Warner, 1986). Sequential mixtures may lead to a semipermanent woody mixed- or single-species crop; this is generally the case when a commodity is produced. In forestry, where the objective is to achieve both an equitable distribution

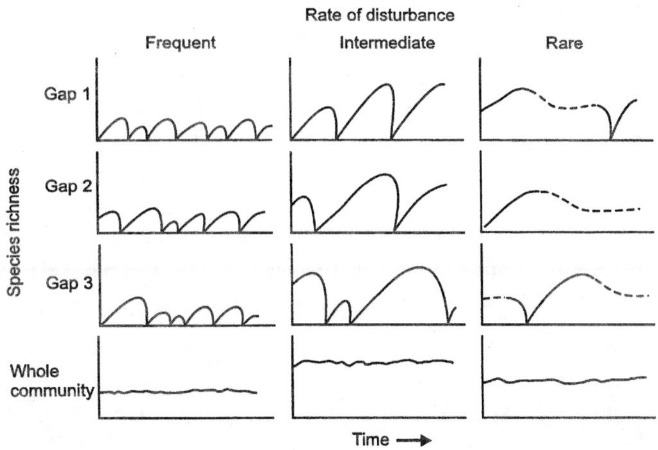


**FIGURE 4.3** Connell's intermediate disturbance hypothesis for maintenance of community species richness. Diagrammatic representation of the time course of species richness in three gaps, and in the community as a whole, at three frequencies of disturbance. Dashed lines indicate the phase of competitive exclusion as a climax is approached (from Begon et al. 1996).

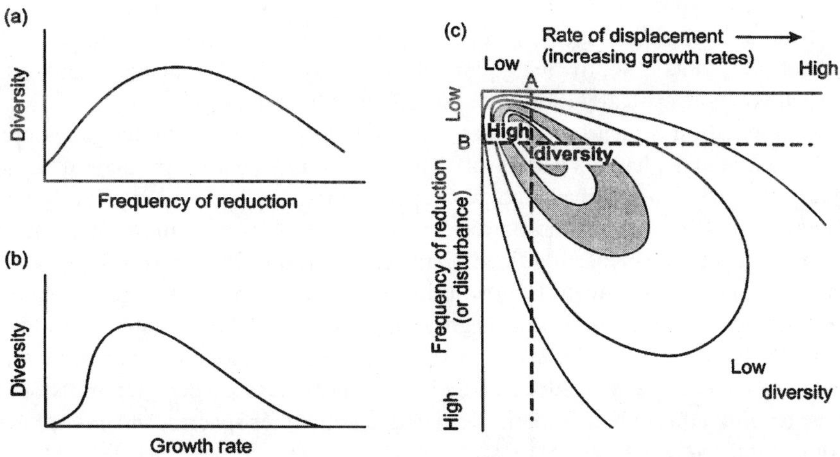


**FIGURE 4.4** Relationships derived from Huston's model of patterns of community species richness. a) Predicted relationship between species richness and frequency of population reduction (equivalent to Connell's model). b) Predicted relationship between species richness and population growth rates in nonequilibrium systems with a low-to-intermediate frequency of population reduction (equivalent to Tilman's resource limitation model). c) Generalized contour map of the relationship between the rate of competitive displacement, equivalent to population growth rate, the frequency of population reduction, and species richness. Graph (a) follows dashed Line A, Graph (b) follows dashed Line B on this map. (modified after Begon et al. 1996).

of use of the land between abutting rural communities and those served by the market, the woody crops are more likely to yield timber and will be harvested at a time that optimizes the profitability of the system as a whole, usually over a somewhat shorter cropping cycle than required for maximum timber productivity.

It is an ecological truism that resources lost through harvesting must be replaced to achieve sustainability. Continuous maintenance of relatively productive successional stages by harvesting will inevitably lead to depletion of soil nutrients, which must be restored either in part by nitrogen fixing plants, and in part or whole by fertilizer or manure application from outside the system. This can be a relatively unprofitable exercise on poor land, and the economics of such systems demand careful review.

### 4.3.2 Resource Utilization

A second consideration is the theoretical maximization of resource utilization that would be achieved by mixtures of plants, or organisms in general, that are resource specialists chosen because of their minimal overlap in resource preference and method of capture. In plants, this criterion applies above and below ground. All organisms experience most mortality during infancy. In plants, progeny mortality soars following flowering, and particularly from seed fall to seedling establishment. It is sometimes claimed that selection is at its most stringent at this stage, and that species have responded by evolving subtle differences in germination and early establishment characteristics, hence microhabitat preferences (Grubb 1977). Whether or not this is so, by being planted as healthy juveniles, plants in cultivation bypass this whole critical period and start effectively on an equal footing.

Maximization is clearly difficult to achieve in a crop that is undergoing continuous successional change, and is difficult to estimate below ground. It is only recently that the theoretical advantage of species mixtures has been rigorously tested (Naeem et al. 1994) and found correct in several respects relating to performance. The extent to which species in mixed plantations exploit different soil depths or differentially exploit soil minerals in solution remains unknown. Farmers frequently experience lower yields from annuals growing close to woody vegetation, though this may partly be due to shading. Trees possess diverse root systems. Some, particularly those of tall stature early in succession (Ewell 1980), develop deep tap roots but the degree to which this enhances their coexistence with other species, and overall yield, is conjectural.

Primary biological productivity is in part related to the area of leaf surface above a given area of ground (Leaf Area Index, LAI). Because photosynthesis in full sunlight uses but a fraction of incident radiation, leaves may be orientated close to perpendicular to incident radiation and still maintain maximum rates. This has the added advantage of reducing heat load, hence evaporative water loss. Horn (1971) showed in theory, which he tested in natural woodland, how the crown geometry of a canopy tree not only influences Leaf Area Density (LAD; the total leaf area within a unit volume of space) within its own crown, but the number and photosynthetic characteristics of the species that can grow beneath it. The spacing of the tallest crowns in natural vegetation influences the amount of lateral light that can penetrate the canopy, and hence the number of leaf layers that can survive beneath (Terborgh 1985). Leaf shape and arrangement and stand canopy architecture are potentially important considerations in the design of mixed-species plantations.

Horn described how woody succession in a temperate forest progresses from species with deep crowns and high LAD, through a stage when several strata, of different species develop; but eventually one of a few shade-tolerant species with a single dense-leaf layer take over the canopy in the absence of disturbance. Horn's own example was succession to beech (*Fagus grandifolia*) dominated forests in New Jersey. Similar examples include the European beech (*Fagus sylvatica*) woods, which were the subject of A. S. Watt's (1947) classic study of forest dynamics. Recently the same overall pattern has been demonstrated in the humid tropics (Ashton 1992, 1998, Ashton and Hall 1992). In dry climates and habitats, trees that are either deciduous or adapted to withstand water stress at individual leaf level by their diffuse crowns and narrow leaves held close to the perpendicular — such as *Eucalyptus, Casuarina,* evergreens, and pines prove most productive. The forest retains high LAI to maturity, and often higher species diversity than mature stands in mesic sites. Fifty-meter tall primary tropical mixed lowland evergreen forests have LAI of about 8, no higher than a well developed oak-maple stand (Oliver and Larson 1996, Waring and Schlesinger 1986). Such forests, when on mesic lower slopes and plains, also support fewer leaf strata below the upper canopy as Horn predicted, (Ashton and Hall 1992) and probably have lower LAI Unlike temperate succession, secondary forest succession on mesic sites in the humid tropics, especially when initiated simultaneously in large patches such as following swidden, early on develops a dense upper canopy and LAI as low as 4 (Ashton 1978). Productivity has not been measured but is likely to be quite low.

The extent to which woody species respond differentially to light conditions below and in the canopy remains contentious. That even closely related species in tropical evergreen forest may differ in light response is clear (Ashton 1995, Ashton and Berlyn, 1992, Ashton et al. 1995, Press et al. 1996), but the extent that this leads to significant differentials in performance in the complex, spatially and temporally variable light conditions of small gaps remains unclear (Ashton 1998, Whitmore and Brown 1996). Certainly, we as yet know nothing concerning those ecophysiological characteristics of agroforestry and horticultural species upon which choices must be made. Wherever there is no shortage of planting stock, enrichment of existing plantations would likely benefit by planting seedlings of several species at each point, and later culling all but the tallest individual.

Canopy structure in artificial mixed plantations will therefore develop along paths determined in part by the choice of species, and in part by site conditions. The tendency is for the number of leaf layers that can be supported, be they on a single crown or in crowns of more than one species vertically stratified, to decline as the stand matures. This will usually be accompanied by a decline in LAI (Horn 1971) and timber yield, and probably in the yield of other useful products (Hart 1980). Traditionally, farmers have retained a broken canopy structure, stratified leaf layers, and a mixture of crop species by selective lopping and felling. This will favor survival of a rich species mixture particularly on fertile, mesic sites, in keeping with the principles of Huston's (1979) general hypothesis of species diversity. Another option is to select upper canopy species with diffuse leaf arrangement. Traditional home garden systems in Asia have sometimes even involved long-lived shade tolerant timber trees such as ebony (*Diospyros*) and *Pericopsis mooniana* in Sri Lanka and hinoki (*Chamaecyparis obtusa*) in Japan. It is under such conditions, where the full range from light-demanding pioneer to shade-tolerant climax species are planted together, that the full advantages of their differences in crown characteristics, as well as height at maturity, can be gained. However, even shade-bearers prove more productive in at least partial direct sunlight, so that shade tolerant fruit trees such as *Garcinia mangostana* (mangosteen) and *Gnetum gnemon* (emping) in Asia are traditionally grown where their crowns are partially exposed to sunlight. Now, owing to the need to intensify use and increase productivity, species selected are increasingly restricted to light-demanding early successional species. Then, there is little to be gained because the physiological characteristics if not the crown architecture of species sharing the same life form will differ little.

Although LAI and net primary production are broadly correlated, it is not necessarily the case that the yield of useful products is so correlated. This will depend on the products chosen. For instance, even shade-bearing fruit trees such as mangosteen and coffee may bear higher yields when planted in partial or full sun; but the partial shade of a tree canopy yields the highest harvest productivity for both cardamom (*Elettaria* spp.) fruit and ginseng roots (*Panax* spp.).

Above ground, or perhaps both above and below ground, there is some, albeit preliminary, empirical evidence that some mixtures can yield more highly than single species crops, particularly of woody species. Rubber underplanted with tea in South China yields more than the two crops grown separately. In a particularly relevant study in the western Himalaya, so-called agrihortisilvicultural systems which are more species-diverse, and include a wider range of growth forms among the crop plants including cereals, rhizomatous and leafy vegetables, fruit and fodder trees, had higher LAI, biological productivity and yield than agrisilviculture in which fuel and fodder trees are grown between fields of herbaceous crops, and agrihorticulture in which vegetables are grown in fruit orchards (Toky et al. 1989). There are also many examples where mixtures have proven less satisfactory than monocultures, at least in the short term and given high inputs of fertilizer or chemical pesticide. The now widespread removal of shade trees in neotropical coffee plantations serves as testimony. A rather detailed study of total economic production, and net value, of coffee and shade trees alone, and coffee, shade and fruit trees showed higher production of coffee and consequently total net value in the simpler mix (Lagemann and Heuveldop 1983). These, and many other such negative examples, are due to the overriding economic value of commodity over subsistence crops (see Raintree and Warner 1986).

## 4.4 ENHANCING STABILITY AND SUSTAINABILITY: PATHOGENS AND SYMBIONTS

Two kinds of interaction between plants growing in mixed-species communities may contribute to ecosystem sustainability and stability. One is the likelihood that there is a negative correlation between host-specific predation and pathogenicity, and the distances between individuals hence population densities of plant hosts, owing to the need for increasingly distant dispersal and searching effort by the predator or pathogen. Crop plants grown in mixture would thereby less frequently experience epidemic disease and mortality, and the mixture as a whole would be more stable than a monoculture both numerically and in yield. The other is symbiotic relationships with microorganisms which facilitate anion uptake, and thereby potentially enhance sustainable productivity. Howe and Westley (1988) have provided a useful review of the theory behind the interactions of plants and herbivores including predators.

### 4.4.1 Enhancing Stability

Janzen (1970) hypothesized that density-dependent mortality from species-specific seed and seedling predators in tree species-rich rain forests would create space around mature trees in which their own species' juveniles would be at a competitive disadvantage compared with those of species not sharing the same predator. Predispersal predation would thereby reduce fecundity of reproductives growing in high densities. Some studies have shown the existence of density-dependent mortality among juveniles in tropical forests (Augspurger 1983a, b, Hubbell et al. 1990, Clark and Clark 1984). Recently Wills et al. (1997) has shown that populations of a majority of the commoner tree species of a rain forest in Panama manifest density-dependent mortality. A classic individual case is that of mahogany and the shoot borer *Hypsiphylla,* which reduces fitness of juveniles, both of which are light demanders. Host specificity is likely more prevalent among pathogens than among predators. On theoretical grounds predation, and possible pathogenicity, will oscillate in intensity owing to the lag in the buildup of predators and parasites following increase of the predator or pathogen (Lotka 1925, Volterra 1926). Wills et al. (1997) obtained strong results once lag-time was taken into account. Further, indigenous host-specific pathogens or predators will be in equilibrium with their hosts or they would have become extinct. Their impact is therefore likely to be less frequently lethal and more difficult to detect. The rapidity with which more or less host specific, lethal, exotic insects and diseases, often insect borne, have spread this century to quite remote populations of chestnut (*Castanea americana*), elm (*Ulmus* spp.), hemlock (*Tsuga canadensis*), and Florida dogwood (*Cornus florida*) in North America gives indication of what indigenous, nonlethal species may achieve in a more subtle way.

A low to intermediate level of seed or juvenile predation, sustained by a browser population limited by top carnivores, may also lead to high species diversity by reducing competition (Terborgh, 1988), but this is not a model that can be simulated in man-made systems!

Mixed-species plantations are seldom so rich that between-plant distances far exceed their crown radii, so that the number of species in the mixture may indeed affect rates of infestation.

The density dependence of pathogens and also herbivores has particular relevance to the current search for multipurpose species, mostly trees with pioneer growth characteristics, often promoted to replace mixed species plantations which may take longer to harvest and are less productive. These species, which are often nitrogen-fixers, are generally principal sources of fodder and fuel, and are managed by coppicing. Both monoculture, alley planting with herbaceous crops, and the permanence of their root stocks make them particularly vulnerable to pests and diseases. This has already been demonstrated in *Leucaena leucocephala,* which has been devastated in the western Pacific by a leaf hopper (Showler 1995). Worwald (1992), in a thoughtful review concluded that the level of mixtures practical in commercial plantations would rarely achieve significant reduction of pest incidence, which would better be controlled by introduction of biological enemies. He nevertheless considered that mixed species plantations have a role in low-input low-output systems such as village forestry and agroforestry.

The diversity of useful tree and shrub species available in arid regions is limited, and the great majority are multipurpose (von Maydell 1983, Palmberg 1986). Pest incidence is less owing to the variability in annual rainfall and growing season, and its influence on predator and pathogen carrying capacity. But it is dangerous to extrapolate such experience to more equable climates.

### 4.4.2 Enhancing Sustainability

Symbiotic microorganisms may enhance the plant's capacity to gain anions. Two kinds of organism are involved.

Mycorrhizae are fungal associates of the roots of most land plants that enhance uptake of phosphates from solution in the soil, either by greatly extending, with their hyphae, the absorptive surface or by emitting secretions that release phosphate into solution from occlusion on sesquioxide molecules (Harley and Smith 1983). Mycorrhizae appear to be almost ubiquitous on the fine roots of tropical trees (Janos 1983, 1987). There is some evidence that mycorrhizae enhance the uptake of nitrogen, calcium, and water. The fungus in turn takes up soluble carbohydrates from the host. Mycorrhizae are most developed and most effective in soil in which exchangeable phosphorus concentrations are low. The capacity of mycorrhizae to take up phosphorus is generally far greater than the rate of diffusion of new phosphate from the substrate and atmosphere, so harvesting of products rich in phosphorus leads to depletion.

Among the several forms of mycorrhizae, ectotrophic mycorrhizae are the most host specific, and many gymnospermous and other temperate trees do not thrive in the absence of certain species. Ectotrophic mycorrhizae are relatively rare in tropical forests, where they are mostly known to occur on trees in the Pinaceae, Dipterocarpaceae, and Leguminosae subfamily Caesalpinoideae, all groups with a tendency to monospecific or family dominance in the forest canopy (Newbery et al. 1997). Although they are recognized to be important in timber plantations, their role in agroforestry remains to be investigated. A number of microorganisms fix atmospheric nitrogen in association with higher plants (Begon et al. 1996). They include Azotobacteriaceae, which fix nitrogen aerobically and are found on leaves and roots; the Rhizobiaceae bacteria which are confined to Leguminosae, where they form nodules on roots and fix nitrogen anaerobically; Enterobacteriaceae which are mainly found in intestinal floras but also rarely on leaves and roots; the Actinomycete

*Frankia* which fixes nitrogen in the root nodules on various plants other than legumes including *Casuarina,* Myricaceae, *Alnus, Ceonothus,* and certain genera in Rosaceae, Rhamnaceae, Ulmaceae, and Ericaceae; and certain blue-green algae which form mutualisms on the leaves and roots of a variety of plants including the superficial roots of some palms, and the leaves of some arboreal Rubiaceae, the herb *Gunnera,* and a variety of cryptogams. The best studied are *Rhizobium* because of their agricultural importance, and *Frankia* (see Bryan, Chapter 3).

From an ecological perspective, in contrast to phosphorus, nitrogen is an element in limitless supply in the atmosphere yet often severely limited in its availability to higher plants. Plants that are able to gain access to atmospheric nitrogen through these mutualisms are therefore initially at a competitive advantage, and may sustain high productivity on soils relatively low in nitrogen. They will eventually enrich the soil around them with their leaf litter which is rich in nitrogen, thus increasing the productivity of neighboring plants which, in certain cases, may actually outcompete them (Begon et al. 1986). Also, whereas mycorrhizae flourish best in acid low-nutrient soils, particularly in the surface organic horizon where present, nitrogen fixation is most effective on relatively fertile soils. Many leguminous trees of the humid tropics lack root nodules.

Farmers appear to have known of the benefits of nitrogen-fixing plants in mixed crop systems since ancient times. The upland farmers of New Guinea, for example, traditionally interplant their herbaceous crops with *Casuarina cunninghamiana,* which also serves as a fuelwood. The use of clover to improve pasture is an ancient practice in northern Europe. The value of multipurpose leguminous trees in traditional agroforestry systems has long been recognized (Von Maydell 1983, Palmberg 1986) (see Bryan, Chapter 3). They have also been widely used in the tropics in regions of fertile soils such as volcanic western Java, parts of Central America and the Andes; but here the risk of epidemic pest infestation is substantial.

## 4.5 CONCLUSIONS

Although the interplanting of nitrogen-fixing plants and the enriched fallow systems used by some cultures in the Amazon (e.g., Posey 1984) are impressive examples, it is rare to find farmers nowadays who choose their species mixture and mixed crop planting design with ecological considerations in mind. Even in the complex agricultural traditions of South Asia, the rather detailed customary protocols for planting home gardens appear to be based on astrological and religious, rather than ecological considerations (Chandrakanth et al. 1990). Currently, the values of commodity crops, often further enhanced by subsidies, so far outweigh those of subsistence crops that mixed cropping systems are becoming simplified in favor of commodity crops whatever the ecological cost. The case of the demise of shade trees in coffee plantations, notwithstanding the consequent increase in soil erosion, again serves as an example. To the farmer, the short-term gain outweighs the long-term risk. In the poorest communities, where population densities of subsistence farmers are close to exceeding the capacity of the land to produce, farmers have no alternative but to increasingly intensify their cropping systems, which may often still involve mixed species plantation, notwithstanding the increasing labor and skill required, and the increasing risk of breakdown through soil deterioration, pestilence, or resource and yield decline involved. Alternative employment opportunities, outside the agricultural sector, are increasingly the only safe solution to rural poverty.

This is not to deny that the ecologist (and the plant breeder and, in the short run, a revival of intercontinental exchange of germplasm) has much to contribute to increasing both productivity and sustainability of mixed plantations. Rather, such innovations must first be found to be economically attractive; and they will need increasing skill on the part of the farmer.

## 4.6 REFERENCES


Ashton, P. M. S. Seedling growth of co-occurring *Shorea* species in the simulated light environments of a rain forest. *For. Ecol. and Manage.* 72(1995): 1–12.

Ashton, P. M. S. and G. P. Berlyn. Leaf adaptations of some *Shorea* species to sun and shade. *New Phytologist* 121(1992): 587–596.

Ashton, P. M. S., C. V. S. Gunatilleke, and I. A. U. N. Gunatilleke, Seedling survival and growth of four *Shorea* species in a Sri Lankan rain forest. *J. of Trop. Ecol.* 11(1995): 263–279.

Ashton, P. M. S., S. Gamage, I. A. U. N. Gunatilleke, and C. V. S. Gunatilleke. Restoration of a Sri Lankan rain forest: using Caribbean pine (*Pinus caribaea*) as a nurse for establishing late successional tree species. *J. of Appl. Ecol.* 34(1997): 915–925.

Ashton, P. M. S., S. Gamage, I. A. U. N. Gunatilleke, and C. V. S. Gunatilleke. Using Caribbean pine to establish mixed plantations: testing effects of pine canopy removal on plantings of rain forest tree species. *For. Ecol. and Manage.* 106(1998): 211–222.

Ashton, P. S. Crown characteristics of tropical trees, in: *Tropical Trees as Living Systems*. P. B. Tomlinson and M. H. Zimmerman, eds. (Cambridge: Cambridge University Press, 1978), 591–615. Cambridge, UK.

Ashton, P. S. The structure and dynamics of tropical rain forest in relation to tree species richness, in: *The Ecology and Silviculture of Mixed Species Forests*. M. J. Kelty, B. C. Larson and C. D. Oliver, eds. Dordrecht, The Netherlands: Kluwer Academic Press, 1992a.

Ashton, P. S. Species richness in plant communities, in: *Conservation Biology: The Theory and Practice of Nature Conservation and Management*. P. L. Fiedler and S. K. Jain, eds. (New York: Chapman and Hall, 1992b), 3–22.

Ashton, P. S. Niche specificity among tropical trees: a question of scales, in: *The Dynamics of Tropical Communities*. D. M. Newbery, N. Brown, and H. H. T. Prins, eds. (Oxford: Blackwell, 1998), 491–514.

Ashton, P. S. and P. Hall. Comparisons of structure among mixed dipterocarp forests of Northwestern Borneo. *J. of Ecol.* 80(1992): 459–481.

Augspurger, C. K. Seed dispersal of the tropical tree, *Platypodium elegans*, and the escape of its seedlings from fungal infection. *J. of Ecol.* 71(1983a): 759–771.

Augspurger, C. K. Offspring recruitment around tropical trees: changes in cohort distance with time. *Oikos* 20(1983b): 189–196.

Begon, M., P. R. Townsend, and J. L. Harper. *Ecology: Individuals, Populations, and Communities*, 3rd ed. Sunderland, MA: Sinauer, 1996.

Boonkird, S. A., E. C. M. Fernandes, and P. K. R. Nair. Forest villages: an agroforestry approach to rehabilitating forest land degraded by shifting cultivation in Thailand. *Agroforestry Syst.* 2(1984): 87–102.

Brandis, D. *Suggestions Regarding Forest Administration in British Burma.* Calcutta: Superintendent of Government Printing, Empire Press, British India, 1876.

Chandrakanth, M. G., J. K. Gilles, V. Gowramma, and M. G. Nagaraja. Temple forests in India's forest development. *Agroforestry Syst.* 11(n.d.): 199–211.

Clark, D. A. and D. B. Clark. Spacing dynamics of a tropical rainforest tree: evaluation of the Janzen-Connell model. *Amer. Naturalist* 124(1984): 769–788.

Condit, R., S. P. Hubbell, J. V. LaFrankie, R. Sukumar, N. Manokaran, R. B. Foster, and P. S. Ashton. Species-area and species-individual relationships for tropical trees: a comparison of three 50-Ha plots. *J. of Ecol.* 84(1996): 549–562.

Conklin, H. C. *Hanúnoo Agriculture: A Report on an Integral System of Shifting Cultivation in the Philippines.* Rome: FAO, 1957.

Connell, J. H. Diversity in tropical rain forests and coral reefs. *Science* 199(1978): 1302–1310.

Davis, T. A. W. and P. W. Richards The vegetation of Moraballi Creek, British Guiana: an ecological study of a limited area of tropical rain forest, Parts I and II. *J. of Ecol.* 21(1933): 350–384; 22(1934): 106–155.

Ewel, J. J. Tropical succession: manifold routes to maturity. *Biotropica* 12, 2 suppl(1980): 2–7.

Ewel, J. J. Designing agricultural ecosystems for the humid tropics. *Annu. Rev. of Ecol. and Syst.* 17(1986): 245–271.

Fernandes, E. C. M., A. Oktingati, and J. Maghambe, The Chagga Homegardens: a multistoried agroforestry cropping system on Mt. Kilimanjaro (Northern Tanzania). *Agroforestry Systems* 2(1984): 73–86.

Gomez-Pompa, A. Tropical deforestation and Maya silviculture: an ecological paradox. *Tulane Stud. in Zool. and Bot.* 1(1987): 19–3.

Grubb, P. J. The maintenance of species—richness in plant communities: the importance of the regeneration niche. *Biol. Rev.* 52(1977): 107–145.

Harley, J. and S. E. Smith. *Mycorrhizal Symbiosis.* New York: Academic Press, 1983.

Hart, R. D. A natural ecosystem analog approach to the design of a successional crop system for tropical forest environments. *Biotropica* 12, 2 Supp.(1980): 73–82.

Horn, H. S. The adaptive geometry of trees. *Princeton Monogr. in Population Biol.* 3. Princeton, NJ: Princeton University Press, 1971.

Howe, H. F. and L. C. Westley. *Ecological Relationships of Plants and Animals.* Oxford, New York: Oxford University Press, 1988.

Hubbell, S. P. Tree dispersion, abundance, and diversity in a tropical dry forest. *Science* 203(1979): 1299–1309.

Hubbell, S. P. A unified theory of biogeography and relative species abundance and its application to tropical rain forests and coral reefs. *Coral Reefs* 16, Supp. S9–S21, 1997.

Hubbell, S. P., R. Condit, and R. J. Foster, Presence and absence of density dependence in a neotropical tree community. *Philos. Trans. of the R. Soc. of London* 330(1990): 269–281.

Huston, M. A general hypothesis of species diversity. *Amer. Naturalist* 113(1979): 81–101.

Jacob, V. J. and W. S. Alles. Kandyan Gardens of Sri Lanka. *Agroforestry Syst.* 5(1987): 123–137.

Janos, D. P. VA mycorrhizae in humid tropical ecosystems, in: *Ecophysiology of VA Mycorrhizal Plants* ed. G. R. Satir. (Boca Raton, FL: CRC Press, 1987), 107–134.

Janos, D. P. 1983. Tropical mycorrhizae, nutrient cycles and plant growth, in: *Tropical Rain*

Janos, D. P. 1983. Tropical mycorrhizae, nutrient cycles and plant growth, in: *Tropical Rain Forest: Ecology and Management,* eds. S. L. Sutton, T. C. Whitmore, and A. C. Chadwick. Blackwell, Oxford, USA.

Janzen, D. H. Herbivores and the number of tree species in tropical forests. *Amer. Naturalist* 104(1970): 501–528.

Lagemann, J. and J. Heuveldop. Characterization and evaluation of agroforestry systems: the case of Acosta Puriscal, Costa Rica. *Agroforestry Syst.* 1(1983): 101–115.

Lotka, A. J. *Elements of Physical Biology.* Baltimore: Williams and Wilkins, 1925.

Michon, G., F. Mary, and J. Bompard. Multistoried agroforestry garden system in West Sumatra, Indonesia. *Agroforestry Syst.* 4(1986): 315–338.

Michon, G., J. Bompard, P. Hecketsweiler, and L. Ducatillon. Tropical forest architectural analysis as applied to agroforests in the humid tropics: the example of traditional village agroforests in Java. *Agroforestry Syst.* 1(1983): 117–129.

Naeem, S., I. L. J. Thompson, S. P. Lawler, J. H. Lawton, J. H., and R. M. Woodfin. Declining biodiversity can alter performance of ecosystems. *Nature* 368(1994): 734–736.

Newbery, D. McC., I. J. Alexander, and J. A. Rother. Phosphorous dynamics in a lowland african rain forest: the influence of ectomycorrhizal trees. *Ecol. Monogr.* 67(1987): 367–409.

Oduol, P. A. The Shamba System: an indigenous system of food production from forest areas in Kenya. *Agroforestry Syst.* 4(1986): 365–373.

Oliver, C. D. and B. C. Larson. *Forest Stand Dynamics.* New York: John Wiley and Sons, 1996.

Palmberg, C. Selection and genetic improvement of indigenous and exotic multipurpose tree species for dry zones. *Agroforestry Syst.* 4(1986): 121–127.

Posey, D. A. Indigenous management of tropical forest ecosystems: the case of the Kayapo Indians of the Brazilian Amazon. *Agroforestry Syst.* 3(1984): 139–158.

Press, M. C., N. D. Brown, M. G. Barker, and S. W. Zipperlen. Photosynthetic response to light in tropical rain forest seedlings, in: *The Ecology of Tropical Tree Seedlings,* UNESCO-MAB Series 17, ed. M. D. Swaine (Carnforth, UK: Parthenon, 1996), 41–54.

Raintree, J. B. and K. Warner. Agroforestry pathways for the intensification of shifting cultivation. *Agroforestry Syst.* 4(1986): 39–54.

Ricklefs, R. E. *Ecology.* 3rd ed. New York: W. H. Freeman, 1990.

Showler, A. T. *Leucaena* psyllid, *Heterophylla cubana* (Homoptera; Psyllidae) in Asia. *Amer. Entomol.* 42(1995): 49–54.

Terborgh, J. The vertical component of plant species diversity in temperature and tropical forests. *Amer. Naturalist* 126(1985): 760–776.

Terborgh, J. The big things that run the world . . . a sequel to E. O. Wilson. *Conserv. Biol.* 2(1988): 402–403.

Tilman, D. Resource competition between planktonic algae: an experimental and theoretical approach. *Ecology* 58(1977): 338–348.

Tilman, D. *Resource Competition and Community Structure.* Princeton, NJ: Princeton University Press, 1982.

Toky, O. P., P. Kumar, and P. K. Khosla. Structure and function of traditional agroforestry systems in the Western Himalaya. I. Biomass and Productivity. *Agroforestry Syst.* 9(1989): 47–70.

Torquebiau, F. Manmade dipterocarp forest in Sumatra. *Agroforestry Syst.* 2(1984): 103–127.

Volterra, V. Variations and fluctuations of the number of individuals in animal species living together. Reprinted in R. N. Chapman, *Animal Ecology,* 1931. McGraw-Hill, New York: 1926.

von Maydell, H. J. Trees and shrubs of the Sahel. Deutsche Gesellschaft für Technische Zusammenarbeit (GTZ). Germany: Eschborn, 1983.

Waring, R. and W. H. Schlesinger. *Ecology and Management of Forested Ecosystems.* (New York: Academic Press, 1986), 340.

Watt, A. S., Pattern and process in the plant community *J. of Ecol.* 35(1947): 1–22.

Whitmore, T. C. and N. D. Brown. Dipterocarp seedling growth in rain forest canopy gaps during six and a half years. *Philos. Trans. of the R. Soc. of London* 351(1996): 1195–1203.

Wills, C., R. B. Condit, and S. P. Hubbell. Strong density and diversity-related effects help to maintain tree species diversity in a neotropical forest. *Proceedings of the U.S. National Academy of Sciences* 94(1997): 1252–1257.

Worwald, T. J. Mixed and pure forest plantations in the tropics and subtropics. *FAO Forestry Paper* 103, 1992.

# 5 Some Applications of Multitrophic Level Ecological Theory to Agroforestry Systems

*Andrew P. Beckerman*[1]

**TABLE OF CONTENTS**



[1] University of Stirling, Department of Biological Sciences, Stirling FK9 4LA, Scotland, UK. a.p.beckerman@stir.ac.uk

0-8493-2206-5/00/$0.00+$.50

## 5.1 INTRODUCTION

Agroforestry systems are structurally complex environments in which a variety of plant-animal interactions take place. These interactions range from competition among plants for light, space, and soil nutrients, to herbivory or parasitism of plants, to predation or parasitism of herbivores. These interactions, and their physiological and behavioral underpinnings, are important because they determine the biological, and therefore economic, output of agroforestry systems.

The primary purpose of this chapter is to illustrate the type of insight that ecological theory can provide to understanding these interactions in agroforestry systems. This chapter introduces basic ecological theory underpinning two classes of interaction prevalent in agroforestry systems: livestock grazing in plantation forests and predator control of herbivore pests.

The theory underlying these interactions can be an integral part of management in agroforestry. The theory provides a predictive template on which to analyze management options and potential outcomes, prior to their implementation. Furthermore, the theory can provide qualitative and quantitative projections of biological detail, such as grazing herbivore stocking levels or the magnitude of herbivore induced plant damage, that can be translated directly in economic and social costs and benefits.

Livestock grazing in plantation forests and the control of herbivore pests by predators both share a common theme regarding the effects of herbivores on the plant resources that dominate agroforestry systems: misplaced emphasis on the animals consuming the plants rather than on the plants. Emphasis in the realm of livestock integration into plantation forests has traditionally been placed on the number of animal units a particular range classification can support (Lewis et al. 1983, 1984, Somarriba and Lega 1991, Anderson et al. 1988). Yet the heart of agroforestry is the investment in the plant resources and it is therefore important to recognize that the dynamics of the plant resources, relative to each other (competition) and to herbivores (herbivory), are essential to the productivity and management of an agroforestry system (Tajuddin 1986, Jaindl and Sharrow 1988, Pearson et al. 1990). (Note that this emphasis and criticism are inappropriate when livestock are the primary or solitary emphasis of the agricultural system.)

The theoretical presentation on grazing introduces a model framework that highlights the challenges and opportunities that multiple plant resources interacting with herbivores represent to an agroforester. The theory summarizes graphically the relationships between herbivore consumption rates, herbivore densities, plant growth rates, plant productivity, and plant density. In doing so, it defines how these relationships potentially define the outcome of competition and herbivory in this type of agroforestry system.

In the case of herbivore pest control, research has also tended to concentrate on the herbivore population level. Perhaps stemming from two classical hypotheses about natural enemies and resource concentration effects (Root 1973, Altieri 1996), much of the traditional treatment of biological control has centered on two trophic level, consumer-resource, predator-prey, or host-parasite theory (Beddington et al. 1978, Hassell 1978, Murdoch et al. 1984, 1985, Briggs 1993, Hochberg 1996). More recent efforts have included plant and or pest genetics (Kimberling and Price 1996) and the concepts of landscape processes including immigration and emigration (Kareiva 1986). However, the bulk of this research is still limited to the dynamics and patterns associated with two trophic levels (but see Price et al. 1980, Duffy et al. 1986, Kennedy 1986, Price and Martinsen 1994, Gutierrez et al. 1990).

We are using this opportunity to introduce the use of a three trophic-level food chain theory as a means of understanding certain types of interactions associated with biological control of pests. The focus will be on how this type of ecological theory transfers focus from predator-prey, host-parasite interactions and the reduction of pest densities, to considerations of plant dynamics and productivity — the principle output of all agroecosystems. The discussion of the theory will be complemented by examples, when available, of where the theory has or would be most appropriate. We will also attempt to integrate, on a conceptual level, an evaluation of the theory into current agroforestry practices.

Again, the primary purpose of this chapter is to describe the type of information ecological theory can contribute to management and decision making in agroforestry systems. It is by no means a complete overview of all of the theory that could be used in addressing interactions and dynamics of agroforestry systems. Instead this chapter offers a sampling of multitrophic level ecological theory and an introduction to models as a means to consider the assumptions and wide range of predictions that theory can contribute to an understanding of how agroforestry systems operate.

## 5.2 AN APPLICATION OF CONSUMER-RESOURCE AND COMPETITION THEORY TO AGROFORESTRY GRAZING SYSTEMS

Grazing herbivores are a common element in many agroforestry systems. During the development of a forest/agricultural product, such as tree species for timber or non-timber products, alternative plant material such as understory grasses and herbaceous material can represent both a problem and an opportunity to an agroforester. Alternative plant material can be a nuisance in terms of competitive interactions with a forest product, particularly during the early stages of stand development and regeneration (Ramakrishnan 1992, Altieri 1996, Ashton et al. 1997). However, additional crops may also present the opportunity to introduce a grazing animal as a second source of income, a tool to manage competitive interactions among plants (Tajuddin 1983, Lewis et al. 1983, 1984, Dangerfield and Harwell 1990, Pearson et al. 1990, Ormazabal 1991, Somarriba and Lega 1991, Louda et al. 1990), or to reduce the potential of fire risks (Anderson et al. 1988). These opportunities, afforded by integrating livestock, can become more attractive as the forest or agricultural products mature and move out of the regeneration phase, particularly in natural (e.g., savannah woodlands, Ponderosa pine [*Pinus ponderosa*] glades) and cultivated/managed plantations where the understory vegetation is maintained as an integral portion of the ecosystem.

This section evaluates the response of a two-plant agroforestry system to the addition of a grazer. It is shown that even though the hypothetical system is ostensibly simple and the models represent some highly intuitive dynamics, they nonetheless offer valuable insight into the role that different types of herbivores can have in productive and unproductive systems; they provide a framework for analyzing grazing systems in agroforestry. Because of the wide array of assumptions about growth and consumption of plant resources that are made possible by real world growth and consumption patterns, the power of this section lies in the framework that is established for qualitatively and quantitatively analyzing the effects of a herbivore feeding on two competing plant resources.

### 5.2.1 Cattle Grazing in a Plantation

As noted above, grazing animals in agroforestry systems can present a problem and an opportunity. Grazing is a problem when the grazers preferentially feed on the primary forest agricultural crops. However, grazers may also represent an economic and biological opportunity if they preferentially consume plants that compete with certain growth phases (e.g., stand development stage; regeneration phase) of the valuable agricultural or forest crops and that alone may not have a significant economic value.

The following discussion is based on the biological scenario of two competing plants and a herbivore that feeds on one of them. For simplicity of presentation, we assume the following: 1) The main production crop is a tree species growing among a grass/herbaceous mixed groundstory; 2) the herbivore feeds primarily on the understory vegetation because the trees are out of reach, are protected by physical or chemical defenses (e.g., thick bark; high concentrations of tannins or phenols) or protected manually by external barriers; 3) the understory vegetation has a negative effect (competition) on tree productivity or quality (e.g., pine plantation in a grassland matrix; acacia woodlands). The method of analysis that follows is not restricted to these assumptions. The approach can be repeated for many other scenarios and variations of these assumptions. In particular, the details of plant competition are minimized in this discussion (see Ashton this volume [Chapter 4]). This aspect of the model could be enhanced substantially to accommodate the wide range of empirical patterns and theory about plant competition (Harper 1977, Tilman 1982, Keddy 1989). However, the models that follow are insufficient to investigate the outcome of a herbivore feeding on both the major crop resource and the competitive groundstory plants as this involves reciprocal and indirect interactions not covered by the following theory (a substantial body of theory does exist on this topic [Holt 1977, Connell 1990, Louda et al. 1990]).

The specific goal of the following analysis is to demonstrate how different levels of herbivory alter the population density and growth rate of a plant species competing with a dominant forest product, as might be found in a plantation forest with herbaceous undergrowth. This is accomplished by asking the following two questions, under an array of different conditions: How effective is the resource at supporting the herbivore stock, and how effective is the herbivore stock at mediating competition between the two plant species?

Intuition would suggest that a herbivore, feeding only on a competitor to a major crop forest product, would reduce the competitive effect of this plant and simultaneously generate an income source via animal production. Presented is a series of graphical models, based on assumptions about the growth form of the consumed plant species, the productivity/growth rate of the consumed plant species, and the functional form of consumption by the herbivore that question the generality of this intuition (Noy-Meir 1975, 1978, Louda et al. 1990, Schmitz and Sinclair 1997). This analysis provides insight into the general principles of plant resistance to herbivory, the type of benefit — to the major forest/agricultural crop — that herbivory provides by reducing competitive interactions between the two plants, and to the economic importance of consumed plant resistance to herbivory.

Louda et al. (1990; see also Noy-Meir 1975, 1978, Schmitz and Sinclair 1997) provide a simple method of analyzing the dynamics of an ecological system in which a herbivore may selectively mediate competitive interactions between two plants. Two considerations are important for developing these models. The first is the functional form of plant growth under differing levels of productivity (Figure 5.1). An important component of the growth form is how much and where density dependent mortality occurs. The magnitude of density dependence, and at what densities the effect occurs in plant growth, is one measure of how resistant to herbivory a plant may be, and how well the consumed plant can maintain a significant positive growth rate in the face of herbivory (resilience). The density range in which the intraspecific effect is strongest is the region where resistance to herbivory is also strongest. In this range of plant density, herbivory will most likely just replace mortality from intraspecific density responses (compensatory mortality; see Louda et al. 1990).

Many forms of plant growth, and the density dependence implicit in it, have been used in models (Noy-Meir 1978). Traditionally, these forms are the product of assumptions about physiology, canopy morphology, and the relative importance of space, light, and nutrient limitation (Noy-Meir 1978) and have rarely been tested explicitly. In fact, density dependence in plants has a fairly long tradition of demonstration without indication of the form of the relationship (Fowler 1997).

The models that follow employ three different growth forms. Noy-Meir (1978) justifies the use of Figures 5.1a and 5.1b in models of crop production and pastures with solid citation of empirical and theoretical studies of plant morphology and physiology. Louda et al. (1990) present Figure 5.1c and offer little justification for it. The

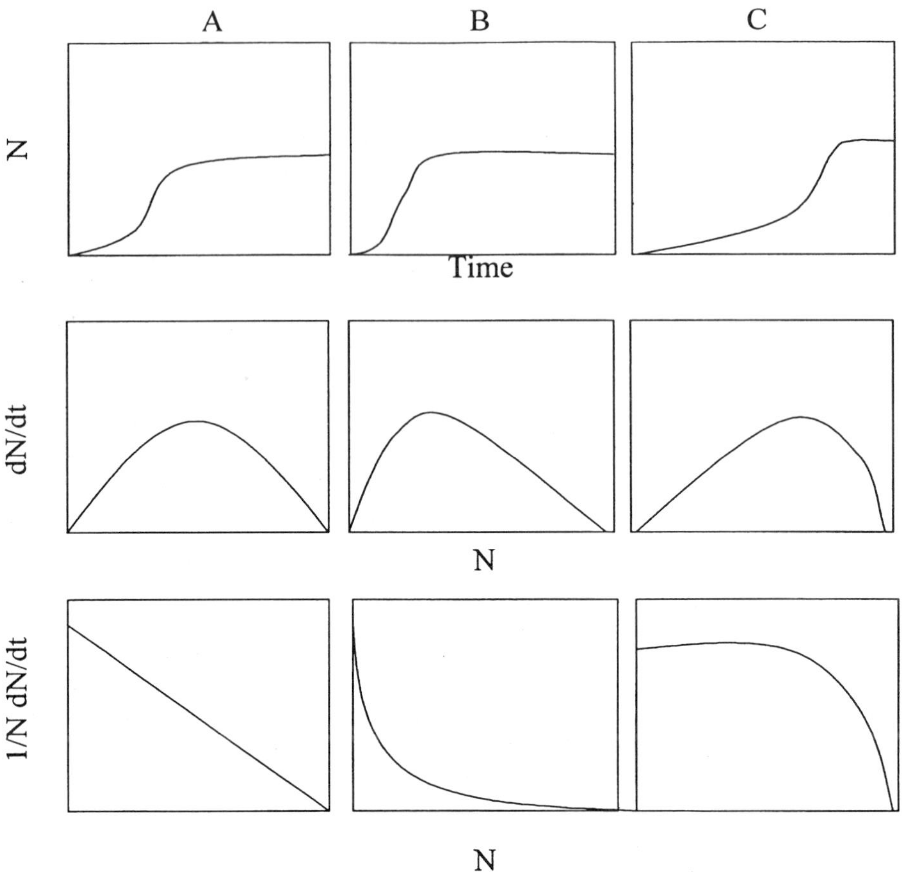


**FIGURE 5.1** The time series, rate, and per capita rate models of population growth used in the analysis of herbivore mediated competitive interactions. Each column refers to a particular growth model: a) linear density dependence; b) nonlinear density dependence, focused at low densities; c) nonlinear density dependence, focused at high densities.

significance of this third form of growth (Figure 5.1c) is that the per capita (per individual) response to density is highest at high plant densities. This phenomenon can generate the strongest resistance to herbivory (see below). However, in the extensive literature reporting on model grazing systems (Noy-Meir 1975, 1978, 1981), there is no use of this growth function.

A frequently used description of plant growth has the form of Figure 5.1b. Noy-Meir (1978) notes that light-limited plant assemblages often behave according to this growth form. Most models resulting in this form of growth generate a function based on light capture efficiency in relation to canopy structure. Figure 5.1a is often used for systems where relative growth rates or biomass is linearly related to the amount of space available for growth, but it has also been used to describe nutrient-limited systems as well (Noy-Meir 1978). Tilman class models (1982, see also Grace and Tilman 1990) describing saturating growth based on nutrient limitation behave similarly to Figure 5.1b at the per capita level. The log (plant characteristic) vs. time plots found in Harper (1977) display a range of forms corresponding to the breadth of models presented here. The steepness of the initial slope in Harper's (1977) graphs correspond to the skewness.

The second consideration is the function describing consumption (Figure 5.2). Three major shapes are commonly discussed in ecological theory: linear (constant consumption as "prey" density increases; Figure 5.2a), inverse density dependent consumption (rate increases at a decreasing rate as prey density increases; Figure 5.2b), and density dependent consumption (rate increases at an increasing rate as prey density increases; Figure 5.2c). There is currently a great deal of debate about the prevalence and implications of the different forms of consumption, how they might arise (Schmitz 1995, Abrams 1997), and the importance of factors such as plant chemistry and structure that are either independent of or tied to plant density (Price et al. 1980, Feeny 1992, Rosenthal and Berenbaum 1992, Futuyma and Keese 1992, Karban et al. 1997). The graphical framework presented here is flexible enough to accommodate a wide array of factors affecting plant growth and herbivore consumption as long as these factors can be integrated into a function of resource consumption.

Following Louda et al. (1990), the models provide detailed predictions about the response of competing resources to herbivory, given a set of assumptions about

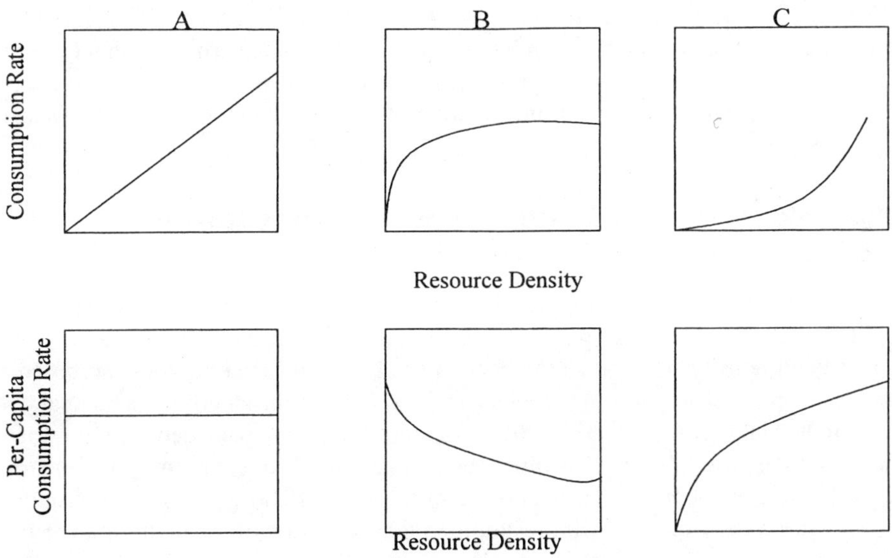


**FIGURE 5.2** Three models of consumption rate and per capita consumption rate vs. resource density: a) linear function response, density independent per capita consumption; b) saturating functional response, inversely density dependent per capita consumption; c) increasing functional response, density dependent per capita consumption.

productivity, growth rates, and consumption rates of the resource consumed by the herbivore. By assuming that reductions in growth rates and biomass of a competing species will have positive effects on the other species, this information is sufficient to predict the competitive repercussions of specialized/managed herbivory in agroforestry situations (see Noy-Meir 1978 for a more mathematical treatment).

The analyses begin with the assumption of logistic plant growth representing linear density dependence. The effects of linear, inverse density dependent, and density dependent consumption by the herbivore on this form of plant growth is reviewed. This protocol is repeated for two additional growth forms representing nonlinear density dependence.

The graphs representing the analyses (Figures 5.3–5.5) have the following components. Following Louda et al. (1990), Two growth curves and two consumption curves are repeatedly shown. The growth curves represent the per capita growth rate of the competitor (Figure 5.1) under high and low productivity conditions (variable r's/intrinsic growth rates in the Louda et al. 1990 presentation) and the consumption curves (Figure 5.2) represent low and high herbivore stocking density (Noy-Meir 1975, 1978, Louda et al. 1990, Schmitz and Sinclair 1997).

Each graph (Figures 5.3–5.5) provides four pieces of information. They each identify a) the magnitude of density/biomass reduction caused by herbivory; b) the subsequent proportional benefit this density/biomass reduction generates by reducing competitive interactions among the plants; c) the risk of herbivore-generated resource extinction (an index of the sustainability of the herbivore component of the agroforestry system); and d) the resistance and resilience of the plants to herbivory.

These factors are important because they represent information necessary to assess the tradeoffs associated with income derived from the herbivore, that derived from reduced competition, and the threat of the herbivore destroying its own resource base. Each graph summarizes this information. We highlight these factors and discuss some of their implications next.

### 5.2.2 Model 1: Linear Density Dependent Plant Growth in the Dominant Competitor

This analysis corresponds to Figures 5.3a–c. Note that all of the graphs describe per capita effects. Figure 5.3a represents linear density dependence (logistic plant growth) and density independent consumption. Under these conditions, herbivores have their greatest impact on the plant resource at low productivity. The shallow slope of the plant growth function allows for large changes in plant density with small changes in herbivore density. The linear per capita growth and consumption function suggest that high stocking densities of herbivores could allow per capita consumption to exceed per capita growth, resulting in the extinction of the plants supporting herbivores under low productivity conditions.

Not only can herbivore stocking densities result in plant extinction, but there is only moderate resistance to herbivory, the resistance decreasing with decreasing productivity. Resistance here is defined by the slope of the production/growth function. The shallower the slope of this function, the easier it is for increases in herbivore den-

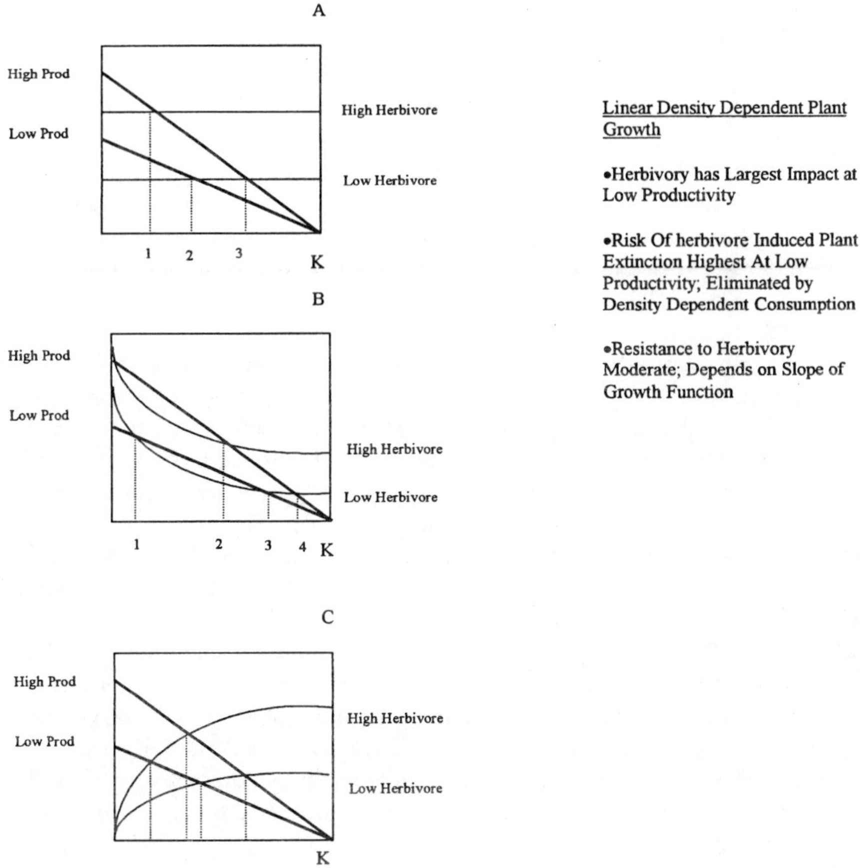


**FIGURE 5.3** The effects of productivity (low and high) and herbivore density (low and high) on a dominant plant competitor that grows logistically (linear density dependence): a) linear function response; points 1, 2, and 3 represent equilibria between growth and consumption and K is the population density achieved when there is no herbivory; b) as in (a) with a inversely density dependent functional response; c) as (a) and (b) with density dependent consumption.

sity to reduce the population size of its plant resources and the harder (slower; lower growth rate) it is for the plant resources to recover from herbivory. The slope of the growth curve represents density dependent reduction in growth from some form of intraspecific competition. The most resistant plants, in terms of growth rates and productivity, are those in which intraspecific competition limits growth faster and more effectively than a reduction in growth caused by an outside factor such as herbivory. Per capita growth rates tied to density in this manner can make herbivory compensatory or simply a replacement of intraspecific plant effects.

To place this in some context, recall that nutrient/resource limitation has been modeled as either Figure 5.1a or 5.1b, space limitation as Figure 5.1a, and light limitation as Figure 5.1b. This suggests that resistance to herbivory, via the plant growth form, is most likely in space and some resource-limited conditions. But in general, the models used to describe growth based on light and resources tend to indicate low levels of resistance or resilience, as found in Figure 5.1b (see Figure 5.4).

Overall, there is very little suggestion from theoretical or empirical work that strong resistance to herbivory (Figure 5.1c, see Figure 5.5) is a result of plant growth form. The form of density dependence is not the only factor that can generate a form of resistance in the plants. Structural and chemical defense mechanisms also represent a form of resistance, which may be independent of, or dependent on, plant and herbivore density (Price et al. 1980, Rosenthal and Berenbaum 1992, Feeny 1992, Futuyma and Keese 1992, Weltzin et al. 1997) and may alter the functional form of plant growth models.

Figure 5.3b retains the linear density dependent growth and adds inversely density dependent consumption. The impact of herbivory is similar to the conditions in Figure 5.3a, with the greatest impact of herbivory, and thus the most benefit from reduced competition, occurring under conditions with low productivity. However, a new situation that arises from this new consumption form is the potential for multiple equilibria between herbivores and plants. Point 1 in Figure 5.3b is an unstable equilibrium, while Point 3 is stable. If the herbivore is added to this system when plant density is in the vicinity of Point 1, but greater than that at Point 1, the plant population will continue to grow, even with the presence of the herbivore (in the region between Points 1 and 3, growth exceeds consumption at all times).

Thus, the potential to limit competition by maintaining the competitor plant density at these low levels will actually result in competitor plant growth and potentially increased competition with the forest product, at least until Point 3, the stable equilibrium between herbivores and the dominant competitor. Alternatively, if the plant density is slightly less than Point 1, any introduction of herbivores will cause the extinction of this plant resource.

The presence of multiple states simply highlights the fact that 1) herbivores will not invariably reduce plant densities, 2) herbivory may not always facilitate reduced competition, and 3) plants and herbivores may not always coexist at the densities they are stocked. Schmitz and Sinclair (1997) review the importance of these types of assumptions in the context of white tail deer (*Odocoileus virginianus*) and North American forests.

Figure 5.3c again maintains the linear density dependent growth assumption but adds density dependent consumption. The general results are similar to those resulting from the assumptions in Figures 5.3a and 5.3b. However, the shape of the consumption function in this scenario results in a relatively larger impact of herbivory as herbivore density increases. Additionally, because per-capita density dependent consumption is defined from the origin of the graph, it becomes impossible for the herbivore to cause the extinction of the resource that supports it.

In summary, the impact of herbivory on a plant that grows logistically (with linear density dependence) is predicted to be largest under low productivity conditions and when the herbivore consumes the plant material in a density dependent manner. These two conditions define the circumstances, in the context of these growth and consumption assumptions, in which herbivores feeding on a competitor to a forest or agricultural crop could have the greatest positive impact on the production of this crop. In general, linear density dependent growth generates moderate resistance to and resilience from herbivory, based on the range of slope possible in the growth function. Density independent and inversely density dependent forms of consumption can cause the extinction of the resource supporting the herbivore.

### 5.2.3 Model 2: Nonlinear Density Dependent Growth with Density Dependence Highest at Low Density

This analysis corresponds to Figures 5.4a–c. In all three panels, density dependence in plant growth is non-linear and strongest at low densities. This corresponds to a logistic model skewed to the right (Figure 5.1a).

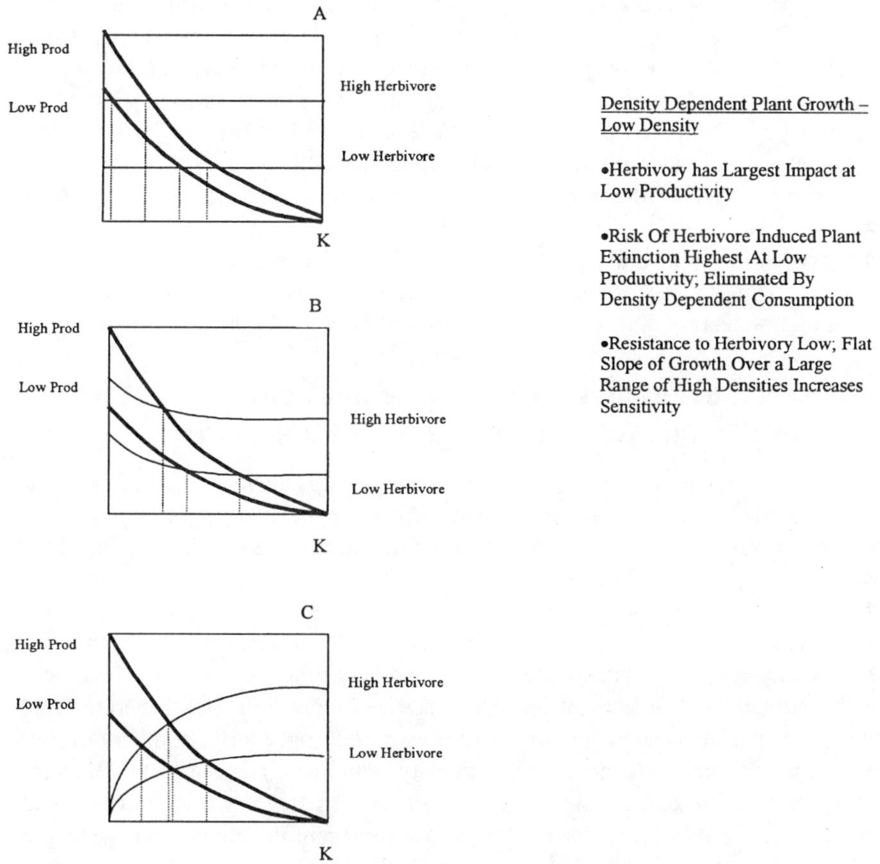


**FIGURE 5.4** The effects of productivity (low and high) and herbivore density (low and high) on a dominant plant competitor that grows with density dependence concentrated at low densities. Dashed lines correspond to equilibria and mark the magnitude of density reductions from K: a) density independent consumption; b) inversely density dependent consumption; c) density dependent consumption.

As above, each graph holds the form of growth constant and varies the form of consumption. This class of scenarios produces similar, general predictions as those produced by linear density dependent growth: the impact of herbivory, and thus the potential for reducing competition between forage plants and the primary forest/agricultural plants, is predicted to be highest at low productivity. However, there are some interesting differences.

In general, the shallowness of the growth functions at moderate to high plant density makes the reduction in competing plant biomass, caused by herbivory, more severe on average than with linear density dependent growth. Inversely density dependent consumption facilitates the largest potential impact of herbivory on plant growth because the shape of these curves causes growth-consumption equilibria to occur at lower plant densities relative to the other consumption functions for a given herbivore density. Because herbivory can have such a large effect on the density of the competitor plant, this scenario predicts the most lucrative return on investing in competitor control for a given stocking density of herbivore. However, the ability of the plant resource to recover and continue to support the herbivores, say in intervening seasons or within a crop rotation, could be hampered by the low growth rates of the competitor plant that ensue over much of the larger densities.

The possibility of the herbivore causing the extinction of the plant resource under this scenario is lowest when the herbivore feeds in a density dependent manner (the risk is eliminated again by the nature of the consumption function), followed by inversely density dependent and then by density independent consumption. A final interesting note is that multiple equilibria are unlikely under this set of assumptions.

### 5.2.4 Model 3: Nonlinear Density Dependent Growth with Density Dependence Highest at High Density

This analysis corresponds to Figures 5.5a–c. It is perhaps the most theoretically interesting scenario, though its applicability is hampered by limited empirical justification of this plant growth form. Here, the density dependence in growth is assumed to be highest at high plant densities. Because of this, the changes in per capita plant growth rate caused by herbivory at high plant densities is compensatory to the changes in per-capita growth generated by the intraspecific density dependent response of the plants. In effect, changes in growth caused by herbivory simply replace the changes that already occur at these densities from intraspecific competition. This generates a form of intrinsic resistance/resilience to herbivory: Increasingly large numbers of herbivores cause very small changes in plant density or biomass (resistance) with increasingly large jumps in plant growth rates (resilience). In essence, this model scenario, as well as Model 1 with high productivity, can represent the concept of grazing as a stimulus to forage production (McNaughton 1985, McNaughton et al. 1988, Pandey and Singh 1992). As a result of this phenomenon, this scenario is the most lucrative from the perspective of herbivore stock.

As in the previous two scenarios, lower productivity leads to a larger impact of herbivory and thus a larger set of potential benefits via reduced competition. However, under all three forms of consumption with this form of growth, the stock-

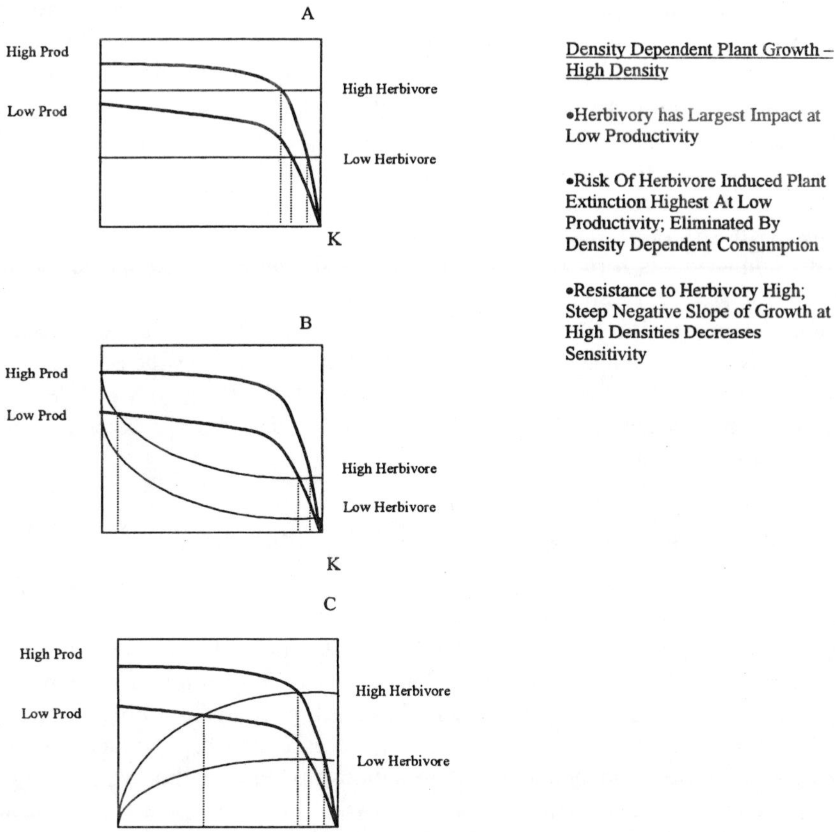


**FIGURE 5.5** The effects of productivity (low and high) and herbivore density (low and high) on a dominant plant competitor that grows with density dependence concentrated at high densities. Dashed lines correspond to equilibria and mark the magnitude of density reductions from K: a) density independent consumption; b) inversely density dependent consumption; c) density dependent consumption.

ing density of the herbivores must be much larger relative to either of the other two scenarios in order to realize a similar competitive response. This is most apparent when this form of growth is paired with inversely density dependent consumption (Figure 5.5b). This particular combination of growth and consumption represents the least effective scenario (out of all nine discussed here) at gaining some benefit from reduced competition. However, it is worth noting that it might be the most significant if the resource consumed by the herbivore was the major forest crop rather than a competitor (i.e., consumption of seedlings).

The maximum potential for herbivore impact under this form of growth occurs when the herbivore consumes the resource in a density dependent manner where high herbivore stocking densities can cause the growth and consumption functions to intersect, in a stable equilibrium, along the flat portion of the growth function (Figure 5.5c.) As described above, the impact of increased herbivory is greatest when the slope of the growth function is shallow; only density dependent consumption can intersect the flat portion of this growth function and generate this effect. With this form of growth, only density independent consumption poses a significant possibility for the herbivore to eat its resource to extinction. Additionally, multiple equilibria are again possible when consumption is inversely density dependent and the implications discussed for Model 1 also apply here.

In summary, this form of growth represents the highest growth rate based on resistance to herbivory and the lowest risk of the herbivore eating its resource to extinction. However, for these reasons, Model 3 also represents the lowest levels of benefit derived from reduced competition. This is not to diminish the potential, however, for benefit derived from grazing. This form of plant growth can tolerate the highest stocking densities with the lowest impact on the plants, a potential benefit under planting conditions.

### 5.2.5 Real Examples and Some Practical Suggestions

A perusal of the agroforestry literature over time reveals more than a handful of examples of livestock use in plantation agroforestry systems. The documentation of livestock use in agroforestry systems occurs most commonly in forests around the world containing the genus *Pinus,* Douglas fir (*Pseudostuga menziesii*), and rubber trees (*Hevea brasiliensis*). Much of this research is limited to describing associations between forest systems and grazing livestock or assessing the productivity of the grazing system itself, independent of the relationship between the livestock and the forest products. The most common assessment asks whether the livestock are healthy and whether there is any damage done to the trees.

This final assessment type, of damage, is interesting given the well-established understanding that understory herbaceous plants can have a detrimental effect on tree growth via competition (Steward et al. 1984). There are a very small number of replicated and controlled experimental studies of the actual effect of livestock grazing on tree species growth and form, either via direct damage or benefits by reduced competition. Three are presented below.

Tajuddin (1986), in a study involving sheep and poultry in Malaysian rubber plantations, found that sheep had a positive influence on the growth of rubber trees over a period of two years. This was accomplished through a combination sheep grazing and the addition of manure during a six- to eight-week period each year. The sheep grazing apparently reduced the impact of 500–600 kg/ha of grasses, forbs, and ferns on the rubber trees.

In a study involving *Pinus elliottii, P. taeda, Trifolium subterraneum,* Pearson et al. (1990) found that cattle had negative or benign effects on these plants. The height and survivorship of both *Pinus* species was less in grazed sites than in lightly grazed (low stocking density and electric wire over seedlings) or ungrazed sites. Lightly grazed and ungrazed sites were not different. Pearson et al. (1990) note that this is not in line with work on Monterey pine and ryegrass in New Zealand where cattle appear to have a net positive effect by reducing competition between the understory and the trees.

Finally, Jaindl and Sharrow (1988) present the results of a long-term study of the effect of sheep grazing on a Douglas fir (*Pseudotsuga canadensis*)-white oak (*Quesrcus alba*) forest in Oregon, U.S. Jaindl and Sharrow examined three site treatments (control, 50% thinning with a 70% reduction in canopy closure, and clear-cut), all followed by douglas fir plantings, crossed by two levels of grazing (control, grazing). The results indicate that site treatment was by far the dominant force in the response of this system over time. The effect of grazing was positive on growth during the first two to five years of the study, but dissipated after that.

Jaindl and Sharrow (1988) also provide a cursory review of the response of Ponderosa pine, western white pine, larch, and Douglas fir to grazing and the reduction of competitor biomass. These studies all indicate a significant positive influence of sheep grazing on the growth of these species: 9–13% increase in Ponderosa pine, 44–56% in western white pine, 38–61% in larch, and 18–26% in Douglas fir.

The three examples above, and the small local studies they cite, indicate the potential for significant increases in growth of the tree species via reduced competition with the herbaceous or grassy understory. Notably, much of this impact occurs during the early stages of tree/stand development. Complementing this indication that the integration of livestock into forestry could be successful is the extensive research on herbicides to control weeds, limit fire potential, and reduce competition (Steward et al. 1984). The same places where herbicides are effective at increasing forest productivity by decreasing competitive interactions should be successful with livestock.

The differences between livestock and herbicides are worth noting, as they highlight the importance of assessing the costs and benefits of competitor control. The costs of the pesticide are in production, accrual, and application, while the benefits are in the increased productivity. Livestock, when not properly managed, can damage the primary tree species rather than helping them. Furthermore, the timing of the grazing rotation and the intensity of the grazing are very important and can limit the benefits. The benefits of grazing-based management, however, are not only in the increased productivity, but in the economic return that the livestock themselves provide. Thus, if the resources and knowledge about grazing and stocking are available (this is more likely than knowledge about herbicides in agroforestry communities), the benefits to this method could outweigh the benefits of the chemical path.

There is obviously a lot of potential to integrate livestock into forestry plantations. Whether it provides a benefit either alone as an alternative source of income or additionally as an enhancement to the tree products via reduced competition has not yet been firmly established. The bulk of the studies on livestock in forestry operations does indicate that livestock represent a financial opportunity in many systems and under a variety of conditions (Dangerfield and Harwell 1990). However, the research has rarely addressed the impact of the livestock on the primary tree products.

How and when increases in forest production can be realized as a result of livestock integration remains a large gap in forestry and agroforestry knowledge. The majority of the research on silvipastoral systems has not addressed this question of mechanism and generally ignores the overall question of whether livestock can benefit the tree species. The theory presented above, which represents a summary of longstanding theory of grazing systems and plant-herbivore interactions, is a framework on which to address this question. The nine scenarios presented above define the extremes of the tradeoff between significant returns from livestock and limited returns from reduced competition to high competitive returns with little direct return from the livestock.

Tests of these models, at a very general level, are not difficult and must at least follow the basic experimental design exemplified in the three examples; treatments/combinations of plant species including grazers must be complemented by a control without the grazers. This type of testing fits easily into an agroforestry system where multiple stands are the norm and the evaluation of the response of trees to grazing represents an on-the-ground optimization procedure for maximizing economic output or other social benefits.

More quantitative tests are also possible. There is little doubt that density dependence occurs in grasses, herbaceous plants, and trees. However, the ability of livestock to provide a benefit to a forestry system via reducing competition among plant species hinges on the form, not the existence, of this density dependence. Furthermore, the timing of density dependence, as in the stand development stage and stage of plant life history (e.g., reproduction/seed production rates, establishment rates, below ground growth rates, above ground productivity), has a large bearing on whether the integration of livestock can be successful.

For example, in *Pinus* plantations, there is suitable quantity of forage for the livestock only in the first few years of stand development and after a thinning procedure. During these times, the competitive interaction between understory and trees is such that the understory vegetation is likely to have a strong impact on growth (early) and possibly reproduction (late). As canopy closure increases, light becomes substantially more limiting to the understory vegetation and the competitive interaction can be reversed.

### 5.2.6 Grazing in Forest Plantations: Summary

The above analyses suggest a few generalizations about selective herbivory mediating a competitive interaction between two plants. In general, herbivory is predicted to be most effective at reducing the biomass of the plant competitor when the plants are growing in a low productivity situation and when density dependence in the competitor's growth is low and or concentrated at low densities. Under these conditions, a unit increase in herbivore density has a much larger effect on plant density and a much smaller effect on plant growth rate than when density dependence generates quite high levels of resistance and resilience to herbivory in the plants. This effect corresponds to the form of growth most commonly defined by physiology and morphology (Noy-Meir 1978).

The potential for multiple equilibria, and in particular of unstable equilibria, in these models prompts caution when introducing herbivore stock to an agroecosystem. They represent a biological and economic risk (extinction and increased competition) for anyone managing a mixed plantation forest with herbivores. For example, in Figure 5.4b, the low density equilibrium presents a paradoxical scenario: Were herbivores to be added to this system at low plant densities, the instability of this point ensures that herbivores will either cause plant extinction (when plant density is to the left of Point X) or that the plants will continue to grow even in the face of the level of herbivory dictated by the herbivore stocking density. As long as the herbivore reproductive schedule lags behind the changes in plant density, the benefits of reduced competition are unlikely under these conditions.

It is important to recognize the graphical analyses, and the predictions they generate, are only effective when the assumptions — in this case the form of plant growth and herbivore consumption — are met. Given this reality check, the theory here represents a set of plausible interactions between two competing plants and a selective herbivore. The applicability of the particular range of insights to any one real-world scenario hinges on that real-world situation matching the assumptions of the different model.

For example, one of the issues raised by this analysis is the possibility that the herbivores introduced in the plantation system might eat their sustaining resource to extinction. While this could result in a loss of herbivore stock, or a transfer of grazing pressure to an alternative resource, this predicted dynamic can be removed by altering a critical assumption about plant growth. The current assumption in the above models stipulates that herbivores can consume all plant biomass and therefore reduce per capita growth to zero. However, the likelihood of, say, cattle consuming the reproductive root mat of a grass understory or the stolons of herbaceous perennial species is fairly low. The possibility of there being some form of growth rate refuge for the plant alters the shape of the growth functions dramatically and changes the type of predictions that emerge from this type of analysis.

Figure 5.6 illustrates a Model 1 growth function augmented by the addition of a growth rate refuge, overlayed by three stocking levels of an herbivore with density independent consumption. Note that this figure is no longer in per capita terms as Models 1–3 were. The significant change in interpretation is that plant extinction is virtually impossible via consumption. Furthermore, there are stable interactions between plant and herbivore at very low plant densities, suggesting that there are conditions under which fairly large stocking levels can be combined with large benefits via reduced competition and a persistent direct return on the herbivore investment. This model exercise demonstrates that realistic knowledge about the plant and herbivore populations — from plant chemistry, architecture, or physiology to refugia or the impact of herbivore trampling — can be easily integrated into the graphical models and analyzed.

Much of this analysis confirms what can be considered common sense from the perspective of the practicing agroforester. These models suggest that knowledge about the competitive relationships between plants and the selectivity of the herbivore is vital to successful management of agroforestry systems integrating grazing animals. What these models do beyond this intuition is create a predictive environment on which to gauge the specific response of an agroforestry system of plant resources to specific types of herbivory. Perhaps it represents too detailed a set of

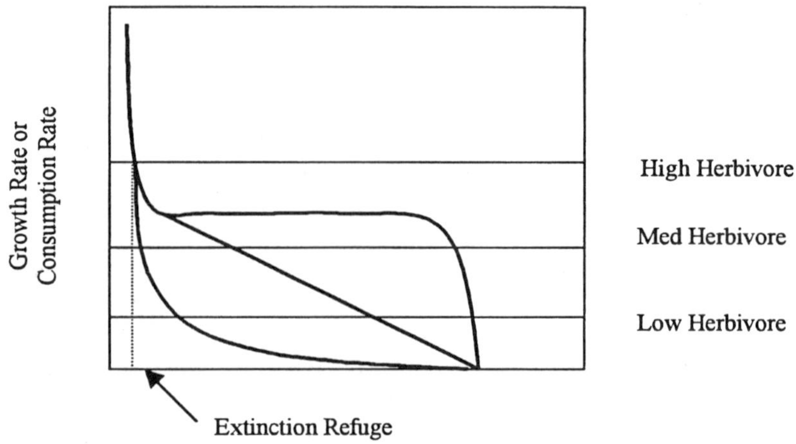


**FIGURE 5.6** The effect of a refuge from herbivory or inaccessible portions of plant material. Even in the face of super high levels of herbivory, the plant is able to support tissue that could result in recovery from herbivore damage. Below the extinction refuge, the plant always has a positive growth rate. In this scenario, it is possible to achieve significant reductions in competitive pressure and significant income from herbivore stocking. The isoclines are easily modified to include functional or physiological processes resulting from herbivory.

information. However, the strength of this detailed information may be in its integration with the detailed economic analyses (Kogan 1986, Dangerfield and Harwell 1990, Somarriba and Lega 1991, Price and Martinsen 1994) that are more and more common in large- and small-scale agroforestry systems integrating grazing animals. Furthermore, given the limited empirical research on at least the qualitative response of tree species to reduced competition, and the vast potential to replace herbicide use with a potentially more lucrative and less chemically intensive method of reducing competitive, fire potential and related issues, the theory here provides an impetus and framework on which to systematically enhance the productivity of the agroforestry systems.

## 5.3 FOOD CHAIN THEORY AND THE RELEVANCE TO PEST CONTROL IN AGROFORESTRY SYSTEMS

One of the most active areas of ecological research related to agriculture, forestry, and agroforestry is biological pest control. Much of the theoretical and empirical research has centered on two species models of predator-prey (Hassell 1978) or host-parasite (Beddington et al. 1978, Hassell 1978, Murdoch et al. 1983, 1984) interactions (but see Gutierrez et al. 1990 and the extensive list of papers reviewed in Schmitz et al. 1998). The majority of the reviews of pest control and pest management are restricted to these basic models and the concepts that surround them et al. 1998). The majority of the reviews of pest control and pest management are restricted to these basic models and the concepts that surround them (Beddington et al. 1978, Murdoch et al. 1983, 1984, Power and Kareiva 1990). (There exists a less mathematical set of models and hypotheses centered around the natural enemies and resource concentration hypotheses [Root 1973, Altieri 1996], but even these have the role of predators on pests as their focus.)

An alternative perspective is presented here, focused on the three trophic level ecological theory, that links the predators of herbivores to plant damage. A focus on three trophic levels has the advantage of highlighting the response of the plants to control of the herbivore pest, rather than that of the pest species to the predator. In this approach, the plant resources are the focal point, as they are in agroforestry systems. The bottom line in agricultural systems is the production of biomass and income from the plants. When balancing the biological and economic inputs to, and outputs from, an agroforestry system, it seems prudent to retain a focus on the plant resource response. For example, a three trophic level approach can be particularly illuminating when confronted by a situation where the reduction in herbivore density has no effect on plant production or when herbivore densities are not impacted by a control predator but there is a significant reduction in plant damage.

The theory of three trophic level terrestrial interactions, centered around herbivore species, has a long and varied history (Hairston et al. 1960, White 1978, Fretwell 1977, Price et al. 1980, Oksanen et al. 1981, Schmitz 1992). In general, the theory makes explicit predictions about how predation minimizes the impact of herbivores feeding on plant resources. This basic research program has bifurcated into research addressing the relationship between specialist herbivores, plant chemistry, host site selection, and predators (Price et al. 1980, Rosenthall and Berenbaum 1992, Altieri 1996) and research addressing the role of predators in mediating the effects of generalist herbivores on their plant resources (Fretwell 1977, Oksanen et al. 1981, Schmitz 1992, Hurd and Eisenberg 1997, Schmitz et al. 1997, Beckerman et al. 1997; see Schmitz et al. 1998 for a review). The remaining portion of this chapter will focus on the generalist herbivore scenario, though some of the conclusions may be relevant to research and applications dealing with specialist feeders.

Recent theory and empirical developments regarding the role of predators in mediating the impact of herbivores on plants (Polis and Winemiller 1996, Schmitz et al. 1997, Beckerman et al. 1997, Schmitz et al. 1998) suggest that both herbivore density and herbivore behavior are important components of the relationship between predator pest control agents and the response of plant communities. The theory-integrating pest behavior in response to the control agent, the resources, and the abiotic environment with the theory on density response in herbivore populations has the capacity to inform agroforesters on how to achieve effective mitigation of plant damage by a pest.

The basic, three trophic level theory posits that the addition of a predator to a system of herbivores feeding on plants will lower the density of the herbivores, thereby reducing the damage on the plants. This simple concept has many possible mechanisms and variations. In the context of reducing pest effects, these mechanisms, not limited to herbivore density reductions, are vital because they dictate the effectiveness with which predators reduce pest impact to suitably low biological and economic levels (Kogan 1986, Carroll et al. 1990, Price and Martines 1994).

For example, predators may act on herbivore prey by reducing the number of some life stage (egg, juvenile, adult) or by altering the behavior and or development of some life stage in a manner that restricts the impact that the herbivore may have on plants. This dichotomy of mechanism produces vastly different pathways by which predators can reduce herbivore impact on plant resources. The effectiveness of the predator, and thus the success of the control program, can hinge on understanding these pathways. Discussed below are four theoretical scenarios that separate density and behavior/development (trait) mediated mechanisms in successful and unsuccessful theoretical mitigation of herbivore impact on plant resources.

### 5.3.1 Scenario 1: Successful Density Mediated Reduction of Herbivore Impacts on Plants

This is perhaps the most common model of indirect interactions between predators and the plants fed upon by the herbivore prey of the predator. Here, predators simply reduce the number of herbivores feeding on the plants, thereby reducing the herbivore impact on the plant biomass. This intuitive model makes a major assumption that is not always made clear. It assumes that per capita consumption by the herbivore cannot increase in response to reduced herbivore numbers. This stipulates that per capita consumption by herbivores is insensitive to changes in the density of the herbivore population or that a consumption response to density changes occurs on a time scale vastly different than the predation events and growth of the plants. If this assumption is relaxed and per capita consumption of plant resources is allowed to increase with lower herbivore densities, a very interesting pattern emerges.

### 5.3.2 Scenario 2: Unsuccessful Density Mediated Reduction of Herbivore Impacts on Plants

If per capita consumption increases with reduced herbivore density, there is a high probability that predation will reduce herbivore density with no reduction in the amount of damage the herbivores inflict upon the vegetation. As herbivore numbers are reduced, the remaining individuals consume proportionately more resources than they would at a higher population density when resource are proportionately less available.

One mechanism through which this could occur is exploitative intraspecific competition where resource limitation (a requirement for competition) is manifest by the supply of food (Schoener 1973, Schmitz 1992, 1993, Chase 1996). Under any level of resource (limiting), the theoretical per capita consumption by any individual herbivore is the amount of resource available divided by the number of herbivores. Reduced herbivore numbers thus results in increased per capita consumption and no net change in the damage on plant resources; fewer individuals means more food per individual.

If supply-limited resource competition drives this scenario, what mechanism could be operating in Scenario 1? If intraspecific competition is occurring, but the herbivores do not respond with an increased per capita consumption, it is plausible that resource acquisition is limited by some other factor than supply. For example, temperature or search time requirements in a complex habitat may limit the time in a day available for acquiring resources. Alternatively, digestive capacity may limit the intake of the resource. If one of these scenarios is the case, then per capita consumption may not increase with changes in herbivore density; the proportional intake of resources is limited more by time or gut capacity than by the supply/availability of resources.

The difference between the supply and time/digestive-limited scenarios provides a way to evaluate the efficacy of predator control before the introduction occurs. By looking at how per capita consumption varies with herbivore density, it may be possible to gauge the success, at the vegetation level, of a successful predator prior to introduction. If per capita consumption increases/decreases with decreasing/increasing herbivore density, predation, or parasitism, operating by reducing density may be ineffective.

There are many other basic situations where predator control would be unsuccessful, such as when a predator's life history and developmental process do not coincide with that of the target herbivore. Alternatively, predators may only consume the most ineffective consumers in a guild of herbivore pests (or the inefficient-sized individuals), leaving behind a suite of highly efficient herbivores that still does significant damage. The above discussion demonstrates that even when the natural history and life history of the predator is well established and synchronized with the prey, it still may not result in successful mitigation of herbivore impact on plants.

Scenario 2 makes it clear that a successful introduction of a predator and the successful consumption of prey may not always translate into successful mitigation of plant damage by herbivores. It reinforces the point that the dynamics of two trophic level models of predator-prey or host-parasite interactions (two-level systems that form the core of much pest control theory; e.g., Hassell 1978; but see Price 1980, Duffy et al. 1986, Kennedy 1986) may not provide sufficient information to successfully explore the options for success and failure in biological control programs.

### 5.3.3 Mitigation of Herbivore Impact by Mechanisms Other Than Density Reduction: General Concepts

Scenarios 1 and 2 assume that interactions between trophic levels are primarily controlled by changes in the density of different organisms. However, this is certainly not the only mechanism by which trophic interactions can occur. An alternative to treating density as the core vector through which direct and indirect interactions occur among three trophic levels is to consider behavior and development (trait vs. density mediation) as the vector of interaction among trophic levels. This distinction, well supported in the aquatic (Huang and Sih 1990, Wellborn et al. 1997) and terrestrial (Hurd and Eisenberg 1990, Schmitz et al. 1997, Beckerman et al. 1997) systems, is significant to biological control efforts in agroforestry systems. (The closest example in the agroecology literature is prey switching by biological control predators, though it is often discussed only in the context of supporting a large complement of density lowering predators (the resource concentration/diversity hypothesis [Altieri 1996]). The presence of trait-mediated interactions between plants and predators suggests that a predator may not need to be effective at killing the prey that damages the plant. Instead, it may only need to be effective at altering herbivore behavior enough to limit its impact on resources.

As described above, trait mediation occurs when a predator alters the behavior or development of the herbivore to a point that limits its impact on the food resource. Trait mediation occurs when the herbivore perceives a risk of predation and subsequently alters its behavior to minimize the risk of mortality due to predation, while at the same time balancing the risk of starvation (McNamara and Houston 1987, Mangel and Clark 1988).

The types of traits that may vary with predation risk are likely specific to different classes of organisms. However, four changes in herbivore behavior represent the basic root mechanisms by which predation risk is manifest. In order for predation risk to impact herbivore growth and development, it must 1) reduce herbivore feeding time and therefore energy intake; 2) force a shift in diet and therefore diet quality and energy intake; 3) force a shift in habitat use and associated diet selection; or 4) alter the timing of oviposition, pupation, or emergence. Any of these effects has the potential consequence of either reducing energy intake or altering life history timing and thereby changing growth rates, fecundity, and offspring survivorship.

Reduced feeding time obviously limits the absolute intake of food. Furthermore, if the herbivore is a generalist of some sort, changes in feeding time may result in a diet shift. If the herbivore chooses a mixed diet in an adaptive fashion (Belovsky 1986), reduced feeding time in the presence of a predator can alter the combinations of food that balance the constraints on foraging relative to the diet chosen under ideal, predator-free conditions (Rothley et al. 1997, Schmitz et al. 1998).

Habitat shifts are the third alternative. In general, a habitat shift represents a choice by the herbivore prey to avoid areas of high risk from predation. This type of response can occur when a generalist feeder samples the habitats containing the food resources differently when predators are not present or when a specialist feeder changes host plants or adds variance to its selection of host plant quality.

The consequences of this type of change include altered diet as a function of the new habitat as well as associated changes in the thermal environment to which the herbivores are exposed. The underappreciated aspect of habitat and diet shifts is that a habitat shift will have a net positive effect on one plant or class of plants and a net negative effect on another; foraging effort will be transferred from the plants with high risk of predation to those with lower risk.

Labeling the fourth alternative a behavioral response requires a degree of sensory perception that may not be present at all stages of the life history (pupae, eggs). However, delayed development and reproduction are well-documented responses to predators and are often tightly tied to habitat shifts and diet choice.

### 5.3.4 Scenario 3: Successful and Unsuccessful Mitigation of Herbivore Impact Via a Trait Mediated Response—Linear Food Chains and Changes in Feeding Time

The conditions under which a predator could successfully mitigate the impact of herbivores without altering their density is fairly straightforward when considering a linear food chain with one resource fed upon by a herbivore that is itself fed upon by a predator. While this may not be a useful example in terms of agroforestry systems where multiple crops are the norm, it is nonetheless important for understanding the impact of specialist pest herbivores and as a baseline for those conditions where multiple resources are available.

Begin by assuming that there is a specialist pest herbivore feeding on an important crop species in the agroforestry system. The success of a predator, one that may not affect density, hinges on the predator instilling a sense of predation risk such that the herbivore no longer acquires sufficient quality or quantity of resources necessary to produce offspring. Starvation induced by the predator is not necessary unless the operative aim of the biocontrol program is the instantaneous removal of the pest.

How the predator generates a risk effect can vary substantially. Time available for feeding may be reduced by a predator that emerges at the time periods of the day normally used by the herbivore for feeding. This may force the herbivore, such as an insect, to feed at a time of day that does not provide the intensity or duration of necessary thermal regimes to acquire sufficient resources (Belovsky and Slade 1986, Schmitz et al. 1997, Beckerman et al. 1997). Alternatively, the predator may be present at all times and the herbivore is either unable to forage at all or is restricted in accessing sufficient resources as local availability declines.

The success of this type of herbivore control has as much to do with the traits of the herbivore (thermal regulation, risk perception) as it does with the hunting mode of the predators. The success of this mode of herbivore control rests completely on an understanding of the natural history and basic behavioral ecology of the predator and the herbivore pest.

The successful control of herbivores by risk effects hinges on resource acquisition by the herbivores being limited by the time they have available for feeding. In fact, if resource acquisition is more limited by the resource supply than time, herbivores can effectively "ignore" the risk-based reduction in feeding time because the resource supply is more limiting. Thus, if resource acquisition is limited more by the supply of the resources, risk effects are not likely to be important.

Another way in which herbivores may not be limited by risk effects is if the timing of predator hunting behavior, or the timing of the introduced predator's life history, is mismatched on the scale of daily feeding cycles or development cycles. If the predator activity is concentrated at times in which herbivore activity is already low, predator-based biocontrol under this situation would be ineffective. Likewise, if development patterns are mismatched, the prey may exceed a size or profitability threshold, causing the predator to have little or no effect on the prey.

### 5.3.5 Scenario 4: Successful and Unsuccessful Mitigation of Herbivore Impact Via a Trait Mediated Response—Multiple Herbivore Resources and Habitat (Diet) Shifts

Scenarios 1–3 are all restricted to relatively linear food chains of introduced predators, herbivore pests, and their plant resources. The lumping together of plant resources into a single functional group generates the consequence that the herbivore pest is assumed to be some kind of specialist. While many insect or small mammal pests are specialists, many are generalists, often consuming the plant material of more than one species. This provides an opportunity for herbivore damage to be transferred, rather than reduced in intensity, and is the basis on which the above theory is expanded to include more realistic ecological systems. Note that host shifts (Futuyma and Keese 1992), a common discussion within research centered on specialist pest species, their hosts and host chemistry, represent a similar type of response to predators.

The classes of response by a generalized feeder to a predator are the same as those described in Scenarios 1–3. The herbivore may be sensitive to mortality and or risk effects generated by the predator. As above, the response depends on whether time or resource supply rate limits resource acquisition by the herbivore.

Begin with a herbivore that has its resource acquisition limited by the supply rate of resources. In this case, assume that the supply rate of both resources limits the herbivore consumption and the herbivore feeds on both resources. If a biocontrol agent is introduced that can inflict mortality on this herbivore, it is likely that there will be a minimal effect on the plant resources. As discussed in Scenario 2, resource intake limited by supply results in per capita increases in consumption; the addition of a predator would have a minimal effect on reducing herbivore damage.

Retaining the same resource limitation assumptions, a predator that presented only a risk would be also be unsuccessful at eliciting an indirect positive plant response. Recall that risk responses at the plant level hinge upon the ability of the predator to alter the available feeding time or feeding schedule of the herbivore. When acquisition of resources by herbivores is limited by resource supply, the herbivores are relatively insensitive to changes in their feeding time or schedule because these factors are less limiting their intake of energy required for development and reproduction.

If resource intake is limited by time in one of the ways described above, there are a variety of situations in which the introduction of a biocontrol predator could be successful. When resource intake is limited by time, the reduction in numbers of herbivores can have a significant impact on the production of biomass. Under these assumptions, individuals remaining in the reduced population have no recourse with which to increase their resource consumption because this would require additional feeding time, which is assumed here to be limiting resource intake. When herbivores feed on multiple resources, the realized benefit to reduced herbivory is proportional to the fraction that each resource represents in the herbivore diet.

If resource intake is limited by time, predation risk, acting via changes in feeding schedules, available feeding time, or feeding locations, can also have profound influences in positive or negative directions. The direction of the influence is contingent upon whether the risk is manifest through vigilance (time reductions) or through the choice of habitat associated with predation events.

If the risk is manifest through a change in feeding time/vigilance, a reduction in plant damage can be generated in two ways. If the predator is wide-ranging and affects feeding time over all plant resources consumed by the herbivore, the reduction in damage will be proportional to the fraction of each resource in the diet, assuming the relative reduction in feeding time is equitable among the resource types. If the effect of the predator is limited to one resource (i.e., the predator itself is some form of a specialist on that resource), then the benefits of reduced feeding time (reduced plant damage) may only be recognized through this resource. However, competitive relationships between the plants and the herbivores, as discussed in Section 2, could greatly complicate this response over time. This scenario is interesting because the efficacy of the predator is limited to one resource. In an agroforestry system of multiple crops, this may offer little benefit to the landowner.

The final alternative is that risk generates a habitat shift in the herbivores rather than a change in feeding time. Here, the herbivores move to resource patches that possess some form of cover or limit the effectiveness of the predator-hunting mode. As a result of the habitat shift, the herbivores by default eat a different diet because the amount of time spent in the different habitats has changed. Consider a herbivore feeding on two resources. If the predator causes the herbivore to spend proportionately more time on Resource 2 to avoid predation on Resource 1, this habitat shift has the intuitive result of reducing impact on one resource, Resource 1, while increasing impact on Resource 2.

This could be beneficial or not, depending on the interest in and value of the plant resources. If both resources are vital to the agroforestry system, a predator generating this type of response has little benefit. However, a habitat shift could cause a herbivore pest to simply move from the part of the plant most useful in the production context to an expendable portion of the plant. As above, both of these alternatives are confounded and compounded by the manner in which competitive interaction between resources or between plant parts is mediated by herbivory (Section 2).

Finally, habitat shifts represent an opportunity if a suitable alternative plant host exists. When planted, and assuming low levels of competitive interaction with primary products, the introduction of a predator could shift the damage to the new expendable crop, reducing the impact on primary products. At the same time, this mechanism might actually increase the species richness of the system, which may or may not be seen as a characteristic worth managing for (Vandermeer 1995).

A potential application of this theory lies in the intercropping of timber forest products and nitrogen fixing plants, where these legumes may offer a food product but also provide a needed source of nitrogen input to the entire ecological system. Here, a habitat shift by insect pests to an intercropped legume crop may accelerate the input of fixed nitrogen to the soil because the insects can speed the transfer of fixed nitrogen from the plant biomass to the soil (frass and insect mortality; Slade and Belovsky 1995). The plausibility and frequency of this type of interaction is unknown, but the theory suggests that patterns of habitat shifts, time budget shifts, and mortality are worth investigation in the context of pest control.

### 5.3.6 Real Examples and Practical Suggestions

Identification of the factors contributing to the success and failure of biological control has been the subject of numerous review papers and theoretical research papers. There appears to be a standard set of characteristics of the environment, the pest species and the predator or parasite introduced for control (Beddington et al. 1978, Hokkanen 1985, Murdoch et al. 1985). These characteristics range from habitat size and heterogeneity of the plant species to the relative growth rates of the pest and predator species to the temperature during and timing of the introduction. The discussion of these characteristics has generally emerged from a definition of success rooted in ecological theory centered on establishing low, stable populations of the pest species (Hassel et al. 1978, Murdoch et al. 1985). However, low levels of pest species are not necessarily an appropriate measure of biological control success from an economic (Hokkanen 1985) or biological perspective (see above). Furthermore, many of these factors have been shown not to be the major predictors of biological control success (Murdoch et al. 1985). A more appropriate measure, from both a biological and economic perspective, may be the amount of damage that the host plants suffer from the pest.

This measure was the justification for presenting the three trophic level ecological theory as an alternative paradigm for measuring the effectiveness of biological control. Based on this theory, the benchmark for success is plant damage — a plant characteristic with distinct economic implications. Furthermore, the theory embodies a range of mechanisms that clearly identifies coarse scale characteristics of ecological systems that could lead to successful reduction in plant damage through biological control.

This theory does not eliminate the usefulness of the standard list of factors usually cited as part of successful biological control, but instead places the factors in the context of a biological mechanism associated with success: a plant biomass perspective. Recall that there are two necessary conditions for successful control: 1) the reduced pest population must not increase per capita consumption; or 2) herbivore population must be forced to switch temporal or spatial habitat use.

These conditions are related to the purported requirement for success (Beddington et al. 1978) that predator/parasite and prey be synchronized with the prey population and be able to respond quickly to the patterns of increase and decrease in the pest population. However, in basing success on plant damage assessment, patterns of success are contingent upon the mitigation of prey damage, not the magnitude and constancy of the prey population reduction. This allows that a small reduction in prey population size can theoretically reduce the amount of plant damage to a level below a predetermined economic threshold. It also suggests that there may be many situations where density reductions could occur, but these density reductions are compensated by a relative increase in consumption or damage by the prey.

These conditions are also linked to the purported need for predator aggregation in areas of high-prey density. The standard justification for this is that stable equilibrium at low pest densities occurs more frequently when this characteristic is included in models. In the context of the tri-trophic theory, recall that a generalist pest species (either multiple host use or a feeding generalist) may be influenced substantially by the localized presence of a predator. When predation or parasitism is confined to a particular season, time of day, or habitat patch type, flexibility in resource use (genetic or behavioral) may result in substantial decreases in pest damage if there is sufficient recourse for the pest species to transfer its feeding/reproduction efforts to different phases of the season, different times of the day or different plant resources. This idea provides a mechanism for the often-stated and little-tested hypothesis about resource diversity having a positive impact on the success of biological control efforts. However, it also illustrates the way in which these conditions could result in a higher concentration of feeding/reproductive effort, for example, by reducing in the variance of host plant use and concentrating damage.

Unfortunately, there is very little field evidence from agricultural, agroforestry, or forestry systems, direct or anecdotal, associated with the three trophic level theory presented above. Most of the reviews and tests of effectiveness are limited to observations or to experimental efforts rooted in the traditional two level predator-prey/host-parasite theory (but see Gutierrez et al. 1990 for perspective; Perfecto 1990, 1991, Gomez and Zamora 1994, Whalen and MacKay 1988, Fonesca 1994, and Horvitz and Schemske 1984 for exceptions). However, agricultural systems, and agroforestry systems in particular, represent some of the best opportunities to systematically test particular aspects of the theory. In particular, the intercropping regimes that characterize most agroforestry systems present an ideal experimental unit for testing for the success of predators and parasites as controlling agents of pest populations.

This requires a fairly straightforward experimental design in which plant damage and production is measured as a function of planting pattern and diversity with: a) no pests or predators; b) with just pests; and c) with pests and predators. In this context, the timing of introductions or emergence patterns, the aggregation and spatial orientation of pest and predators, the synchrony or asynchrony of growth rate responses, and in general the mitigation of plant damage become a function of the habitat structure (e.g., climate affecting time vs. supply-based resource limitation) and the behavior and physiology of the pest and predator species.

This standard experimental design is subject to very basic statistical analysis of variance (ANOVA) and is easily integrated with predetermined economic thresholds of damage. Furthermore, the evaluation can be executed in any type of ecological system. As the number of tests increases geographically and according to different types of agroforestry systems, a database will emerge that itself can be subject to analysis (Hokkanen 1985).

### 5.3.7 Conclusions Regarding the Use of Predators to Control Herbivore Pests in Agroforestry Systems

Scenarios 1–4 have centered on two plausible assumptions about what limits resource intake in herbivore species. The conditions in which the introduction of a predator would be successful at reducing herbivore damage on plants when the herbivore is limited by resource supply or time available to feed is outlined. Also outlined are the conditions under which the reduction of herbivore density would or would not be successful, and under what conditions density reductions would not be necessary for effective reduction in herbivore damage.

Table 5.1 summarizes this information. The most general conclusion is that predator control is unlikely to be effective in situations where herbivore resource intake is limited by resource supply rather than time available for feeding. Interestingly, this general concept applies equally well to predator-herbivore-plant systems as to parasite-host-plant systems: When resource supply limits resource intake instead of some temporal factor, per capita compensation by the herbivore or pest is likely to override the effects of predation, predation risk, or parasitism. The consequence of this is that the success of biological control in agroforestry potentially hinges on the distribution of resource supply limited intake on geographic, local, and microhabitat scales. The success of biological control is likely, therefore, to be correlated with climate, patterns of plant architecture affecting microclimate and perhaps fertilization techniques, all of which affect supply rate and consumption of plants by herbivores. These correlations suggest an opportunity for manipulating the thermal and resource environment experienced by herbivores in agroforestry systems. Mixing of crops and trees can be used to alter the thermal and resource supply conditions in a manner that could positively affect pest control efforts (see Vandermeer 1995 and Altieri 1995 for a review and the current status of this discussion).

There are three additional messages from this overview. First, the traditional theory of biocontrol centers around the stability of two trophic level systems of predator-prey or host-parasite where the metric of success is the maintenance of herbivore/pest populations at a low density. This focus, while useful in many contexts, is not focused on the primary product of an agroforestry system, the plant resources. The previous analysis demonstrates that using three trophic level theory and focusing on the response of the plant species to changes in herbivore density or behavior can illuminate a wide range of scenarios leading to both successful and unsuccessful control of herbivore pests.

**TABLE 5.1**
**The response of plants to a predator feeding on a herbivore when predators mediate herbivore density or behavior and herbivore resource intake is limited by supply or time.**

| | Supply Limited | | Time Limited | |
|---|---|---|---|---|
| | Single Resource | Multiple Resource | Single Resource | Multiple Resource |
| Density Mediated | ∅ | ∅ | + | + |
| Trait Mediated | ∅ | ∅ | + | Diet or Habitat Shift |

Given this range of predictions, based on clear sets of assumptions about herbivore feeding and predator hunting, the second message is that the application of predator introductions to mediate plant damage by herbivores is contingent on biotic and abiotic characteristics of the agroforestry system. While this is a relatively obvious point, it has not been the central theme in research and application of biological control. It is clear from the above analysis that climate and resource supply dynamics, the factors that determine how resource intake by a consumer is limited (i.e., feeding time vs. resource supply), are critical pieces of information for gauging when a predator will be successful at reducing herbivore impact. This focus on plant productivity and climate is a substantially different focus and demonstrates how alternative modes of plant damage reduction can be realized. It suggests that there may be regional and organismal biases in achieving these reductions due to climates and choice of predator and offers a series of testable hypotheses about why.

Finally, the success and failure of biocontrol efforts in agroforestry systems will rest on making decisions informed by natural history and using experimental designs to implement and monitor the efforts. By basing initial decisions on natural history, it is possible to use the above theory, or any other, to gain some prior indication of the outcome of the introduction. This works because the theory makes predictions based on assumptions that are derived directly from the natural history. The use of experimental designs to implement and monitor such introductions accomplishes two things. It ensures that an investment is made in more than one option (i.e., alternative predators or manipulations, each with its own predicted effect) and that there is a reference (control) point against which to compare natural changes and changes caused by the implementation of the biocontrol program.

## 5.4 SUMMARY AND CONCLUDING POINTS

This chapter presents two pieces of theory that are relevant to agroforestry systems. The first focuses on grazing animals in mixed plantation forestry and the importance of specialized consumers among competing plants. The second focuses on predator control of plant damage by herbivore pests. As noted in the introduction, this is by no means an exhaustive list of ecological theory involving multiple trophic levels that can be applied to agroforestry systems. In fact, agroforestry systems are potentially some of the more complex ecological units, as they often combine plant resources that do not normally associate with one another. There is a great deal of theory that can inform agroforesters on the interactions among soil nutrients and micro-organisms, the plant production that dominates the agroforestry system, and the consumers and predators that rely in some direct or indirect manner on these plants and climate.

The two core examples presented highlight the importance of clear and plausible assumptions informed by natural history. In both the grazing case and the biocontrol case, there were a number of plausible assumptions that lead to vastly different outcomes, some successful in the context of the applied system and others not. The asumptions were rooted in the natural history of the environment (e.g., temperature and vegetation structure affecting feeding time and sensitivity to predators) and the organisms (e.g., the relationship between herbivory rates and plant growth rates in the context of grazing). This alone does not validate the models and the theory, but instead makes the predictions dependent on real and accessible information about the ecological system. When this dependency is satisfied, the value of the theory is that it can demarcate clear alternative outcomes to a given type of investment in an agroforestry system.

Does agroforestry need ecological theory of this type? The primary purpose of this chapter is to describe, through example, the type of information ecological theory can contribute to management and decision making in agroforestry systems. The type of explicit qualitative and quantitative information available from considering ecological theory can complement the increasingly complex and explicit economic and social analyses associated with temperate and tropical agroforestry systems. In fact, the strength of the ecological theory lies in making links with social and economic theory as a guide to "experimental" evaluations of alternative agroforestry techniques that promote ecological, social and economic ends.

## 5.5 REFERENCES

Abrams, P. A. Dynamics and interactions in food webs with adaptive foragers, in: *Food Webs: Integration of Patterns and Dynamics,* G. Polis, and K. O. Winemiller. eds. (New York: Chapman Hall, 1997), 101–125.

Altieri, M. A. *Agroecology: The Science of Sustainable Agriculture.* 2nd ed. Boulder, CO: Westview Press, 1995.

Anderson, G. W., R. W. Moore, and P. J. Jenkins. The integration of pasture, livestock, and widely-spaced pine in South West Western Australia. *Agroforestry Syst.* 6(1988): 195–211.

Ashton, P. M. S., S. Gamage, I. A. U. N. Gunatilleke, and C. V. S. Gunatilleke. Restoration of a Sri Lankan rain forest: Using Caribbean pine (*Pinus caribaea*) as a nurse for establishing late successional tree species. *J. of Appl. Ecol.* 34(1997): 915–925.

Beckerman, A. P., M. Uriarte, and O. J. Schmitz. Experimental evidence for a behavior-mediated trophic cascade in a terrestrial food chain. *Proceedings of the National Academy of Sciences, U.S.* 94(1999): 10735–10738.

Beddington, J. R., C. A. Free, and J. H. Lawton. Characteristics of successful natural enemies in models of biological control of insect pests. *Nature* 273(1978): 523–519.

Belovsky, G. E. Optimal foraging and community structure: implications for a guild of generalist herbivores. *Oecologia* 70(1986): 35–52.

Belovsky, G. E. and J. B. Slade. Time budgets of grassland herbivores: body size similarities. *Oecologia* 70(1986): 53–62.

Briggs, C. Competition among parasitoid species on a stage structured host and its effects on host suppression. *Amer. Naturalist* 141(1994): 372–397.

Carroll, C. R., J. H. Vandermeer, and P. M. Rosset. *Agroecology.* New York: McGraw-Hill Publishing, 1990.

Connell, J. H. Apparent vs. "real" competition in plants, in: *Perspectives on Plant Competition,* J. B. Grace and D. Tilman, eds. (San Diego: Academic Press, 1990), 126–140.

Crawley, M. J. *Plant Ecol.* Oxford: Blackwell Science, Ltd., 1997.

Dangerfield, C. W. and R. L. Harwell. An analysis of a silvopastoral system for the marginal land in the Southeast United States. *Agroforestry Syst.* 10(1990): 187–197.

Dangerfield, C. W. and R. L. Harwell. An analysis of a silvopastoral system for the marginal land in the Southeast United States. *Agroforestry Syst.* 10(1990): 187–197.

Duffy, S. S. and K. A. Bloem. Plant defense-herbivore-parasite interaction and biological control, in: *Ecological Theory and Integrated Pest Management Practice,* M. Kogan, ed. (New York: John Wiley and Sons, 1986), 97–112.

Feeny, P. The evolution of chemical ecology: contributions from the study of herbivorous insects, in: *Herbivores: Their Interactions With Secondary Plant Metabolites. 2nd ed. Vol. II, Ecological and Evolutionary Processes,* G. A. Rosenthall, M. R. Berenbaum, eds. (San Diego: Academic Press, 1992), 35–46.

Fonesca, C. R. Herbivory and the long-lived levels of an Amazonian ant-tree. *J. of Ecol.* 82(1994): 833–842.

Fretwell, S. D. The regulation of plant communities by the food chains that exploit them. *Perspect. in Biol. and Med.* 20(1977): 169–185.

Futuyma, D. J. and M. C. Keese. Evolution and coevolution of plants and phytophagous arthropods, in: *Herbivores: Their Interactions with Secondary Plant Metabolites. 2nd ed. Vol. II, Ecological and Evolutionary Processes,* G. A. Rosenthall and M. R. Berenbaum, eds. (San Diego: Academic Press, 1992), 85–98.

Grace, J. B. On the relationship between plant traits and competitive ability, in: *Perspectives on Plant Competition,* J. B. Grace and D. Tilman, eds. (San Diego: Academic Press, 1990), 90–113.

Gutierrez, A. P., K. S. Hagen, and C. K. Ellis. Evaluating the impact of natural enemies: a multitrophic perspective, in: *Critical Issues in Biological Control,* M. Mackauer, L. E. Ehler, and J. Roland, eds. (Andover, U.K.: Intercept, 1990), 81–109.

Hairston, N. G., F. E. Smith, and L. B. Slobodkin. Community structure population control and competition. *Amer. Naturalist* 94(1960): 421–425.

Harper, J. L. *Population Biology of Plants.* London: Academic Press, 1997.

Hassell, M. P. *The Dynamics of Arthropod Predator Prey Systems.* Princeton, NJ: Princeton University Press, 1978.

Hochberg, M. E. Consequences for host population levels of increasing natural enemy species richness in classical biological control. *Amer. Naturalist* 147(1996): 307–318.

Hokkannen, H. M. T. Success in classical biological control. *Crit. Rev. in Plant Sci.* 3(1985): 35–72.

Holling, C. S. The components of predation as revealed by a study of small mammal predation of the European pine sawfly. *Canadian Entomol.* 91(1959): 293–320.

Holt, R. D. Predation, apparent competition, and the structure of prey communities. *Theor. Population Biol.* 12(1977): 179–229.

Horvitz, C. C. and D. W. Schemske. Effects of ants and an ant-tended herbivore on seed production of a neotropical herb. *Ecology* 65(1984): 1369–1378.

Huang, C. and A. Sih. Experimental studies on behaviorally mediated, indirect interactions in a three trophic level system. *Ecology* 71(1990): 1515–1522.

Hurd, L. E. and R. M. Eisenberg. Arthropod community responses to manipulation of a bitrophic predator guild. *Ecology* 71(1990): 2107–2114.

Jaindl, R. G. and S. H. Sharrow. Oak/Douglas Fir/sheep: A three crop silvopastoral system. *Agroforestry Syst.* 6(1988): 147–152.

Karban, R., A. A. Agrawa, and M. Mangel. The benefits of induced defense against herbivory. *Ecology* 78(1997): 1351–1355.

Kareiva, P. Trivial movement and foraging by crop colonizers, in: *Ecological Theory and Integrated Pest Management Practice,* M. Kogan, ed. (New York: John Wiley and Sons, 1986), 25–46.

Keddy, P. A. *Competition.* New York: Chapman Hall, 1989.

Kennedy, G. G. Plant-plant-pathogen-insect interactions, in: *Ecological Theory and Integrated Pest Management Practice,* M. Kogan, ed. (New York: John Wiley and Sons, 1986), 47–59.

Kimberling, D. N. and P. W. Price. Competition, leaf morphology, and host clone effects on leaf galling grape phylloxer (Homoptera: Phylloxeridae). *Environ. Entom.* 25(1996): 1147–1153.

Kogan, M. *Ecological Theory and Integrated Pest Management Practice.* New York: John Wiley and Sons, 1986.

Lewis, C. E., G. W. Burton, W. G. Monson, and W. C. McCormick. Integration of pines, pastures, and cattle in south Georgia, USA. *Agroforestry Syst.* 1(1983): 277–297.

Lewis, C. E., G. W. Burton, W. G. Monson, and W. C. McCormick. Integration of pines and pastures for hay and grazing. *Agroforestry Syst.* 2(1984): 31–41.

Louda, S. M., K. H. Keeler, and R. D. Holt. Herbivore influences on plant performance and competitive interactions, in: *Perspectives on Plant Competion,* J. B. Grace and D. Tilman, eds. (San Diego: Academic Press, 1990), 167–180.

Mangel, M. and C. W. Clark. Towards a unified foraging theory. *Ecology* 67(1986): 1127–1138.

McNamara, J. M. and A. I. Houston. Starvation and predation as actors limiting population size. *Ecology* 68(1987): 1515–1519.

McNaughton, S. J. Ecology of grazing systems in the Serengeti. *Ecol. Monogr.* 55(1985): 259–294.

McNaughton, S. J., R. W. Reuss, and S. W. Seagle. Large mammals and process dynamics in African ecosystems. *Bioscience* 38(1988): 794–800.

Murdoch, W., J. D. Reeve, C. B. Huffaker, and C. E. Kennett. Biological control of olive scale and its relevance to ecological theory. *Amer. Naturalist* 123(1984): 371–392.

Murdoch, W. W., J. Chesson, and P. Chesson. Biological control in theory and practice. *Amer. Naturalist* 125(1985): 344–366.

Noy-Meir, I. Stability of grazing systems: an application of predator-prey graphs. *J. of Ecol.* 63(1975): 459–481.

Noy-Meir, I. Stability in simple grazing models: Effects of explicit functions. *J. of Theor. Biol.* 71(1978): 347–380.

Oksanen, L., S. D. Fretwell, J. Arruda, and P. Niemela. Exploitation ecosystems in gradients of primary productivity. *Amer. Naturalist* 118(1981): 240–262.

Ormazabal, C. S. Silvopastoral systems in arid and semiarid zones of northern Chile. *Agroforestry Syst.* 14(1991): 207–217.

Pandey, C. B. and J. S. Singh. Rainfall and grazing effects on net primary production in a tropical savannah, India. *Ecology* 73(1992): 2007–2021.

Pearson, H. A., V. C. Baldwin, and J. P. Barnett. Cattle grazing and pine survival and growth in subterranean clover pasture. *Agroforestry Syst.* 10(1990): 161–168.

Perfecto, I. Indirect and direct effects in a tropical agroecosystem: The maize-pest-ant system in Nicaragua. *Ecology* 71(1990): 2125–2134.

Perfecto, I. Ants (Hymenoptera: Formicidae) as natural control agents of pests in irrigated maize in Nicaragua. *J. of Econ. Entom.* 84(1991): 65–70.

Polis, G., and K. Winemiller. *Food Webs: Integration of Patterns and Dynamics.* New York: Chapman Hall, 1996.

Power, A. and P. Kareiva. Herbivorous insects in agroecosystems. in: *Agroecology,* C. R. Carroll, J. H. Vandermeer, and P. M. Rosset, eds. (McGraw-New York: McGraw-Hill Publishing, 1909), 123–145.

Price, P. W. and G. D. Martinsen. Biological pest control. *Biomass and Bioenergy* 6(1994): 93–101.

Price, P. W., C. E. Bouton, P. Gross, B. A. McPheron, J. N. Thompson, and A. E. Weis. Interactions among three trophic levels: influence of plants on interactions between insect herbivores and natural enemies. *Annu. Rev. of Ecol. and Syst.* 11(1980): 41–65.

Ramakrishnan, P. S. Shifting agriculture and sustainable development: an interdisciplinary study from northeast India. *Man and the Biosphere Series, Vol. 10.* Paris: UNESCO, and Carnforth, U.K.: The Parthenon Publishing Group, 1992.

Root, R. B. Organization of a plant-arthropod association in simple and diverse habitats: the fauna of collards (Brassica oleracea). *Ecol. Monog.* 43(1973): 95–124.

Schmitz, O. J. Exploitation in model food webs. *Theor. Population Biol.* 41(1992): 161–183.

Schmitz, O. J. Functional responses of optimal consumers and the potential for regulation of resource populations. *Wild. Res.* 22(1995): 101–113.

Schmitz, O. J. and A. R. E. Sinclair. Rethinking the role of deer in forest ecosystem dynamics, in: *The Science of Overabundance: Deer Ecology and Population Management,* W. J. McShea, H. B. Underwood, and J. H. Rappole, eds. (Washington, DC: Smithsonian Institution Press, 1997), 127–146.

Schmitz, O. J., A. P. Beckerman, and K. M. O'Brien. Behaviorally mediated trophic cascades: effects of predation risk on food web interactions. *Ecology* 78(1997): 1388–1399.

Schmitz, O. J., P. A. Hämback, and A. P. Beckerman. Trophic cascades in terrestrial ecosystems. *Amer. Naturalist,* in press.

Schoener, T. W. Population growth regulated by energy or time: some simple representations. *Theor. Population Biol.* 4(1973): 56–84.

Slade, J. B. and G. E. Belovsky. The effect of grasshopper herbivory on nitrogen cycling in a prairie environment. *Bull. of the Ecol. Soc. of Amer.* 76 Supp. Part 2 (1995): 246–247.

Somarriba, E. and, F. Lega. Cattle grazing under Pinus caribea. 1. Evaluation of farm historical data on stand age and animal stocking rate. *Agroforestry Syst.* 13(1991): 177–185.

Steward, R. E., L. L. Gross, and B. H. Honkala. *Effects of Competitive Vegetation on Forest Trees: A Bibliography with Abstracts.* USDA Forest Service. General Technical Report WO 43, 1984.

Tajuddin, I. Integration of animals in rubber plantations. *Agroforestry Syst.* 4(1986): 55–66.

Tilman, D. *Resource Competition and Community Structure.* Princeton, NJ: Princeton University Press, 1982.

Vandermeer, J. The ecological basis of alternative agriculture. *Annu. Rev. of Ecol. and Syst.* 26(1995): 201–224.

Wellborn, G. A., D. K. Skelly, and E. E. Werner. Mechanism creating community structure across a freshwater gradient. *Annu. Rev. of Ecol. and Syst.* 27(1996): 337–363.

Weltzin, J. F., S. Archer, and R. K. Heitschmidt. Small mammal regulation of vegetation structure in a temperature savannah. *Ecology* 78(1997): 751–763.

Whalen, M. A. and D. A. MacKay. Patterns of ant and herbivore activity on five understory euphorbiaceous saplings in submontane Papua, New Guinea. *Biotropica* 20(1988): 294–300.

White, T. C. R. The importance of a relative shortage of food in animal ecology. *Oecologia* 33(1978): 71–86.

# *Part II*

## *Pattern and Process of Forests as Templates for Silvicultural Applications in Agroforestry Systems*

# Part II Introduction

The second part of this book covers the pattern and processes of how forests and trees grow in space and time. It has a more synecological focus than the first part emphasizing the importance of understanding native forest dynamics and using this knowledge for silvicultural applications in agroforestry systems. The first three chapters (6, 7, and 8) in this part of the book are reflections of the first two chapters (2 and 3) in the first part of the book. However, Part II has more of a whole system emphasis, and a concentration on describing ecological pattern of plant species distribution in relation to the site factors of water and climate (Chapter 6), soils and nutrient use (Chapter 7), and their agroforestry application (Chapter 8). Chapters 9 and 10 build upon these chapters (6, 7, and 8) to provide a stand dynamics framework as a silvicultural model for application in agroforestry systems. In particular these two chapters emphasize the importance of facilitative versus competitive forces underlying the successional dynamics of tree-mixtures.

# 6 Water and Climatic Relationships in Agroforestry

*David M. Smith*[1]

## TABLE OF CONTENTS



## 6.1 INTRODUCTION

Agroforestry is conducted by making choices of combinations of species that will grow and interact in ways that provide the desired products and other benefits of management. The different species can coexist if there is an appropriate partitioning of resources between them (Caldwell and Pearcy 1994). The ideal condition is one in which the leaves and roots of different plants or groups of plants do not function simultaneously in the same units of space above or below ground. The partitioning can be thought of as spatial or temporal although the boundaries between the two kinds are usually blurred enough that absolutely perfect distinctions can only seldom be made.

The physiological functions of plant tissues can take place only if they contain available liquid water. They halt if water becomes unavailable because of (1) low temperature, (2) inadequate supply of oxygen to roots, or (3) simply an absolute lack of water. Photosynthesis takes place if leaves are sufficiently supplied with water, carbon dioxide, and the appropriate wavelengths of light (Ledig 1983, Kimmins 1987).

[1] Yale University School of Forestry and Environmental Studies, New Haven, CT 06511; Tel.: (203) 248-1077; Fax: (203) 432-3809

0-8493-2206-5/00/$0.00+$.50

In agroforestry, the basic objective is to match plants of different species with the specific kinds of microenvironments to which each species is adapted. Trees are tall, they cast shade, and can be very effective in root competition. Because of this, the tree component plays the chief role in determining the characteristics of these microenvironments in agroforestry systems (Reifsnyder and Darnhofer 1989). Figure 6.1 is a schematic illustration that shows how the shade and roots of one tree create different microenvironments at one time during a day. These microenvironments can be thought of as components of the growing space which are the units of space above and below the ground surface that can be occupied by functioning leaves or roots.

The effects of solar radiation that result from the slanting rays of the sun change during the course of a day so shadows shorten and lengthen. Even in the tropics, the sun is directly overhead only for comparatively short periods of time. Therefore, the microclimatic patterns depicted in Figure 6.1 keep changing during the course of a day as well as with the seasons. To *thrive* in a particular microenvironment a plant must be well-adapted to the existing range of changing light conditions which operate day after day; to *endure,* the plant must be able to survive some extremely unfavorable condition that may exist for only a brief period of time.

In the shadows of tree crowns, the light received is filtered in different ways. Most of the photosynthetically useful radiation received by subordinate plants is that of sunflecks that comes through gaps between leaves and stems. The sunflecks have a central area of direct sunlight surrounded by a zone of scattered solar radiation that comes from all points in the sky. Some light is transmitted through leaves but varying portions of this are in the green wavelengths that are not used in photosynthesis. Sometimes filtered light is called "green shade" although this is is an accurate term only for the light transmitted through leaves.

Diffuse solar radiation is that which has been scattered during its passage through the atmosphere. The short blue wavelengths are scattered more than the long red wavelengths. Diffuse radiation comes most abundantly from the part of the blue sky that is directly overhead, although the scattering is such that all illuminated surfaces receive some diffuse radiation. The vertical lines showing the effect of a tree crown on diffuse radiation oversimplify the situation and signify only that spaces directly beneath tree crowns receive far less diffuse radiation than those directly beneath the sky. Spaces that are in shadows but are open to the sky above can be thought of as being in "blue shade" which is much more favorable to photosynthesis than "green shade." Such spaces are also shielded from most of the intense heating effects of direct solar radiation.

These heating effects are most significant in a boundary zone at the ground surface because most solar energy passes through the atmosphere without much alteration and is first absorbed (or reflected) by whatever solid or liquid surface it hits first. The boundary zone consists of the laminar layer of air above the surface of the soil and an evaporative layer below it.

The air in the laminar layer is so tightly held by friction that airborne heat, water vapor, and other substances move into and out of it very sluggishly and mainly by the vertical movement of individual molecules. The turbulent layer above is one in which wind-driven turbulence moves heat and substances much more rapidly in large, chaotic clumps of molecules.

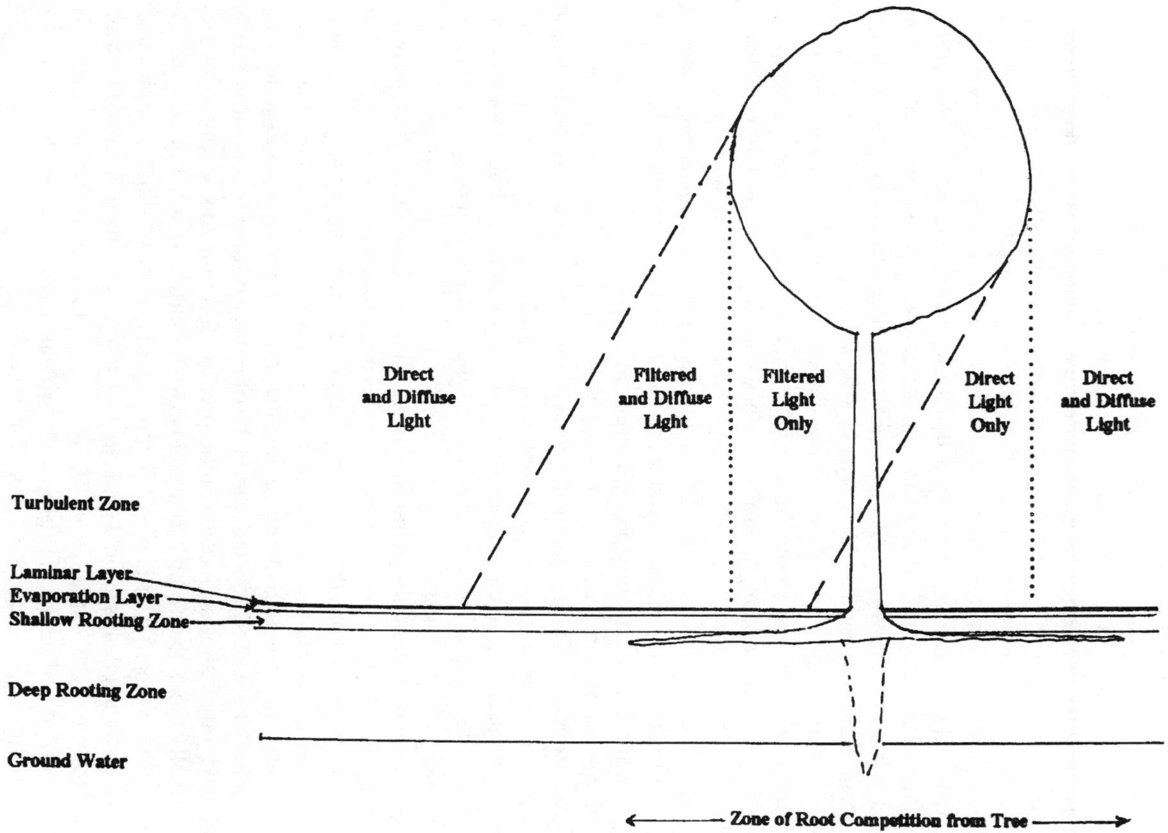


FIGURE 6.1 The above- and below-ground microenvironmental zones which comprise the growing space available to the plants of agroforestry. The sun shines from the right side of the sketch.

The functional significance of the soil evaporation layer is that water can escape from it by direct evaporation as distinct from transpiration (Leyton 1983). It is difficult but not impossible for plant roots to exploit this stratum. It includes the leaf litter and other nonliving organic layers that may be present and the **capillary fringe** of the mineral soil. The capillary fringe is the zone in which surface tension effects can bring liquid water to the surface. The main point is that direct evaporation can take place in these uppermost layers of soil, especially in unshaded places.

The boundary layer is also that in which many exchange processes take place. Water from the soil and carbon dioxide from the air are united by solar energy in photosynthesis. Water vapor escapes to the atmosphere from living tissues and nonliving surface materials. The momentum of atmospheric wind is so dampened by the frictional drag of the solid surfaces that the lowermost molecule of air is tightly held to those surfaces where it can only vibrate. When it is exposed to direct solar radiation, the boundary layer becomes hot and heats the air above by turbulent or convective movement of heated air upward. During the night the surface materials radiate energy to the sky and can become the coldest layers of the site.

The most important *temporal* microenvironmental variations in light conditions are those caused by changes in sun elevation during days and seasons as well as the difference between day and night. However, they can vary widely in space depending on the effect of different kinds of shade on the amount of quality of light, as well as on topographic differences in the degree and aspect of slope and aspect (relation to compass direction). Variations in water are more commonly temporal because of seasonal climatic factors, notably precipitation and temperature.

The deep rooting zone is that from which water can be moved upward only by roots and transpiration. The ground water zone is any layer in which water is held at the saturation level by impediments to drainage. The boundary between it and the deep rooting zone may fluctuate depending on changes in these impediments. The water in the ground water zone is available to the roots of plants only if there is sufficient oxygen for root respiration.

Precipitation wets the soil and transpiration ordinarily dries it from the top downward. When water falls on the soil it seeps down through the pore space until it has wetted all particle surfaces to their capacity to retain liquid water by surface tension. The wetted soil volume is then said to be at **field capacity,** which is the maximum amount of water that can be retained against the force of gravity. Liquid water cannot move below the wetted stratum although some may diffuse in the vapor phase. The reason why transpiration losses tend to proceed from the top downward is that there are usually more absorbing rootlets close to the surface than farther down. If it has rained recently enough, it is possible to have a layer of soil that is depleted of water below a stratum that is at field capacity and above one not yet depleted by transpiration. However, the exploitation of soil water proceeds somewhat unevenly because most of it is done by short-lived unicellular mycorrhizal mycelia and root hairs that proliferate temporarily in units of moist soil.

The exploration of the soil by fine roots is so flexible and efficient that it must be presumed that trees will exploit any unit of soil space that is available to them if they are not dormant and their woody roots come close to that space (Leyton 1983).

Truly deep-rooted species exploit all available soil levels but their seedlings may suffer while they are penetrating the upper soil levels to which their herbaceous associates are better adapted. On the other hand, these herbaceous associates can be in competition with the trees except at times when the trees are dormant or do not have roots that extend into space occupied by the herbaceous plants.

The weight of evidence, such as that of Ewel et al. (1982), Jonsson et al. (1988), and Dhynai et al. (1990) suggests that the plants of any community that are rooted in the soil have feeder roots in the upper strata. This means that it must be presumed that the components of any agroforestry system can compete for water in the upper strata of soil unless they are dormant or have failed to extend horizontally into some part of this stratum.

The notion that tree roots extend only as far horizontally as the branches of a tree is simply not true. The roots, which do not have to support themselves mechanically, usually extend much farther than the branches (Figure 6.1). They may also have intraspecific grafts with neighboring trees such that the species involved may, especially on dry sites, dominate the growing space even when the crowns of the trees do not touch. Sometimes the grafts can cause the root system of a cut tree to survive and supply water to some adjacent tree that remains standing. Reliance on sprout regeneration of trees is advantageous in agroforestry systems because it can allow one to avoid times when small trees grown from seed must compete for moisture with shorter-lived species that have roots only in the shallow-rooting zone.

Because of these effects, insidious root competition between trees for water and nutrients can be as effective as the more obvious competition for light between the foliar crowns of plants. Because of root-grafting or sprouting, the cutting of trees does not necessarily eliminate root competition. The root systems of sprouting species can remain alive through repeated cuttings of the tops and in almost permanent contact with the deep rooting stratum. With species that form functional intraspecific root-grafts the neighbors of a cut tree may simply take over its root system.

Sometimes the only way of eliminating root competition is by physically removing the roots from the soil or separating them from their foliar sources of carbohydrate by cutting them off in trenching operations. The burning of dense accumulations of dead wood, such as old logs, lying on the surface can maintain high temperatures in the soil long enough to kill roots (Weaver 1951). Virtually all of the herbicides that kill woody plants do so by killing roots.

Some species may require freedom from water loss not only from root competition but also from that induced by direct evaporation from the capillary fringe layer. This is especially true during the early stages of the life of small seedlings that are planted or germinate in place naturally. This combination of soil-moisture conditions can be produced by eliminating roots from the zones of side shade shown in Figure 6.1. Direct solar radiation can cause seedlings to die of loss of water from the capillary fringe and lethal surface temperatures even in climates with frequent rains during growing seasons. Lethal temperatures can develop on initially wet surfaces after an hour or two of exposure to direct sunlight (Smith 1951). Some shade-tolerant species of plants may require or function best in partial shade throughout their lives.

Soil moisture conditions in the deep-rooting zone are much more stable. In fact, if that zone is continually supplied with water, any root competition between plants rooted in it would be only for chemical nutrients.

## 6.2 TEMPORAL PATTERNS OF SOIL MOISTURE

The time patterns of availability of soil moisture are far from perfectly related to the timing of precipitation. There can be important lag effects resulting from storage of water in the soil. If the rooting zone is very deep, supplies of water may even be sufficient to supply deep-rooted plants throughout a long dry season. The storage capacity depends not only on the depth of the rooting stratum but also on the texture of the soil. A given volume of coarse sand can store very much less than one rich in microscopic clay particles. In cold climates much water can be stored in snow or ice and become available when the soil thaws. Interception by foliage withholds much water from the soil and can even negate the effect of small showers. Trees are particularly important in losses of water to interception and transpiration because they have large amounts of foliage which is also so high above the ground that it is very exposed to the drying effects of sun and wind. The availability of water can be increased by irrigation or the creation of terraces or other manipulations of the soil surface to concentrate water in localized spots.

Supplies of soil moisture can be affected by the rates at which plants transpire water. Within any genetic group, even within a species, there can be much inherent variation in such attributes as the amount of water stress that will cause stomates to close and reduce transpiration. There are fundamental differences in photosynthesis and water use between C3, C4, and CAM (Crassularian Acid Metabolism) species groups (Tieszen, 1983). Almost all tree species and many other plants are C3 species. Some grasses (including maize and sorghum), sedges, and some understory species conduct C4 photosynthesis and need less water for each unit of sugar produced than is the case with C3 plants. CAM plants, such as pineapple, are most efficient and, in nature, even more highly adapted to arid conditions. They open their stomates at night to accumulate $CO_2$ and close them during the day when photosynthesis takes place, but are not adapted to shaded conditions.

Trees can have some special effects on distribution of soil water that are not shown in Figure 6.1. Their crowns can intercept significant amounts of precipitation and lose it by direct evaporation. Some of the interception is temporary because water runs down the branches as stemflow and becomes concentrated around the bases of the trunks of the trees. In foggy places, small cloud droplets coalesce into large drops on leaf surfaces and fall to the ground as fog-drip, which is a major supplement to rain and snow on terrain that often projects into the clouds. Fog-drip does not collect in rain gauges in the open and is, therefore, not counted in conventional measurements of precipitation.

## 6.3 DORMANCY

Dormancy is the main defense of plants against seasons of drought or cold when soil water is not available to them. For annual plants, the seed is the dormant state. With most herbaceous perennials, such as grasses, the tops die and the roots go into varying degrees of dormancy. Deciduous trees shed their leaves. The leaves of some evergreen plants may undergo biochemical changes by which their chlorophyll changes and they go into a kind of resting stage. The seed or spore stage is the ultimate in dormancy. However, all dormant tissues continue to respire and lose water either slowly or very slowly.

The seed is the epitome of the dormant plant. It can be released from dormancy when the grower wants it to germinate either in the field or in a nursery, provided that there are no photoperiodic requirements that regulate the season of germination. This phenomenon provides much flexibility in the timing of the growth of annual plants which is dictated by the weather. Seasonal climatic effects can be anticipated but cannot be controlled except by irrigation or drainage.

The evergreen habit has two very different manifestations with gradations between the two. Where water is always available, as on many soils of rain-forest climates, the evergreen plants are capable of conducting photosynthesis except at night. At the other extreme are the evergreens of soils or climates that are only seasonally or sporadically warm and moist enough for photosynthesis. Evergreens of such relatively unfavorable climates can initiate photosynthesis promptly and opportunistically without having to construct new foliage. The evergreen species of both categories may include angiosperms and gymnosperms. There are, for example, both evergreen and deciduous oaks.

## 6.4 PATTERNS OF TEMPORAL PARTITIONING OF GROWING SPACE

The terminology used by Huxley (1983) and Nair (1990) may be adapted to distinguish between various patterns of partitioning of growing space. These are shown in Figure 6.2, with lines that indicate when the particular plants are photosynthesizing. The thick lines denote upper-stratum plants that are arborescent or woody plants and the thin lines lower-stratum plants that are herbaceous or annual.

**Coincident** describes the case in which plants grow simultaneously. Plants that are **concomitant** all start together, but one is grown continuously and another for only part of the time. For example, a deep-rooted tree crop may be functioning all the time because of an ample water supply while plants in the shallow-rooting zone run out of water and must die or become dormant. **Intermittent** refers to seasonal climates in which all of the plants involved become dormant at some time during the year but not necessarily simultaneously. The term **continuous** refers to situations in which some plants are growing all the time, even though some component may be dormant or absent part of the time. **Separate** denotes shifting-cultivation systems in which the growing of woody plants and herbaceous ones is not attempted simultaneously.

Some of the following terminology is based on Huck (1983) but most on Huxley (1983). In their use of growing space, plants of different species can be **complementary, overlapping,** or **competitive (concurrent).** Combinations of species that grow actively at the same time are concurrent; they can become compet-

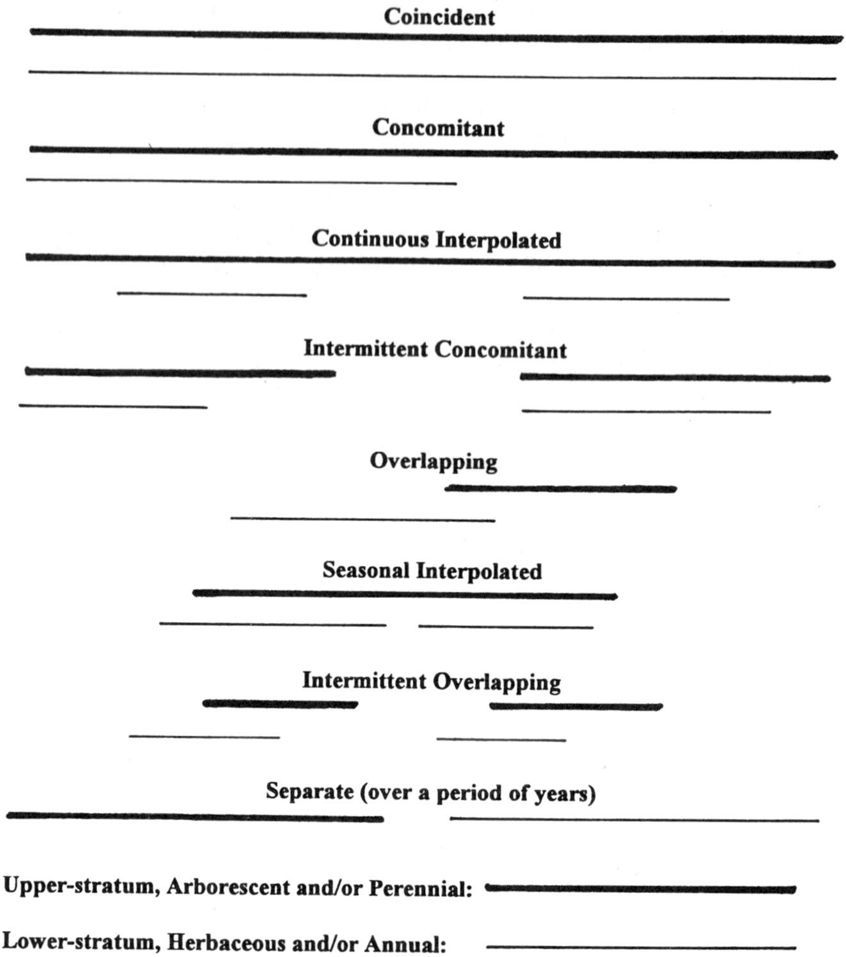


FIGURE 6.2 Temporal or seasonal patterns of interaction between the upper- and lower-stratum plants of agroforestry systems.

itive if both share the same growing space simultaneously. Overlapping species grow actively only part of the year but interact with each other only part of that time. Complementary combinations are those that do not use the same growing space at the same time.

## 6.5 COMPETITIVE EXCLUSION

Sometimes grasses or shrubs can be established beneath trees in ways that cause them to prevent the establishment of more trees. Where closed stands of trees would otherwise dominate the vegetation, this kind of agroforestry combination can be thought of as induced savanna vegetation. This illustrates the application of the principle of **competitive exclusion;** if a unit of growing space is truly and completely filled with plants, no other plants can invade.

## 6.6 CLIMATIC DIFFERENCES AND EFFECTS

The times during which the various species components of agroforestry systems can function are governed by seasonal climatic patterns (Nair 1989). These different climates are often categorized by the kind of natural vegetation that is adapted to them. The Koeppen classification of climates (Figure 6.3) is the classic one that is based on such vegetation indicators (Griffiths and Driscoll 1982). The relationship between agroforestry and each climate is described and discussed in the following paragraphs; the Koeppen letter designations for each climate are given in parentheses. Most climates that are too dry for natural trees or too cold for partially shaded annual crop plants are omitted.

**Hot, wet (Af)**—the completely aseasonal tropical rain forest of broad-leaved evergreens. There is no dormant season. Water is continuously available except for short periods when the surface layers of soil may be temporarily depleted, especially where there is no shade from trees. The variety of species that can be employed in these climates is wider than in any other. Most annual herbaceous crop plants can be started at any time of the year. Complicated agroforestry systems with multi-layer vegetation can be developed if close attention is given to the partitioning of light. Because competition for water is not serious, alley cropping is feasible, provided there are continuing efforts to prevent too much shading of subordinate crops. While there are virtually no problems with shortages of water there can be serious ones with leaching of nutrients, pan formation, and other phenomena associated with excessive precipitation. These climates are found in exceedingly well-watered places such as parts of the western Amazon Basin, central Congo, Sri Lanka, Malaysia, Indonesia, Colombia, and Panama.

Liyanage et al. (1984) have described an example of the variety of plants that can be grown beneath overstories of coconut palms in southwestern Sri Lanka. The fruits include banana, pineapple, and passion fruit. Then there are the spice plants such as chillies, tumeric, ginger, and betel. Alternatives among export crops are black pepper, cacao, cinnamon, cloves, coffee, and nutmeg. No cereals are listed but tubers that can be grown include cassava, sweet potato, taro, and yams. Soybeans, peanuts, and other legumes can be grown in the areas that are somewhat less rainy than the rest. Many other tree species could be grown in upper strata.

**Hot with short dry season (Am)**—a seasonal tropical climate with forests having varying proportions of deciduous and evergreen broad-leaves. The periods of dormancy vary in length but are much shorter for the tree layers than for those of herbaceous plants. In most cases, agroforestry systems like those of the Af climates with multi-layered vegetation can be developed. Examples of areas with these

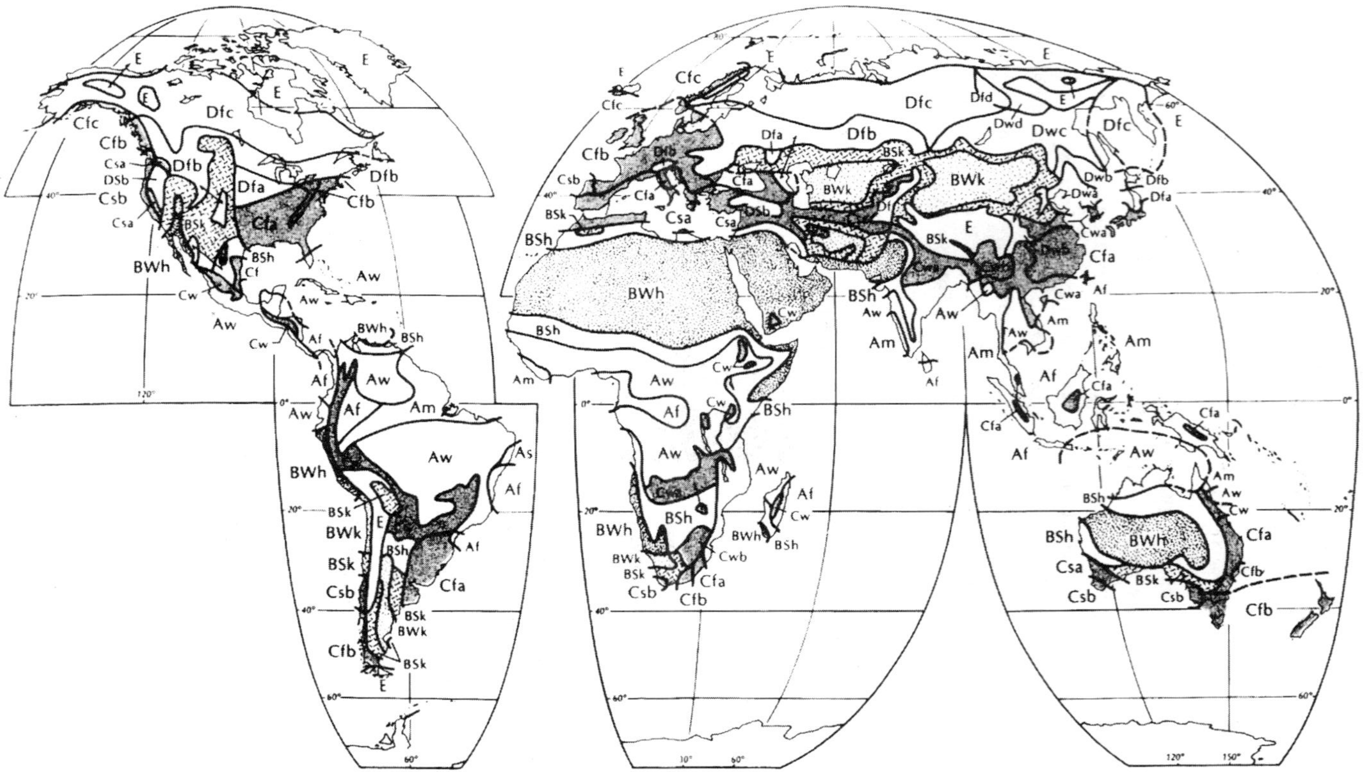


**FIGURE 6.3** The Koeppen Classification of world climates and precipitation regimes. (From a map made by the Cartographic Laboratory of Texas A & M University, published by Griffiths and Driscoll in 1982, and used by permission.)

climates are most of the Amazon Basin, the Philippines, southern India, Burma, and Liberia. In southern Thailand, upper-stratum trees grown in this zone include *Hevea brasilensis, Melia azedrach, Dipterocarpus alatus,* and *Casurina junquiana* (Watanabe, et al., 1988).

**Hot with dry and wet seasons (Aw′ and Aw″)**—a tropical monsoonal climate with pronounced dry seasons. The timing of rainy seasons is commonly determined by the advance and retreat of monsoonal rain belts into the continents. At their maximum intrusion in the Aw′ climates, there is one wet season. In areas of Aw″ climates affected first by the advance of the rain belts and later by their retreat, there are two peaks of precipitation with moderately moist periods between the two and then a definite dry season. The natural vegetation of such areas are usually savannas with the different layers limited to widely varying numbers of scattered trees with grass beneath. Tree species include *Tectona grandis, Eucalyptus camalduensis, Albizia procera, Acacia albida, Gliricidia sepium, Leucaena leucocephala, Dalberghia sissoo, Acacia nilotica, Cordia abyssinica, and Ziziphus spina-christi* (Miehe 1986, Poschen 1986). Many species of pines, such as those native to southern Mexico, are adapted to these climates. Typically the grasses grow vigorously during the rainy season after which the tops die and the roots go into dormancy. A wide variety of annual herbaceous crops, including maize, sorghum, and other cereals, can be grown in these climates. Such climates are found around the edges of the Congo and Amazon basins, the Caribbean area including southern parts of Mexico and Florida, parts of Ecuador and Thailand, as well as the north coast of Australia.

**Dry steppe and dry desert (B)**—trees do not naturally grow in these climates. However, agroforestry can be practiced at oases, on river banks, under irrigation, and in other places where there are supplies of water that are not controlled by the local climate. These climates exist in and adjacent to the Sahara, Sonora, Kalahari, Atacama, and Australian deserts, as well as Pakistan and the vast grasslands of the interiors of continents. Adaptable species are much like those of the Aw climates.

**Mid-latitude, uniform precipitation, hot (Cfa) or warm (Cfb) summer**—complex broad-leaved deciduous forest with some conifers, especially on dry sites. A dormant season is induced by winter cold. During the summer the partitioning of light is almost as intricate as that of the moist tropics. It is possible to maintain multilayered vegetation. Annual herbaceous plants can be grown during the warm season if there is sufficient light. The natural vegetation of deciduous forests of these climates often has a flora of herbaceous perennials that blooms and grows most actively during the spring before the trees come into leaf.

A very wide variety of agroforestry combinations of species are theoretically feasible. Among the combinations of trees and crops grown in these climates are the fast-growing *Pawlonia* trees with rice in Asia. Silvopastoral practices may include such trees as walnuts, oaks, or chestnuts over grasses. These distinctly seasonal climates exist in broad areas of the eastern coasts of Asia and North America, as well as New Zealand and eastern Australia, eastern South Africa, Uruguay, southern Brazil, southeastern Alaska, southern Chile, and some western coasts of Europe.

**Mid-latitude, dry winter, hot or warm summer (Cw)**—drought-resistant broadleaves and conifers, often with grass understory; trees grow well in the moister parts but may be vulnerable to winter frost damage. Examples of such climates exist in Arizona, New Mexico, highland Mexico, Pakistan, and northern India. Although more conductive to silvopastoral activity than to agroforestry with annual food crops, irrigation greatly increases latitude for use of both tree and crop components. Among the tree species used in Pakistan are *Eucalyptus camalduensis, Albizia procera, Morus alba, Leucaena leucocephala, Dalberghia sissoo,* and *Acacia nilotica* (Akbar et al. 1990). There are important beneficial effects of windbreaks in reducing evapotranspiration in these climates.

**Mid-latitude, wet winter, dry and hot or warm summer (Cs)**—all vegetation drought-resistant; grassland, savanna, or closed forest, depending on how much water is stored from winter precipitation. These are the Mediterranean climates which exist around the Mediterranean Sea, southeastern Australia, and generally on the west coasts of continents. Sometimes the growing seasons come during winter, spring or autumn rather than summer. In cloudy coastal fringes adjacent to cold waters fog-drip may suffice to prevent summer drought as in the redwood and Sitka spruce-western hemlock belts of the western coast of North America.

The agroforestry tree species of Mediterranean climates include olive, cork oak, *Pinus radiata, Eucalyptus globulus,* and various species of fruit trees. If it is very dry in the summer, the herbaceous crops are likely to be limited to grasses and cereals. Under more favorable conditions a wide variety of vegetables can be grown.

At low elevations in parts of California, deciduous oaks tend to occupy soils where there are dependable supplies of soil water during spring and early summer. The opportunistic evergreen oaks often occupy drier soils with grass where there are sporadic periods when it is moist enough and warm enough for photosynthesis, mainly during the cooler parts of the year.

**High latitude, uniform precipitation, hot (Dfa) or warm (Dfb) summer**—varies from grassland to savanna ("prairie"); suitable for silvopastoral husbandry—mid-continent areas of the northern U.S., Canadian prairie provinces, Ukraine, central Russia, and northern Japan. Dfa climates are limited to the Northern Hemisphere. The trees species employed here are often oaks or pines.

**High latitude, dry summer (Dsb)**—savanna of grass and drought-resistant shrubs and conifers such as at the lower elevations of eastern Washington, where silvopastoral systems are often employed.

**High latitude, dry winter (Dw)**—forests of either conifers or fire-following deciduous species in eastern Siberia and northeast China. Perhaps suitable for silvopastoral systems.

**Highland climates**—both climate and vegetation vary widely, depending on effects of temperature-reduction associated with high elevation and orographic effects of terrain on precipitation. The vegetation or ground cover can range from tropical rain forest to glaciers. In other words, almost any of the other climatic regimes can prevail and can be identified as such. Temperature effects depend much more on altitude than latitude. Seasonal precipitation regimes will be essentially the same as at lower elevations but with very important exceptions depending on the direction of rain-bearing winds. Precipitation will be increased on windward slopes and reduced, sometimes to very small amounts on leeward slopes. There is much fog-drip in zones where clouds intersect the land surface.

The necessity of putting lines on maps obscures the point that the boundaries are bands of transition rather than true lines. Furthermore they are not even in the same places in every week or every year. The meteorological processes that cause precipitation are especially complex and highly mobile. Most seasonal climates are the result of the continual shifting of rainy belts and changes of the sources of humid air.

The classifications are also based on average conditions. The fact that it normally rains during June in Belize does not mean there will never be drought in Belize in June. For this reason, the annual plants chosen for use in agroforestry should be ones adapted to the variability of rainfall in the locality and not merely to average conditions. However, if more than one annual species is customarily used there is a collective adaptability in which the loss of one species is compensated by the survival of another.

It should be noted that the direct relationship between plants and water climate depends on soil moisture and not on the basically indirect connection with precipitation. Artificial irrigation can cancel the effects of precipitation. Water is also continuously available on most river floodplains even with rivers that flow through deserts. In fact, the combinations of vegetation used in agroforestry can be the way of imposing some semblance of order on the chaos of natural vegetation that often proliferates on alluvial sites. The storage of water in snow can prolong the supply of available soil water long into a dry season.

On the other hand, a soil on which poorly aerated, ponded water persists for long periods may be so deficient in available water that any semblance of agriculture is impossible and only specialized tree species can grow. An excessively drained sand may be too droughty even in high rainfall areas and the same is true with the poor moisture storage of thin soils on bedrock.

Climatic classifications can be useful in selecting exotic species for use and in understanding the time schedules of agroforestry. Species moved to the wrong zones may suffer not only from extremes of temperature or soil moisture but also from fungi and other pests to which they are not climatically adapted. For example, moving species from west coasts to east coasts is often counterproductive as is moving species from summer dry to summer wet climates. *Pinus caribaea* thrives in summer-wet climates while *Pinus radiata* is a species which is winter-wet and summer-dry.

## 6.7 ROOT PHENOMENA

Far less is known about the development of root systems than about the display of foliage. Dissection by digging can be difficult and roots are classic cases of phenomena that cannot be studied without disturbing them. This is especially true of the unicellular mychorrizal mycelia and root hairs which actually do most of the absorption of water and nutrients. There is evidence that these microscopic structures are very short-lived and that they extend temporarily into soil spaces that have water or nutrients and then die when they have exploited the resources there (Huck 1983). This kind of explorator exploitation appears to be governed mainly by need for nutrients, at least where supplies of water are ample (Fitter 1994).

However, the main consideration involving roots and water is that the woody perennial members of agroforestry systems are more deeply rooted than the herbaceous plants. Therefore, the perennials normally can function for longer periods. There are exceptions, such as the spring flora of some temperate deciduous forests. In these, herbaceous perennials of the ground flora develop leaves as soon as it becomes warm enough at the level of the forest floor; the plants then complete most of their annual development before the trees are fully in leaf. Some of the ring-porous deciduous trees are especially slow in developing leaves and casting shade because they must first develop the large and very specialized vessels in the new xylem before they can supply water to the new leaves.

Even though established trees are usually deep-rooted, they are not when they are young. They can be in severe competition with shrubs and herbaceous species when the trees' roots are still confined to the boundary layer and the shallow-rooting zone. This poses especially critical problems in agroforestry systems when new trees are being started. Sometimes the problems can be mitigated by depending on planting trees that already have roots, especially roots long enough to extend below the soil evaporation layer. If new trees arise from the stumps or roots of cut trees, they do not have to depend on moisture in the shallow-rooting zone.

Storage of water by soils can allow roots to function for significant periods after the cessation of precipitation. If the soil is sufficiently warm, porous, and deep, the supply of water for trees and other deep-rooted plants may endure though months without rain and photosynthesis continues.

Perhaps no species is more suitable for use in agroforestry in seasonal parts of the tropics than *Acacia albidum* which bears its leaves during the dry season but is leafless in the rainy season when grasses and other herbaceous plants can grow beneath it. Its natural habitat is in gallery forests of central African river flood plains although it is widely planted in savanna regions (Miehe 1986). Its remarkable adaptation is probably due to a deep-rooted growth habit that enables it to maintain contact with permanent water tables such as those which exist in its natural habitat. It is not clear what survival value lies in being leafless during rainy seasons. It can be pollarded to provide forage during dry seasons. *Gliricidia sepium* of Central America has the same kind of seasonal behavior.

## 6.8 RELATIONSHIP TO CHOICES OF SPECIES

The trees and other woody perennials used in agroforestry are more sensitive to seasonal moisture regimes and other ecological factors than the herbaceous components. While they may have dormancy mechanisms, such as the deciduous habit, various above-ground parts of the plants must respire and transpire just enough to remain alive. The herbaceous perennials, which are next most sensitive, can retire into deep dormant conditions in which only the roots remain alive. The seed is the dormant stage of annual plants; seeds can be stored in the soil, on standing dead plants, or by artificial means until conditions are appropriate for germination.

Artificial irrigation greatly increases the latitude for choice of species and the same is true of moist but well-aerated soils along the banks of streams. On the other hand, physiological dryness limits choices to rather specialized species on the wet and poorly aerated soils of various kinds of wetlands. Thin or sandy soils with little moisture-storage capacity also require choices of drought-resistant species.

The safest choices of agroforestry species are likely to come from the native vegetation, which has a history of adaptation to local precipitation regimes. If exotics are introduced, it is best that they be genetic material from similar climates. There is, however, much latitude for use of exotics. It is greatest with those herbaceous annuals that can grow to harvestable size during short growing seasons. Some tree species are well-known for their adaptability to a wide variety of climatic conditions. Most of them are adapted to periods of moisture deficiency. Among them are *Eucalyptus camalduensis, Pinus radiata, Leucaena leucocephala, Tectona grandis, Acacia albidum,* and *Gliricida sepium.*

The effect of moisture regimes is not limited to that on soil water; they may also influence the degree of safety from attack by fungi and other pests.

## 6.9 REFERENCES

Akbar, G., M. Amhad, S. Rafique, and K. N. Babar. Effect of trees on the yield of wheat crop. *Agroforestry Syst.* 11 (1990): 1–10.

Caldwell, M. M. and R. W. Pearcy, eds. *Exploitation of Environmental Heterogeneity by Plants. Ecophysiological Processes Above- and Below-Ground.* (San Diego: Academic Press, 1994), 429.

Dhyani, S. K., P. Narain, and R. K. Singh. Studies on root distribution of five multipurpose tree species in Doon Valley, India. *Agroforestry Syst.* 12 (1990): 149–161.

Ewel, J., F. Bendict, C. Berish, B. Brown, S. Gleissman, M. Amador, R. Bermudez, A. Martinez, R. Miranda, and S. Price. Leaf area, light transmission, roots and leaf damage in nine tropical plant communities. *Agro-Ecosystems* 7 (1982): 305–326.

Fitter, A. H. Architecture and biomass allocation as components of the plastic response to soil heterogeneity, in: *Exploitation of Environmental Heterogeneity by Plants. Ecophysiological Processes Above- and Below-Ground,* M. M. Caldwell and R. W. Pearcy, eds. (San Diego: Academic Press, 1994), 305–323.

Griffiths, J. F. and D. M. Driscoll. *Survey of Climatology.* (Columbus, OH: C. E. Merrill Publishing Co., 1982), 358.

Huck, M. G. Root distribution, growth, and activity with reference to agroforestry, in: *Plant Research and Agroforestry,* P. A. Huxley, ed. (Nairobi, Kenya: International Council for Research in Agroforestry, 1983), 528–542.

Huxley, P. A. Comments on agroforestry classifications: with special reference to plant aspects, in: *Plant Research and Agroforestry,* P. A. Huxley, ed. (Nairobi, Kenya: International Council for Research in Agroforestry, 1983), 161–171.

Jonsson, K., L. Fidjeland, J. A. Maghembe, and P. Högberg. The vertical distribution of roots of five tree species and maize in Morogoro, Tanzania. *Agroforestry Syst.* 6 (1988): 63–69.

Kimmins, J. P. *Forest Ecology.* (New York: Macmillan Publishing Co., 1987), 531.

Ledig, F. T. The influence of genotype and environment on dry matter distribution in plants, in: *Plant Research and Agroforestry,* P. A. Huxley, ed. (Nairobi, Kenya: International Council for Research in Agroforestry, 1983), 427–454.

Leyton, L. Crop water use: principles and some considerations for agroforestry, in: *Plant Research and Agroforestry,* P. A. Huxley, ed. (Nairobi, Kenya: International Council for Research in Agroforestry, 1983), 379–400.

Liyanage, M. de S., K. G. Tejwani, and P. K. R. Nair. Intercropping under coconuts in Sri Lanka. *Agroforestry Syst.* 2 (1984): 215–228.

Miehe, S. *Acacia albidum* and other multipurpose trees on the fur farmlands in Jebel Marra highlands, Western Darfur, Sudan. *Agroforestry Syst.* 4 (1986): 89–119.

Nair, P. K. R. *Agroforestry Systems in the Tropics.* (Boston: Kluwer Academic Press, 1989), 664.

Nair, P. K. R. Classification of agroforestry systems, in: *Agroforestry: Classification and Management,* K. G. MacDicken and N. T. Vergara, eds. (New York: John Wiley and Sons, 1990), 31–57.

Poschen, P. An evaluation of the *Acacia albida*-based agroforestry practices in the Hararghe highlands of Eastern Ethiopia. *Agroforestry Syst.* 4 (1986): 129–143.

Reifsnyder, W. E. and T. O. Darnhofer, eds. *Meteorology and Agroforestry.* (Nairobi, Kenya: International Council for Research in Agroforestry, 1989), 546.

Smith, D. M. Influence of seedbed conditions on the regeneration of eastern white pine. *Conn. Agric. Exp. Sta. Bull.* 545. (1951), 61.

Tieszen, L. L. Photosynthetic systems: implications for agroforestry. in: *Plant Research and Agroforestry,* P. A. Huxley, ed. (Nairobi, Kenya: International Council for Research in Agroforestry, 1983), 323–346.

Watanabe, H., P. Sahunalu, and C. Khemnark. Combinations of trees and crops in the taungya method as applied in Thailand. *Agroforestry Syst.* 6 (1988): 109–177.

Weaver, H. Fire as an ecological factor in the southwestern ponderosa pine forests. *J. of For.* 49 (1951): 93–98.

# 7 Nutrient Cycling and Nutrient Use Efficiency in Agroforestry Systems

*Florencia Montagnini[1], Carl F. Jordan,[2] and Rodrigo Matta Machado[3]*

## TABLE OF CONTENTS



[1] Centro Agronómico Tropical de Investigación y Enseñanza (CATIE) 7170 Turrialba, Costa Rica

[2] University of Georgia, Department of Ecology, Athens, GA 30602, U.S.

[3] University Federal de Minas Gerais, Instituto de Ciencias Biológicas, Belo Horizonte, M.G. 31270.010, Brazil

0-8493-2206-5/00/$0.00+$.50

## 7.1 INTRODUCTION

In humid tropical regions, soil nutrients are key factors influencing plant productivity and long-term sustainability of a production system. It has long been maintained that nutrient losses and increased weed invasion are the principal contributors to abandonment of fields of annual grains or root crops after two to four years of use (Watters 1971, Van Wambeke 1992, Bandy et al. 1993). Increased weed infestation of agricultural fields after nutrient loss results from the greater ability of weeds to take up scarce nutrients relative to most annual or root crops, thus out-competing these plants in the use of nutrients and other resources. Many "weed" species are perennials and woody species, whose ability to compete depends on a variety of mechanisms associated with roots, woody stems, and litter production. A key to increasing the sustainability of agricultural systems is to incorporate species that have these nutrient-conserving capabilities, while at the same time producing a crop that has economic and/or subsistence value. Many indigenous peoples have devised such systems through trial and error, and this chapter presents examples from the Kayapó in Brazil, and from other traditional systems from Asia and Africa.

The importance of nitrogen-fixing species in agroforestry systems is widely recognized. In this chapter, we discuss this and other effects of woody species on soil fertility, the interactions between species of woody plants, and the significance for design of agroforestry systems. The concept of nutrient use efficiency by plants is presented as a potential tool to aid in system design and management. Finally, we stress the need for initial economic incentives to allow for system establishment and to increase the adoptability of sound agroforestry techniques by local farmers.

## 7.2 NUTRIENT DYNAMICS IN AGROFORESTRY SYSTEMS

### 7.2.1 EXAMPLES OF TRADITIONAL SUSTAINABLE AGROFORESTRY

Shifting agriculture has been practiced in the tropics for many centuries. Today, shifting (also called "swidden," "slash-and-burn") agriculture is the predominant land-use practice on about 30% of the arable soils of the world and provides sustenance for an estimated 300 million of the world's poorest people (Andriesse and Schelhaas 1987). Traditional shifting agriculture uses long forest fallows between short periods of farming. Long fallows make the traditional technique sustainable but also require extensive amounts of land. When land is scarce, farmers shorten forest fallows and lengthen agricultural periods, resulting in soil nutrient depletion, reduced crop yields, and increased weed invasion.

Similar patterns are reported in the tropics worldwide, in spite of differences in ecological and socioeconomic conditions. For example, the "jhum" cycle is a system of shifting agriculture widely practiced by about 30 million people in the hilly, subtropical region of northeast India (Ramakrishnan 1992). A variety of versions of jhum agriculture are practiced by the different tribes in cycles that range from 10 to 30 years. However the shortening of the jhum cycle to four to five years during recent times has led to concerns about the resulting soil degradation and yield declines in a pattern much like that experienced in lowland humid regions of Latin America and Africa.

#### 7.2.1.1 An Example from the Kayapó in Brazil

There are numerous examples of agricultural practices of indigenous peoples, or of colonists such as the Amazonian caboclos, who use trees to maintain or restore soil fertility without chemical fertilizers (Anderson 1990, Gómez-Pompa and Kaus 1990, Subler and Uhl 1990, Balée 1992, Lescure et al. 1992, Nations 1992, Jordan 1995). One well-documented example is that of the Kayapó in Brazil (Posey 1982). The Kayapó live today on a 2.5-million-hectare reserve in the Xingú River Basin in the Amazon region. Although the Kayapó are nomads for part of the year, cultivation of plants for food and medicine is an important part of their culture. Cultivation begins with the clearing of a circular field. Trees are felled so that the fallen stems radiate outward, and the bulk of the forest canopy biomass ends up near the perimeter of the circle. Root crops such as sweet potatoes or yams (*Dioscorea* spp.), taro (*Colocasia esculenta*), and manioc (*Manihot esculenta*) are planted in open corridors left between the fallen trees. The crops are already rooted and growing before burning occurs.

Burning is carefully managed. Tribal elders agree upon an appropriate day when winds are minimal and the fields will burn thoroughly but not too quickly. The farmers begin burning the piles of dried debris one at a time. A protracted burn minimizes the heat, so that the root crops will lose their green tops but not their viability. These preburn crops are given a head start on weeds that will establish in the ash.

Papaya, bananas, cotton, urucú (*Bixa orellana*), and tobacco, which all require a high quantity of nutrients, are planted on the outer margins of the field, where ash concentrations are highest. A few weeks after the burn, the farmers gather up unburned sticks and limbs for a second fire. In the resulting piles of ash, other high-nutrient requiring plants like beans, squash, and melons are planted.

The fields of the Kayapó last for many years. Sweet potato and yam bear in fields that are four or five years old. Bananas and urucú, and domesticated varieties of a large vine-like plant called "kupa" commonly continue to bear edible leaves and stalks for eight to 12 years, and some fields that are 40 years old still yield edible kupa.

Many plants useful to the Kayapó establish naturally in the old fields. Some of these spontaneously colonizing plants have important medicinal values, and others provide seeds, berries, and roots for food. Some of the colonizing plants bear fruits that make excellent fish bait. Others attract animals or birds. The animals drawn to the leafy and bushy plants in these sites are easier to hunt than those inhabiting the canopy of the high forest.

Because the Kayapó understand this process and take advantage of the species that sequentially occupy a site, they do not need to continually seek new forest to cut and burn. After many years, when an old site develops into a closed forest, it can be cut and used again, with no long-term degradation of the site.

Kayapó practices contrast with the shifting cultivation in the Amazon carried out by colonists from southern and northeastern Brazil. The latter depend mainly upon crops such as corn, rice, and cassava which grow well for only two or three years. When yields decline, the colonists abandon the fields and clear new forest.

### 7.2.1.2 Homegardens: Traditional Low-Scale, Low-Input Agroforestry

Homegardens are systems for the production of subsistence crops for the gardener and family, with or without the addition of cash crops. They can be located immediately surrounding the home or slightly further away, but still near the residential area. It is claimed that homegardens originated in prehistoric times when hunters and gatherers accidentally dispersed seed of highly valued fruit trees close to their homes (Soemarwoto 1987). In the near east, homegardens are documented in paintings dated 3,000 years B.C. and the practice continues in modern times. There are several well documented examples from Java, and they are also common in other parts of Indonesia, and in Malaysia, India, and in countries of Africa and Latin America. They hold great species diversity with many life forms varying from climbers to tall trees and vines, creating a forest-like, multistory canopy structure. The canopies of most homegardens consist of two to five layers. Usually there are no rows, blocks, or definite planting distances among components. Chemical fertilizers are generally not used; dung, household wastes and pruning residues are used instead. The use of species with antipest properties is also a widespread practice that decreases the need for chemical pesticides (Fernandes et al. 1989, Michon et al. 1989). They generally have stable yields and great variety of products, allowing continuous or repeated harvests during the year under a low-input system.

In west Java, the average size of homegardens is $< 0.1$ ha, with an average of 19 and 24 species per garden in the dry and wet seasons, respectively. Size of home gardens decreases with altitude with highest number of species occurring at 500–1000 m. Poor people tend to grow more staples, vegetables, and fruits; well-off people tend to grow more ornamentals and high-economic value cash crops. More subsistence crops are grown in remote areas; more cash crops are grown near cities. Culture and tradition influence composition: e.g., more medicinal plants are found in west Java, while tobacco and coffee are more commonly grown in Muslim districts of southern Ethiopia; animals are found in most gardens but pigs are not found in Muslim homegardens; in west Java with intense rains fishponds are usually present (Soemarwoto 1987).

Homegardens can be sustainable production systems, however this is true under low-input and low-yield conditions. For example, the homegardens of the Chagga, in Mt. Kilimanjaro (Tanzania) represent ecologically sustainable land-use systems, but their productivity is relatively low and needs to be increased if they are expected to support larger populations (Fernandes et al. 1989). Migration of youngsters to urban areas has disrupted the traditional transmission of the knowledge and experience required for the successful management and perpetuation of the complex multicropping system. Availability of fertilizers has decreased the need for organic manures, thus greatly reducing labor inputs in homegardens and therefore reducing nutrient recycling processes. If homegardens are to be used for raising the standard of living of people to satisfactory levels, the question arises whether the yield and the income can be significantly increased without sacrificing their sustainability.

### 7.2.2 Nutrient Mobilization and Losses in Shifting Agriculture

Scientists have long believed that sharp decreases in growth of annuals like corn and rice after two or three years of shifting cultivation were due to nutrient losses such as leaching of calcium and potassium and volatilization of N. However, results from the Man and the Biosphere project at San Carlos de Río Negro in Venezuela (Jordan 1989) suggested that during the first two to three years of cultivation of a cleared forest, only a very small proportion of the nutrient stocks was actually lost through leaching (Figures 7.1, 7.2). The decrease in production was instead due to binding of formerly labile phosphorus by iron and aluminum in the mineral soil, thus rendering the P unavailable to crop plants.

In undisturbed forests of the region, P appears to be readily available to the trees (Jordan 1989). Even after cutting and burning the forest, and during the first few years of cultivation, most of the soil P was kept mobile because it was chelated to Fe and Al by organic acids leached from decomposing organic matter on the soil surface. After the site was cut and burned, and as the humus and litter gradually disappeared during the three years of cultivation, liberation of organic acids decreased and an increasing proportion of P was bound in the soil. By the end of the third year, all of the humus, and most of the tree trunks were gone. The study concluded that lack of labile P caused diminishing crop productivity, and that the conversion of P from labile to bound states was due to the disappearance of humus and organic matter on the soil surface.

Despite binding of P in the soil, secondary successional vegetation invaded the site, and by the fifth year stocks of Ca, K, Mg, and N began to decline (Figures 7.1, 7.2). There were no detectable changes in total ecosystem stocks of P during the experiment. However, there was an increase in P in biomass as the successional forest became established. Apparently, successional plant species were able to take up bound P from the soil. This P was probably in Fe- and Al-bound fractions that were unavailable to the crop plants.

What is the mechanism through which woody vegetation is able to take up the P unavailable to crops? One possibility is excretion of piscidic acid[1] from the roots (Ae et al. 1990), or leaching of citric and malic acids from the decomposing leaf litter of the trees (Han 1989). These organic acids can replace the P bound by Fe and Al in the clay, liberating the P and rendering it soluble and readily available for uptake. Annual plants may not have these P recycling mechanisms, or crop plants bred for productivity may have lost the capability of taking up these forms of P (Chapin 1980, 1983).

---

[1] Piscidic acid = (p-hydroxy benzyl) tartaric acid.

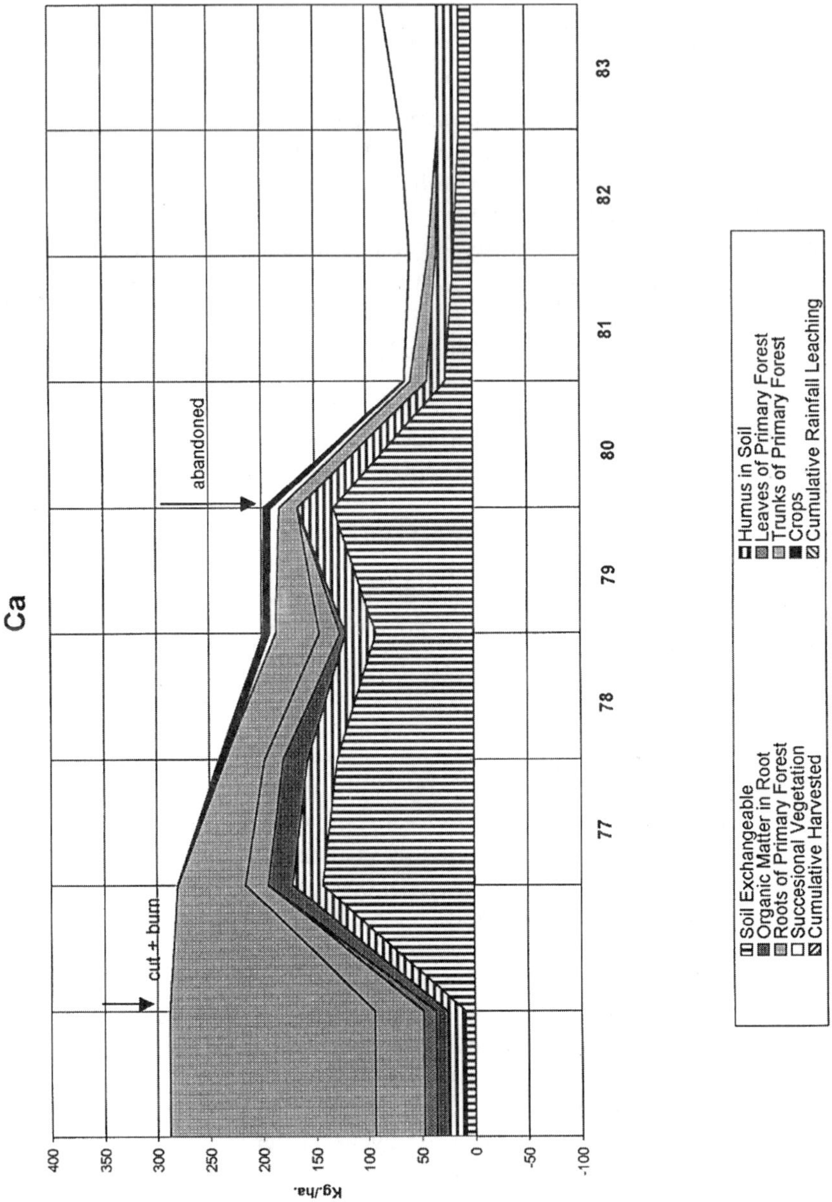
Ca
Kg./ha.
400
350
300
250
200
150
100
50
0
-50
-100
77
78
79
80
81
82
83
cut + burn
abandoned
Soil Exchangeable
Organic Matter in Root
Roots of Primary Forest
Succesional Vegetation
Cumulative Harvested
Humus in Soil
Leaves of Primary Forest
Trunks of Primary Forest
Crops
Cumulative Rainfall Leaching

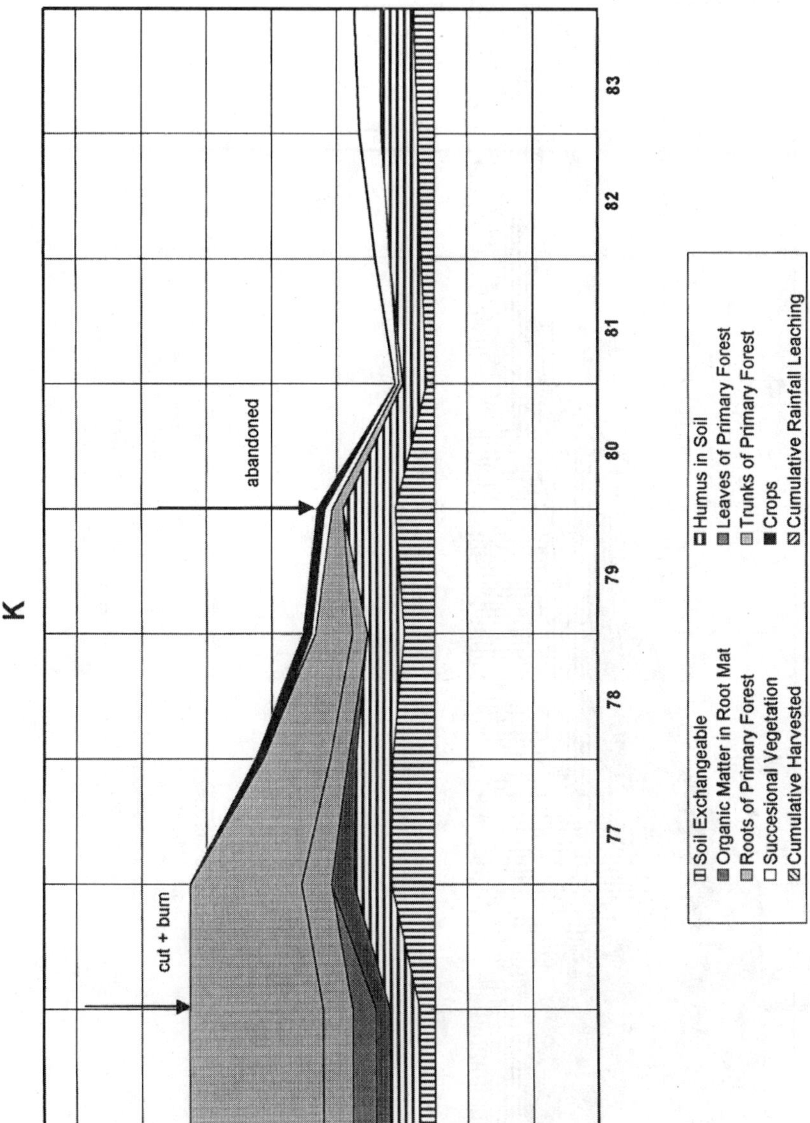


**FIGURE 7.1** Stocks and cumulative losses of calcium and potassium as a function of time in the experimental plot at San Carlos de Rio Negro, Venezuela.

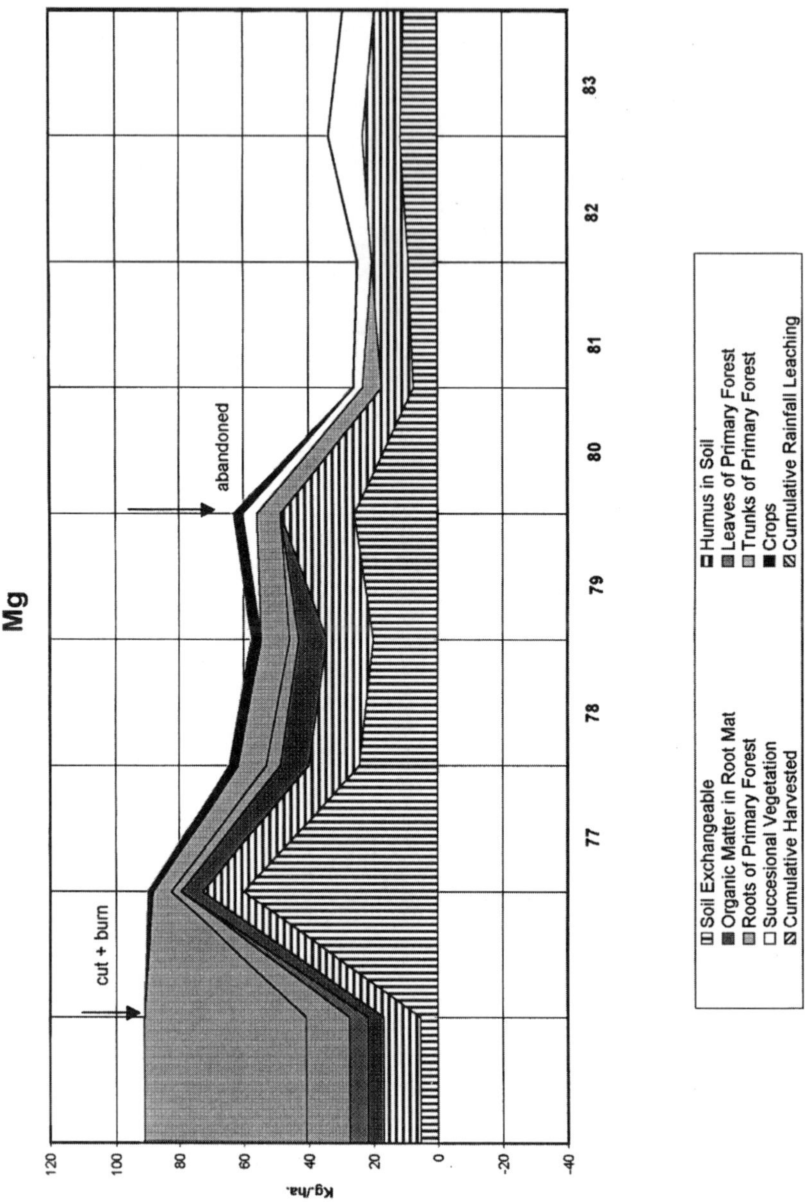
Mg
Kg./ha.
120
100
80
60
40
20
0
-20
-40
77
78
79
80
81
82
83
cut + burn
abandoned
Soil Exchangeable
Organic Matter in Root Mat
Roots of Primary Forest
Succesional Vegetation
Cumulative Harvested
Humus in Soil
Leaves of Primary Forest
Trunks of Primary Forest
Crops
Cumulative Rainfall Leaching

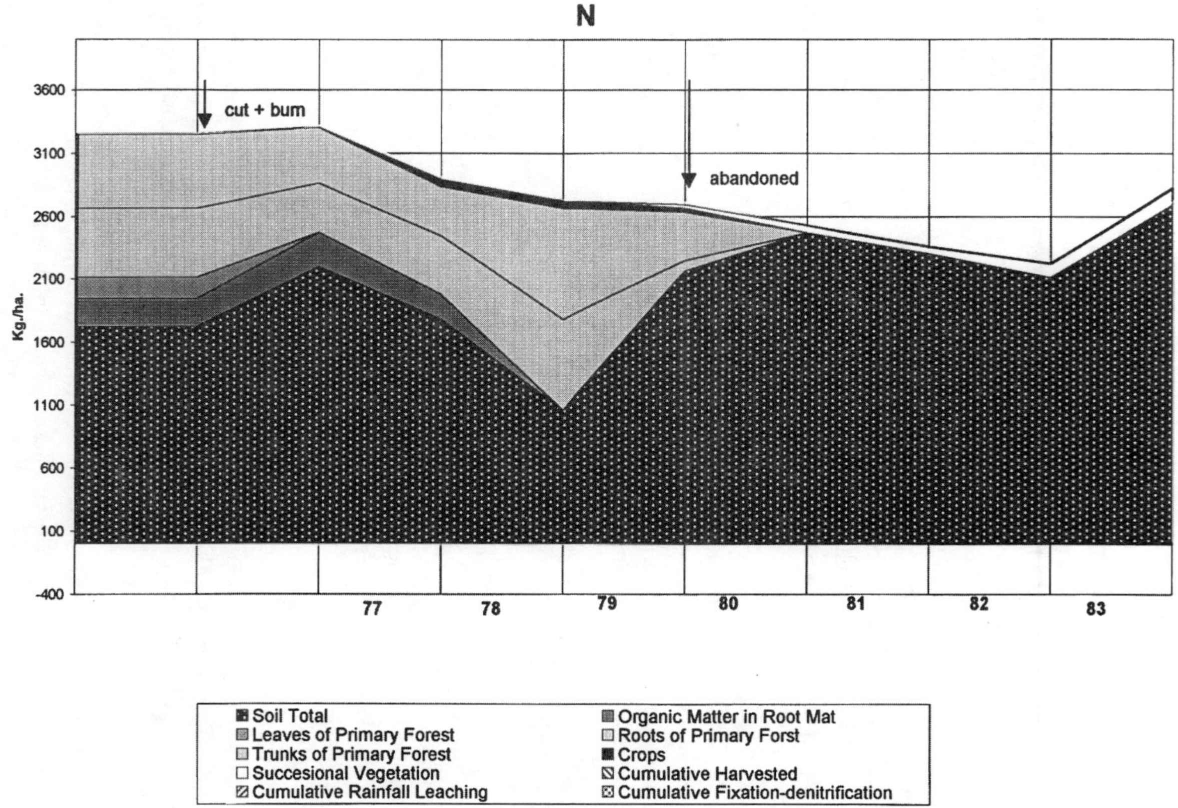


**FIGURE 7.2** Stocks and cumulative losses of magnesium and nitrogen as a function of time in the experimental plot at San Carlos de Rio Negro, Venezuela. Keys as for Figure 7.1.

Mobilization of bound P by woody plants accounts for sustainability of agroforestry systems such as those of the Kayapó. The tree crops are established soon after clearing and burning, before the tree trunks, branches, and organic matter in the upper soil horizons are completely decomposed. Because the new trees are already producing litter before the remains of the old ones disappear, the production of organic acids is not interrupted, and the P is kept in a labile state.

This suggests that it would be desirable to start an agroforestry system immediately after clearing a fallow or primary forest. For example, establishment of rubber trees (*Hevea brasiliensis*) in the Brazilian Amazon region has been highly successful under the partially opened canopy of a secondary forest (Mesquita 1995). At the time of planting, a thick layer of litter and humus on the forest floor supplied nutrients, improved the microclimate, prevented erosion, and was an energy source for soil microorganisms that improved the physical and chemical properties of the soil.

Unfortunately, many agroforestry systems are established on sites that have been cropped or kept in pasture for many years. In such sites, nutrient status of the soils is low. Figures 7.1 and 7.2 show that in the experimental plots at San Carlos de Río Negro, nutrient loss continued even after cultivation was abandoned after three years, and nutrient recovery did not begin until year five. Apparently in this system with high rainfall (3600 mm/year) and low cation exchange capacity, it took a couple of years for fallow vegetation to cover the site and reestablish nutrient cycling mechanisms that lead to recovery. By the time the reversal began, nutrient stocks were only a fraction of those in the ecosystem when cultivation started.

### 7.2.3 Managed Forest Fallows

Improved fallows have been proposed as a management alternative to shifting cultivation in the tropics (Nair 1990, Kass et al. 1993). Traditionally shifting cultivators have encouraged the presence of certain tree or herb species in fallows to restore site fertility, suppress weeds, and increase economic yields. Several types of traditional "enriched" fallows have been described, including those techniques involving planting or tending selective species for fruit, fuelwood, or timber in fallow fields or secondary forests which are maintained by local populations over long periods of time, for local consumption, for markets, or for both (Kass et al. 1993, Denevan at al. 1984, Padoch et al. 1985, Padoch and De Jong 1987, Raintree and Warner 1986, Unruh 1990, Deal Amo and Ramos 1993, Sips 1993). Some of these systems produce crops for local consumption and for a regional market, providing substantial cash income for many farmers (Padoch et al. 1985).

The use of managed fallows based on a single species is quite widespread in the Americas, occurring from subtropical areas of Brazil to highland regions of Central America (Kass et al. 1993, Sips 1993). These systems include both biologically and economically enriched fallows. Some of the economically enriched fallows of the Amazon require more intensive management by which certain trees are protected during clearing or planted during the cropping period and maintained during the fallow (Padoch and De Jong 1987). In the planted fallows, one or more species with biological or economic value are introduced to shorten the fallow regeneration period or increase its economic value (Szott et al. 1991, Vergara 1987).

For example, in an experimental fallow system in the Peruvian Amazon, selected soil-improving tree species were planted in abandoned shifting agriculture fields (Szott et al. 1991). The species planted were acid-tolerant woody legumes, *Cajanus cajan* and *Inga edulis.* Two years after planting, the total ecosystem levels of Mg and Ca declined, while total N and K increased, and the levels of P increased after 4.5 years. In these experiences, weed control was achieved more rapidly with herbaceous species *(Pueraria phaseoloides, Desmodium ovalifolium);* however, good suppression of weeds was also eventually obtained by the wood legumes.

Experiences such as those described by Szott et al. (1991) are relevant in tropical regions worldwide where land becomes scarce and the fallows are not long enough to restore soils to their productive capacity. For example, *Gliricidia sepium,* a N-fixing tree native from the neotropics and broadly used there as shade for coffee, for living fences, and as a common component of homegardens, is currently a popular fallow tree in some parts of lowland rain-forest regions of western Nigeria (Adejuwon and Adesina 1990). Its poles are staked as support for training yam vines, and as the species propagates vegetatively, the staked poles coppice within a short time and at the end of the cropping cycle they become part of the fallow vegetation. Compared to natural fallows in which the development of trees is random, the progress of the cultivated fallows of *G. sepium* leads to greater organic matter build up and increases in nitrate-nitrogen and potassium concentrations in the soils. Other leguminous trees that are also becoming popular in the region such as *Leucaena leucocephala* or *Derris indica* could also be used with similar results (Adejuwon and Adesina 1990).

Other improved fallow systems rely more on the introduction of valuable species in the fallow period, in combination with other species that have soil restoring capacity. For example, in a traditional shifting cultivation system in the lowlands of Papua New Guinea, the cropping cycle is usually 18 months, consisting on mixed food crop gardens with yams, bananas, taro, sugar cane, and some fruit trees. The fallow cycles last up to 30 years. Robusta coffee, a cash crop component, has been added to the fallows in some areas since the 1950s. The coffee is interplanted with *Leucaena leucocephala* as shade, and food crops are planted in the establishment phase. This seems to be a promising enriched fallow alternative, and the local fallow gardeners appear to be willing to take on such innovations (Allen 1985).

### 7.2.4 Nutrient Dynamics in More Recent Agroforestry: Alley Cropping

In the mid-1980s, considerable research focused on the development of low-input technologies for sustainable food production by small-holder farmers. Results from this research suggest that in situations where planted fallows are not feasible because land is scarce, techniques such as alley cropping and the application of mulch (green covers) may become practical alternatives (Kang and Wilson 1987, Kang et al. 1990).

In alley cropping, annual crops are grown between hedgerows of preferably N-fixing leguminous shrubs and trees, which are periodically pruned to prevent shading of companion crops. The prunings can then be used as mulch and green manure to improve soil fertility and produce high-quality fodder. Alley cropping is regarded as an improved bush-fallow system with the following potential advantages:
1) Cropping and fallow phases are combined; 2) Cropping periods are longer, and land is used more intensively; 3) Soil fertility is effectively maintained with the use of species selected for that purpose; and 4) The need for external inputs is reduced (Kang and Wilson 1987).

In areas of Nigeria and in other forest-savanna transition regions of Africa with nonacid soils, results of field and on-farm trials have shown that alley cropping with corn, cowpea, rice and cassava between hedgerows of *Leucaena leuocephala* and *Gliricidia sepium* allowed higher levels of crop production than monocultures (Kang and Wilson 1987). Cowpea and rice showed no significant increase in yield in comparison with monocultures, but when planted in alley cropping with *Leucaena,* they did not respond to N fertilization, indicating that the N supply from prunings was adequate, and additional fertilizers were not necessary.

When used in regions of low soil fertility, the addition of green manure from the trees grown in the hedgerows can significantly increase the yield of the crops grown in the alleys. For example, in western Kenya, experiments carried on red, acid soils showed that maize and bean yields were higher when grown in alley cropping with *Leucaena leucocephala, Cajanus cajan,* and *Sesbania sesban.* These responses were still significant in the third testing season (Onim et al. 1990).

However, in some instances although the tree prunings can add substantial amounts of nutrients and organic matter to the soil, these quantities may not be enough to reach the levels required by the associated crops. For example in experiments conducted in Ibadan, Nigeria, prunings from *Gliricidia sepium* hedgerows yielded the highest N while prunings from *Cassia* spp. yielded the highest organic matter. N supplementation was needed to optimize the yield of maize, with higher amounts required in *Flemingia* alleys than in *Gliricidia* or *Cassia* alleys (Yamoha et al. 1986).

Agroforestry techniques such as alley cropping are better able to increase available stocks of nutrients than monocultures. The following examples illustrate the contributions of this agroforestry system to P and N dynamics. Figure 7.3 compares total labile P in a replicated alley cropping system (*Albizia julibrissin* as the hedge, *Sorghum bicolor* as the grain crop) on an Ultisol in the state of Georgia, U.S., with plots without alleys that otherwise had received the same treatment. Both alley-cropped and monocultured sorghum were previously green-manured in summer with velvet bean (*Mucuna deeringiana*) and in winter with crimson clover (*Trifolium incarnatum*). There was a gradual increase in labile P between the two systems, but even by the third year, sorghum production in the agroforestry system was not greater than that in the monoculture (Matta-Machado and Jordan 1995).

This study was carried out on soil degraded by a century of cotton and soybean

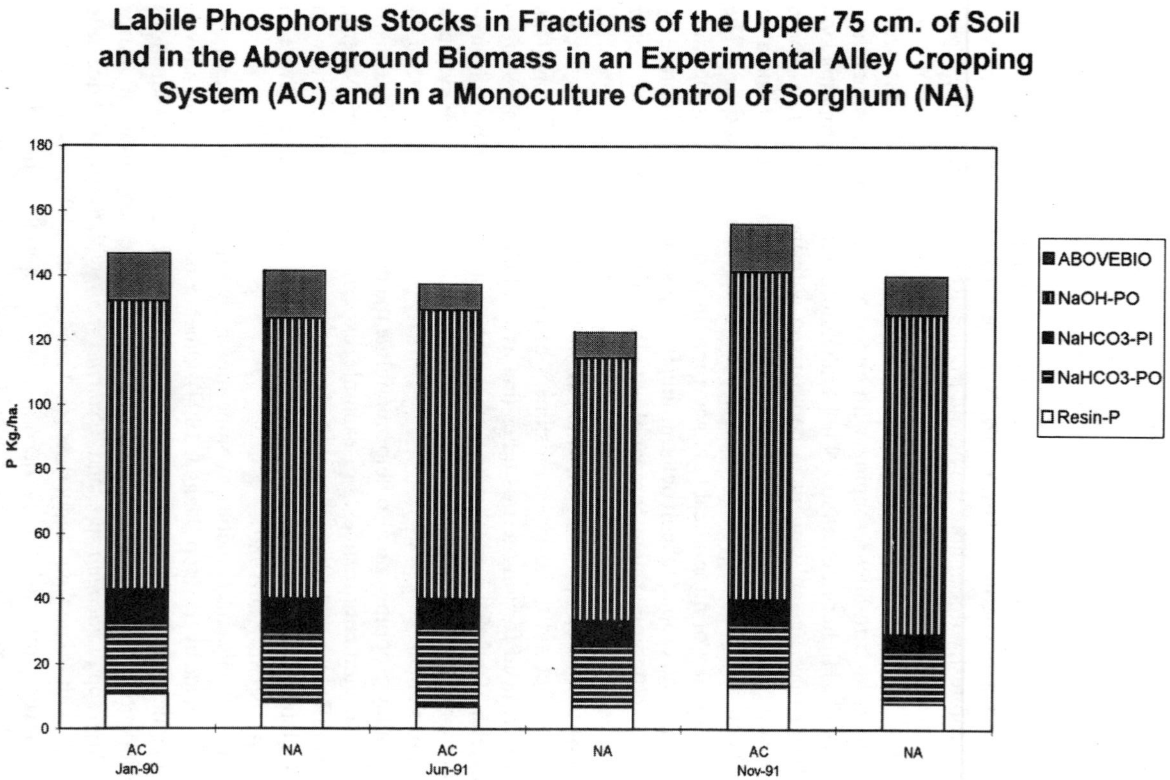


**FIGURE 7.3** Labile phosphorus stocks in fractions of the upper 75 cm of soil and in the aboveground biomass in an experimental alley cropping system (AC), and in a monoculture control of sorghum (NA). There is a trend of gradually increasing differences between the two systems throughout the three-year experimental period. Due to large variability, differences were not significant except between those fractions labeled "a" and "b." PO is organic phosphorus, PI is inorganic phosphorus, NaOH and $NaHCO_3$ are the extractants. ABOVEBIO is aboveground biomass.

farming. While agroforestry can increase nutrient stocks compared to nonagroforestry plots, this increase comes slowly. A better strategy, if it is affordable, might be to initially fertilize the agroforestry system to enable faster production. An even better strategy, if a suitable site is available, is to start the system following a fallow. When an agroforestry system begins with high levels of nutrients, the system has a higher chance of maintaining itself.

Research at the Tropical Agricultural Research and Training Center (CATIE) in Turrialba, Costa Rica, has addressed the role of N in alley cropping annual crops with trees used in agroforestry throughout the humid tropics. Results of long-term experiments with maize grown with *Erythrina poeppigiana* and *Gliricidia sepium* hedgerows showed that after seven years, maize productivity and N uptake were more than twice as high in alley cropping with either species than in monoculture (Haggar et al. 1993). Higher rates of soil N mineralization in the alley cropping systems led to faster maize establishment in comparison with the single crop. These higher rates of soil N mineralization resulted from the build up of readily mineralizable organic N compounds in the soil after seven years of tree mulch application. The long-term accumulation of mineralizable N was more important than the synchrony of mulch N release and crop uptake in determining the higher yields and N uptake by maize in the alley cropping system compared with the sole crop. In spite of higher yields, total recovery of much N by the maize in a single year was only about 10 kg/ha, with most of it taken up during the first two months following planting; however, this initial effect led to faster establishment of maize in the alley crop. This again points to the need for long-term studies that assess the benefits of alley cropping to crop yields. This also serves to stress that early results may not be outstanding, and that other external inputs may be necessary initially to ensure the system's successful establishment.

The use of alley cropping can also help ameliorate the detrimental effects caused by certain soil management practices. For example, experiments in Ibadan, Nigeria, have shown overall nutrient declines in soils through a four-year cropping cycle, with the highest depletion of soil nutrients in plow-till systems and the least severe in *Leucaena*-based systems (Lal 1989).

In alley cropping systems sometimes competition between trees and crops can significantly reduce yields by the crop (Haggar 1994). Other research on nonacid, alluvial soils in Yurimaguas, in the Peruvian Amazon, showed that rice yield reductions from light competition were evident up to 1.5 meters from the hedgerows of all tree species tried (*Inga* spp., *Leucaena* spp., and *Erythrina* spp.), with the greatest decrease in yield found with *Leucaena* hedgerows (Salazar et al. 1993). Weed control was better achieved with the slowly decomposing *Inga* mulch, but rice yields (at a distance $>$ 1.5 m from the hedgerows) were higher with *Leucaena* and *Erythrina*.

Species choice strongly influences the success and overall applicability of alley cropping techniques. For example, many plant species contain allelochemicals that suppress weeds and other plants. Allelopathic interactions are useful when the suppressed plants are considered weeds, but care must be exercised when the mulch is applied to crop plants (Regnier and Janke 1990). Finally, on very acid soils with high aluminum saturation, some of the tree species preferred for alley cropping, such as *Leucaena* spp. and *Gliricidia* spp., do not grow well, and they have to be replaced by other species more adapted to those conditions (e.g., *Calliandra* spp., *Cassia* spp., *Inga* spp., *Flemingea* spp., or *Paraserianthes* spp.).

### 7.2.5 Modified Traditional Systems: Trees Used for Shade of Perennial Crops

At CATIE, Costa Rica, research has long intended to modify the predominant traditional systems of the region — coffee or cacao with shade trees — to improve their productivity and sustainability. This research started with examination of nutrient cycling variables in the existing systems in the region and was followed by experimental systems where nutrient cycling could be examined in controlled experiments. For example, on a farm near CATIE, Beer (1988) compared the annual nutrient return in litter fall and prunings in systems of coffee with *Erythrina poeppigiana* (poró), and coffee with poró and *Cordia alliodora* (laurel). Both trees are common in agroforestry systems with perennial crops in Latin America. The total annual input of litter fall plus pruning residues was similar in both systems. Total annual litter fall input from poró was less than half in association with laurel than without laurel, but litter fall from laurel compensated for the reduced litter fall from poró. In addition, the inclusion of laurel with poró and coffee resulted in a more even distribution of annual nutrient input. Annual inputs of Ca and Mg in litter fall and prunings were larger in the system including laurel than in the system with poró alone. There were no differences in the total input of N or P between the two systems, and the system including laurel had a smaller annual input of K. In spite of these differences among the two systems, the amounts of nutrients recycled by the associated trees reached the recommended levels of fertilizer required for coffee production in both cases.

Which system to choose will depend on the most limiting nutrients in each case. In the previous example, the system with laurel was preferred by many farmers because, apart from a more even nutrient input throughout the year, the value of the laurel timber added an economic incentive. In addition, laurel is a self-pruning species while poró has to be pruned to increase biomass recycling; the system including laurel is therefore less costly than the system with poró alone. Results of experiments at CATIE confirmed these findings. Recommendations were also drawn to modify the traditional systems in order to take advantage of the nutrient cycling benefits from the shade trees and to consider labor availability and timber value (Fassbender et al. 1991).

Other research on similar traditional systems have also demonstrated the importance of nutrient cycling by shade trees and perennial crops. In Ocumare de la Costa, Venezuela, Aranguren et al. (1982) concluded that shade trees of cacao plantations contributed about half of the total annual litter fall. The rate of N transfer to the soil via litter fall was 321 kg/ha. N output with the harvest of cacao pods was 45 kg/ha, with approximately 20 kg returned to the field after processing from pod shells. The authors concluded that the net harvest output of N could be compensated by inputs of N in shade-tree leaf litter from species of *Inga* and *Erythrina*.

### 7.2.6 Agroforestry in Semiarid Environments

The semiarid tropics cover an area of about 20 million $km^2$ and are inhabited by about 700 million people, nearly half of them in India (Vandenbeldt 1990). It covers most of western, eastern, and south-central Africa; most of India, northeastern Burma, northeastern Thailand, and northern Australia; large parts of eastern and central South America, and western parts of Central America. In all, about 34% of the land area of 48 countries is included in the semiarid tropics. They are characterized by low standards of living and low and erratic crop yields due to defficient and strongly seasonal and undependable precipitation (Swindale 1982, Vandenbeldt 1990). The climate is characterized by high atmospheric water demand; high mean annual temperatures (>18°C); and low, variable annual rainfall (400–1900 mm). The climate of most of the semiarid tropics is monsoonal, with over 90% of the rainfall occurring in the period of April to October in the northern hemisphere and October to April in the southern hemisphere.

In arid and semiarid environments, agroforestry systems help to provide greater insurance against weather abnormalities (Swaminathan 1987). Some multiple purpose trees commonly used in agroforestry such as species of *Acacia, Prosopis,* and *Casuarina,* can grow well in arid areas and can be combined with grain crops. Perennial shrubs such as *Sesbania grandiflora* and *Cajanus cajan* are also promising for producing food, fodder, and fuelwood. In semiarid regions, browsing from shrubs and trees, more resistant to dry periods than herbaceous forage, can represent as much of 20–25% of the total intake for livestock (Le Houérou 1987). This provides stability and productivity to livestock and is a main source of income of farmers.

In Africa, shrubland, woodland (open dry forest with more than 50% canopy cover), and savanna (annual or perennial grass or herb cover with <50% canopy cover) cover about 10 million $km^2$, or about 35% of the continent (Le Houérou 1987). Trees, shrubs and palms have always played important ecological and economic roles in semiarid Africa. In ancient Egypt, certain species of trees and palms (e.g., *Balanites aegyptica, Phoenix dactylifera*) were worshipped as sacred trees (von Maydell 1987). The baobab (*Adansonia digitata*) and *Acacia albida* are other trees of great significance to people in Saharan, Sahelian, and Sudan savannas. *A. albida,* a N-fixing tree widespread in arid and semiarid Africa, keeps its foliage through the dry season and only sheds its leaves at the beginning of the growing season; therefore there is no competition for water or nutrients with the associated crops. In fields where *A. albida* grows naturally at densities of 20–40 trees/ha, the most common intercrops are with millet, sorghum, maize, and peanuts. The pods of *A. albida* are good fodder during the dry season, and the tree also produces good firewood, thorny fencing material, tannins, gum, and bee forage (von Maydell 1987).

In India, the extent of the arid and semiarid region is about 300,000 $km^2$, mostly in the northwest (Shankarnarayan 1989). The *Prosopis cineraria* based systems of semiarid India are similar to the *A. albida* systems of semiarid Africa. *P. cineraria,* also a N-fixer, is an important source of animal feed, fuel, and timber. As *A. albida, P. cineraria* has a very deep tap root and it can be lopped at a young age (about eight years) with yields of 40–70 kg of fuelwood, 20–30 kg of leaves and 5 kg of pods per year (Singh 1987). In fields with relatively low densities of *P. cineraria* (about 120 trees per ha, depending on soil type and rainfall), it is usually intercropped with millet and legumes. Improved soil fertility and higher moisture content have been found in *P. cineraria* intercropped systems, along with higher grain yield and forage biomass production (Singh 1987).

In semiarid ecosystems, soils under tree canopies apparently have greater levels of organic matter, calcium, magnesium, potassium, and phosphorus than those in open grass (Belsky et al. 1993, Campbell et al. 1994, Isichei and Moughalu 1992, Kellman 1989). However, in these systems the direct effects of trees cannot be easily distinguished from other, indirect effects, because higher fertility underneath the trees is often associated with a more favorable microenvironment (Belsky et al. 1993). Additionally, because the trees in savanna ecosystems have generally been in place for a relatively long time before crops are interplanted and soils are sampled, it is hard to determine cause and effect relationships between the presence of the trees and improved soil conditions under their canopies.

### 7.2.7 The Role of Animals: Agrosilvopastoral Systems

Agrosilvopastoral systems—the combination of timber, fuelwood, or fruit trees with animals, with or without crops—are practiced at many scales. A large-scale system may include timber plantations with grazing to control weeds and to obtain a more immediate return from the sale of animal products. Cattle raising can also complement subsistence agriculture, with animals integrated in home gardens or in systems of fodder production to feed animals in stables. In farms where alley cropping is practiced, animal manure can be added to mulch to contribute rapidly available nutrients to crops. Mulches of manure can also serve as a nutrient source for microbial decomposers, speeding up the decomposition of plant materials. In some regions, the incorporation of trees—especially MPTS—can change cattle raising from an inefficient use of land to a more ecologically and economically feasible activity. The incorporation of trees can improve system productivity either by increasing pasture yields or through the production of tree fodder from leaves and fruits (Gill et al. 1990, Cobbina 1994/1995).

Though the presence of animals in a tree plantation may accelerate nutrient cycling, if the animal load is too high, soil compaction may affect tree growth (Montagnini 1992). It is often difficult to separate the effects of soil compaction and nutrient recycling by animals from the impacts of trees and forage grasses or legumes. In experimental systems in the Atlantic region of Costa Rica, a fast-growing leguminous tree species (*Erythrina berteroana*) was introduced in native grass pastures. Over a three-year study period, soil organic C increased in two nongrazing treatments, and soil bulk density decreased in the control areas with no grazing and no trees (Cooperband and Logan 1993). Changes in soil pH, exchangeable cations, Al and P levels were not clearly a result of the impacts associated with grazing, or of changes associated with site preparation.

Other experiments on eroded marginal lands in subtropical India compared soil erosion in agrosilvopastoral systems with *Leucaena leucocephala* (used for fuelwood and fodder) and *Pennisetum purpureum* (used for fodder) with the traditional rain-fed crop sequence of *Sesamum indicum* followed b *Brassica napus* (Grewal et al. 1994). Results suggested that the agrosilvopastoral systems were more effective in conserving soil than the traditional cropping system. However, animal grazing did not have an influence on soil variables because the animals were kept outside of the system.

Agrosilvopastoral systems require careful management to take advantage of nutrient cycling and to avoid soil compaction and trampling by cattle. This can be achieved at a small scale where manual recycling of manure, managing and rotating pastures, and hauling fodder to feed animals in stables or enclosures are more feasible. However, these practices are labor intensive and will only be adopted when the benefits and returns are attractive to the farmers.

## 7.3 THE EFFECTS OF TREES ON SOIL FERTILITY

Agroforestry systems have often been regarded as having protecting functions on physical and chemical properties of soil. It can be argued that erosion control can be achieved through the use of soil covers, independently of the presence of trees. A number of herbaceous species were used in early attempts to improve fallows in tropical areas, but in the more seasonal climates, the herbaceous crops with their shallow roots were not able to sustain the dry season very well. Some herbaceous, leguminous species such as *Pueraria phaseoloides* and *Centrosema pubescens* are still used with success as green covers of commercial tree plantations (rubber, cacao) in several tropical humid regions. In the 1970s and 1980s, as selective herbicides became more accessible and the "minimum tillage" practices became popular in tropical regions, the use of cover crops such as *Pueraria phaseoloides* and *Mucuna utilis* became widespread as *in situ* mulches. However, small shrubs such as *Crotalaria* spp. and *Cajanus cajan* were found to be more adequate than herbaceous species because they could stand drier periods and competed less with the associated herbaceous crops. Tree or shrub species that stand pruning and have high rates of organic matter production and good nutrient cycling abilities can be combined in agroforestry systems with agricultural crops or cattle, with the added advantage of yielding products such as fuelwood, timber, fruits, or fodder.

In addition, in agroforestry all system components including trees, crops, and their litter can contribute to nutrient cycling and soil protection. For example, a summary of erosion rates under tropical forest, tree crops and some agroforestry systems is given in Table 7.1. If the rates of soil loss shown are considered low (<2 tons/ha/yr), moderate (2–10 tons/ha/yr) and high (>10 tons/ha/yr), then the lowest erosion rates correspond to natural rain forest, forest fallows, multistory tree gardens, and undisturbed forest plantations; intermediate or moderate to high rates are found for cropping periods of shifting agriculture and taungya systems; and the highest rates are recorded for tree plantation crops and for forest plantations, both under intense management. The wide ranges shown by the data indicate the importance of management rather than the intrinsic nature of the systems. Additionally, it can be seen that the highest erosion rates were recorded for the two intensively managed systems in which there was no surface cover (Young 1989).

**TABLE 7.1**
**Rates of erosion in tropical forest and tree crop systems (Wiersum 1984, in Young 1989).**

| Land use system | Erosion (ton/ha/yr) |
|---|---|
| Multistory tree gardens | 0.01–0.14 |
| Natural rain forest | 0.03–6.16 |
| Fallow period of shifting cultivation | 0.05–7.40 |
| Undisturbed forest plantations | 0.02–6.20 |
| Tree crops with cover crop or mulch | 0.10–5.60 |
| Cropping period of shifting cultivation | 0.40–70.05 |
| Cultivation period of taungya system | 0.63–17.37 |
| Tree crops, clean weeded | 1.20–182.90 |
| Forest plantations, burned or litter removed | 5.92–104.80 |

A key to the success of agroforestry systems is the choice of fast-growing wood components with positive impacts on soil properties (Sanchez et al. 1985, Nair 1989, Young 1989, Montagnini 1992). A number of publications compile characteristics, uses, and properties of Multiple-Purpose Tree Species (MPTS) for their application worldwide (e.g., NAS 1979, NAS 1980, Glover and Adams 1990, Lantican and Taylor 1991). The majority of the MPTS lists and databases include information on the role of trees on soil fertility or nutrient cycling. These guides emphasize the N-fixing ability and nodulating status of leguminous trees and actynorrhizal plants with few statements on their potential effect on other ecosystem nutrients.

Several indigenous and exotic tree species show potential for improving chemical, biological, and physical characteristics of soil, and could be advantageous in agroforestry combinations. For example, out of about 30 economically valuable tree species tested in experimental conditions in humid lowland regions of Costa Rica, Brazil, and Argentina, nearly half had a positive influence on soil total N, organic matter, and/or exchangeable cations and P, in comparison with nearby pasture (Table 7.2). In several cases the values of the parameters tested under the tree canopies were close to those found in adjacent young secondary forests. Most of the other species tested did not appear to change soil conditions substantially, and they could be used on non-degraded soils (Table 7.2).

A number of constraints limit the broad utilization of results, such as those shown in Table 7.2, as a single criterion for species selection for agroforestry. Primarily, the results of standard soil fertility tests used in agriculture may not always reveal the soil's productive potential because they do not include all chemical forms of nutrients available for plant uptake. For example, although mineral N ($NO_3^-$ and $NH_4^+$) makes up less than 10% of the total soil N pool, it is the form of N available to plants. Data on N mineralization rates under tree species are frequently not available because their estimation requires time-consuming laboratory or field incubation of soil samples. Evaluating the effects of trees on soil P availability is even more

**TABLE 7.2**
**Topsoil chemical characteristics in pure stands of 24 indigenous tree species at La Selva, Costa Rica; Porto Seguro, Bahia, Brazil; and Misiones, Argentina.**

| Site/Tree species | pH | C (%) | N (%) | P (cmol.kg $^{-1}$) | K | Ca | Mg |
|---|---|---|---|---|---|---|---|
| **a. La Selva, Costa Rica[1]** | | | | | | | |
| *Stryphnodendron microstachyum* | 5.4ab | 3.42ab | 0.29b | 5.6a | 0.27a | 0.45a | 0.63ab |
| *Vochysia ferruginea* | 5.4ab | 3.76a | 0.32a | 7.1a | 0.22a | 0.73a | 0.61ab |
| *Vochysia guatemalensis* | 5.3ab | 3.13ab | 0.29b | 5.2a | 0.11a | 0.25a | 0.37ab |
| *Hyeronima alchorneoides* | 5.1b | 2.96c | 0.22b | 1.5b | 0.09a | 0.31b | 0.21b |
| **Abandoned pasture** | 5.3ab | 2.73c | 0.22b | 4.9a | 0.19a | 0.32b | 0.27b |
| **Secondary forest** | 5.3ab | 4.33a | 0.33a | 3.6b | 0.17a | 0.68a | 0.55ab |
| **b. Porto Seguro, Bahia, Brazil[2]** | | | | | | | |
| **N-fixing leguminous species:** | | | | | | | |
| *Bowdichia virgilioides* | 4.9 | 1.98def | 0.16def | 1.32def | 0.06bcd | 1.35bc | 0.39de |
| *Centrolobium minus* | 4.6 | 1.87efg | 0.16def | 1.19efg | 0.05fgh | 0.53hi | 0.21i |
| *Centrolobium robustum* | 4.5 | 1.65ij | 0.13f | 1.07fgh | 0.05fgh | 0.40i | 0.16i |
| *Inga affinis* | 4.9 | 2.10cde | 0.18cd | 3.64a | 0.07bcd | 0.76gh | 0.49bc |
| *Parapiptadenia pterosperma* | 4.9 | 2.38ab | 0.20bc | 0.78ij | 0.08b | 1.40bc | 0.60a |
| *Pithecellobium elegans* | 4.8 | 1.67hij | 0.15ef | 0.59kl | 0.05efg | 0.79gh | 0.40de |
| *Platymenia foliolosa* | 4.7 | 2.08cde | 0.18bcd | 0.13m | 0.05efg | 1.05cde | 0.42cd |
| **Non-N fixing leguminous species:** | | | | | | | |
| *Arapatiella psilophylla* | 4.7 | 1.94def | 0.18bcd | 1.45de | 0.06bcd | 0.38i | 0.37de |
| *Caesalpinia echinata* | 5.1 | 2.41a | 0.17cde | 1.54de | 0.07bcd | 1.17bcd | 0.39de |
| *Cassia spp.* | 4.7 | 1.94def | 0.16def | 1.40def | 0.07bcd | 0.56hi | 0.34de |
| *Copaifera luscens* | 5.0 | 2.02cde | 0.17cde | 0.63jk | 0.06cde | 1.15bcd | 0.34de |
| *Dimorphandra jorgei* | 4.9 | 1.97def | 0.19bc | 0.97ghi | 0.03j | 0.98def | 0.32efg |
| *Hymenaea aurea* | 4.4 | 2.00def | 0.16def | 2.03c | 0.06bcd | 0.26i | 0.24hi |
| *Macrolobium latifolium* | 4.7 | 1.90efg | 0.16def | 0.67jk | 0.04hij | 0.36i | 0.25fg |
| **Of other families:** | | | | | | | |
| *Bombax macrophyllum* | 4.8 | 1.78ghi | 0.13f | 1.42de | 0.06bcd | 0.84efg | 0.33ef |
| *Buchenavia grandis* | 4.6 | 2.06cde | 0.14f | 2.09c | 0.06bcd | 0.80fg | 0.33ef |
| *Eschweilera ovata* | 5.3 | 1.82fgh | 0.31a | 0.58kl | 0.11a | 1.38bc | 0.53ab |
| *Lecythis pisonis* | 5.3 | 1.99def | 0.18bcd | 0.23lm | 0.04ghi | 1.46b | 0.32ef |
| *Licania hypoleuca* | 5.0 | 1.63j | 0.14f | 1.61d | 0.07bcd | 1.31bcd | 0.35de |
| *Pradosia lactescens* | 4.9 | 2.15bcd | 0.18bcd | 0.81ij | 0.05fgh | 0.84efg | 0.24gh |
| **Primary forest** | 4.9 | 1.99def | 0.15ef | 0.96hi | 0.08bc | 1.23bcd | 0.36de |
| **Secondary forest** | 5.1 | 2.15abc | 0.22b | 2.46b | 0.07bcd | 2.20a | 0.62[a] |
| **c. Misiones, Argentina[3]** | | | | | | | |
| *Balfourodendron riedelianum* | 5.8 | 2.6b | 0.34ab | n.d. | 0.55bc | 7.1bc | 1.7c |
| *Bastardiopsis densiflora* | 7.1 | 6.3a | 0.65a | n.d. | 1.28a | 20.4a | 3.4ab |
| *Cordia trichotoma* | 6.4 | 4.0ab | 0.46ab | n.d. | 0.79b | 13.6ab | 2.6abc |
| *Enterolobium contortisiliquum* | 6.1 | 3.4ab | 0.39ab | n.d. | 0.67b | 8.7bc | 3.5a |
| *Ocotea puberula* | 6.1 | 4.4ab | 0.59a | 6.09a | 1.11a | 17.3a | 4.7a |
| **Grass control** | 5.8 | 2.2b | 0.0.27b | n.d. | 0.26c | 6.3c | 2.4bc |

**Sources:** [1] Montagnini and Mendelsohn (1996), [2] Montagnini et al. (1994), [3] Fernández et al. (1995).

**Note:** For each site, differences among means are statistically significant when followed by different letters ($p < 0.05$). n.d.: not detected difficult, although P release from litter and its uptake by test crops can give an indirect indication of impacts on soil P availability.

The impacts of trees on soil fertility depend on nutrient recycling characteristics such as litter chemistry and decomposition rates. Tree litter can be used as mulch with different outcomes: A fast mulch decomposition rate may accelerate the growth of associated crops on poor soils, while in other cases a more persistent litter may provide a steady source of nutrients and a better soil cover year-round. In the example shown in Table 7.2, high rates of litter fall and slower decomposition resulted in high litter accumulation and high soil organic matter under *V. ferruginea,* making this species well-suited for protecting soils against erosion. In contrast, litter from *V. guatemalensis* may be especially important for Ca and Mg recycling (Montagnini et al. 1993). Although the litter of *H. alchorneoides* was less abundant than the other three species, it had a relatively faster decomposition rate and higher nutrient content. These characteristics promoted fast nutrient recycling, especially of N, Ca, Mg, K, and P.

Additional factors influencing nutrient release from litter are its polyphenol and lignin content, as both compounds lower the quality of plant materials. The polyphenol to N ratio, for example, may serve as an index for short-term immobilization patterns observed for legumes with relatively high polyphenol content, and the lignin plus polyphenol to N ratio may serve as an index for longer-term release patterns (Palm 1995). The timing of nutrient release by trees is also important so that nutrients will be supplied in synchrony with crop needs (Palm 1995). Yet as discussed in the previous section, synchronizing nutrient supply may not be as important as the long-term buildup of tree litter mulch and soil organic matter in influencing crop yields and, equally important, in maintaining soil productivity.

It has been argued that the ability of agroforestry systems to enhance nutrient availability is very limited on infertile soils compared to systems on fertile soils, although they can indeed play an important role in reducing nutrient losses in both situations (Szott et al. 1991). Litter production and quantities of nutrients recycled in litter are greater on fertile than infertile soils; however, use of prunings to accelerate nutrient fluxes may increase plant productivity on infertile soils (Szott et al. 1991).

Inclusion of woody components in a production system can provide benefits from the tree products themselves (timber, fuelwood, leaf mulches, and other tree products) and from their potential ecological advantages, especially their nutrient cycling abilities. The choice of a tree species will often depend on whether both productive and ecological advantages can be achieved in the same system, and in some cases one prevailing function may be desired.

## 7.4 NUTRIENT USE EFFICIENCY AND SPECIES CHOICE

The nutrient use efficiency (NUE) concept has been employed to describe the differential ability of tree species to accumulate organic matter in relation to nutrients taken up from the soil. NUE has been defined on different scales of space and time. At the plant population and community levels, NUE is generally defined as the amount of biomass produced per unit of nutrient taken up (Grubb 1989, Binkley et al. 1992, Medina 1995). Since direct measurements of nutrient absorption are not usually made for mature tree stands, leaf litter fall nutrients are generally used as an estimate of annual nutrient uptake (Vitousek 1984, Grubb 1989, Binkley et al. 1992). Ideally, efforts should be made to measure total production by stems, branches, and roots and to consider nutrient uptake plus losses to herbivory and leaching of nutrients (Grubb 1989). Differences in nutrient cycling and efficiency of use may result from an ability to use various nutrient conserving mechanisms, from physiological to mutualistic interactions (Chapin 1980, 1983).

It is also important to consider the relationship between the recycling ability of the species and its potential short and long-term impacts on soil nutrient amelioration. A "nutrient cycling index" (NCI), taken as the inverse of NUE, i.e., the amount of nutrients in annual litter fall/annual tree biomass production, has been used to assess suitability of tree species for agroforestry combinations. For example, Fassbender, et al., (1991) found that the P recycling index was about six times higher in combinations of cacao with *Erythrina poeppigiana* than with *Cordia alliodora* in agroforestry systems in Turrialba, Costa Rica. *Cordia,* a timber species, accumulated much P in stem biomass, while *Erythrina,* a shade tree with good nutrient cycling properties, produced large amounts of leaves and branches resulting in greater P recycling.

When put in context with nutrient recycling characteristics of a species, NUE can indicate appropriate system design and management to maintain productivity and recover or conserve nutrients over the long term. The ability of a species to produce large amounts of biomass with less nutrients may be an important consideration in choosing species for degraded, nutrient-poor sites.

### 7.4.1 Applicability of NUE in System Design and Management

From the 24 tree species shown in Table 7.2, data on tree productivity, leaf litter fall and litter chemistry of eight species in Bahia, Brazil, and four species at La Selva, Costa Rica, were used to calculate NUE values (Table 7.3). Because the species were part of a forestry project with fast-growing timber trees, NUE values were calculated as the annual stem biomass increments/nutrients in annual leaf litter fall (Montagnini 1995). There were not enough data on litter fall or productivity for the other species shown in Table 7.2. At both Bahia and La Selva, the highest efficiencies were for K and P, and the lowest were for N, Ca, and Mg (Table 7.3).

Results from Bahia suggest that *B. macrophyllum* and *P. foliolosa,* with overall high NUE values, would grow well on relatively nutrient-poor soils, and thus could be good alternatives for reforestation of degraded sites following the abandonment from agriculture and pasture that is frequent in the region. *B. macrophyllum* tended to accumulate high amounts of litter under its canopy while *P. foliolosa* stands had relatively high amounts of organic matter and total N in the topsoil in comparison with adjacent areas of secondary forest (Montagnini et al. 1994). These features indicate that these species may be well-suited for soil rehabilitation, including increasing soil organic matter content and protecting against soil erosion. Also at Bahia, species

**TABLE 7.3**
**Nutrient use efficiencies (NUE) (Megagrams of annual stem biomass increment/kg nutrient in annual litter fall).**

| Porto Seguro, Bahia, Brazil: | N | Ca | Mg | K | P |
|---|---|---|---|---|---|
| **N-fixing leguminous species:** | | | | | |
| *Centrolobium robustum* | 0.05 | 0.05 | 0.37 | 0.87 | 1.74 |
| *Platymenia foliolosa* | 0.08 | 0.34 | 7.00 | 3.51 | 4.66 |
| **Non N-fixing leguminous species:** | | | | | |
| *Caesalpinia echinata* | 0.04 | 0.03 | 0.54 | 0.78 | 1.81 |
| *Hymenaea aurea* | 0.06 | 0.09 | 0.45 | 0.64 | 1.91 |
| **Species of other families:** | | | | | |
| *Bombax macrophyllum* | 0.36 | 0.20 | 0.88 | 20.70 | 20.70 |
| *Buchenavia grandis* | 0.05 | 0.04 | 0.38 | 0.68 | 1.70 |
| *Eschweilera ovata* | 0.12 | 0.10 | 0.44 | 0.96 | 5.73 |
| *Lecythis pisonis* | 0.12 | 0.11 | 0.89 | 1.58 | 3.56 |
| **La Selva, Costa Rica:** | **N** | **Ca** | **Mg** | **K** | **P** |
| *Stryphnodendron mycrostachyum* | 0.08 | 0.06 | 0.35 | 0.63 | 0.53 |
| *Vochysia ferruginea* | 0.08 | 0.03 | 0.26 | 0.38 | 0.56 |
| *Vochysia guatemalensis* | 0.16 | 0.05 | 0.29 | 0.84 | 1.07 |
| *Hyeronima alchorneoides* | 0.09 | 0.03 | 0.15 | 0.22 | 0.45 |

such as *B. grandis* and *H. aurea,* with overall lower NUE values, would be most appropriate for agroforestry combinations where crops could benefit from nutrient recycling from litter.

At La Selva, *V. ferruginea* showed comparatively low efficiency values for all the nutrients considered, confirming the beneficial role of this species in recycling organic matter and positively impacting soil fertility as shown in Table 7.2. The relatively low efficiency (high recycling) of N and P found for *S. microstachyum* and *H. alchorneoides* was also shown in experiments where maize grown with mulch of these species grew better and absorbed more N and P than with mulch of other species (Montagnini et al. 1993).

In projects that aim to recover soil nutrients in degraded sites, species with high nutrient use efficiency should be combined in time or space with species with low use efficiency. However, NUE values alone may not be enough to assess the role of a tree species on ecosystem nutrients. For example, in spite of high NUE values, a tree species may pose high demands on soil nutrients over the long term. In humid tropical regions nutrients are expected to be critical factors influencing tree productivity, while in regions with a marked dry season water use efficiency rather than NUE would be a more important factor influencing species choice and system design. Still other ecological adaptations of the species (e.g., light use efficiency, root architecture, resistance to pests and diseases) could be more important in selecting species for agroforestry combinations.

## 7.5 ADOPTABILITY OF AGROFORESTRY SYSTEMS

The introduction of trees in a production system will only be beneficial if competition between trees and crops for resources is minimized while positive effects on soil fertility are enhanced: Examples of such systems include parklands with N-fixing species such as *Faidherbia albida* in the Sahel, and some sequential systems of relay intercropping and improved fallows (Sanchez 1995). Apart from their beneficial effects on soils, tree species with rapid canopy closure may decrease the growth of weeds, thus the canopy characteristics of the trees will affect their suitability for interplanting with annual crops and will dictate the management practices required when used in agroforestry systems. Finally, decisions to change any system must take into account the objectives that the change is seeking to achieve; thus species selection for agroforestry must be based on several factors apart from their positive influence on soils and crop yields (Wood 1990). Furthermore, species choices for forestry and agroforestry systems are mandated by local people's preferences, which in turn depend on seedling availability, official incentives, and markets.

Choice of suitable crop species is also important for the success of alley cropping systems. Maize and rice, for example, are more light demanding than beans or cassava and thus more affected by shading by trees. The height of pruning and the width of alleys can be adjusted to avoid excessive competition between crops and trees. In experiments at CATIE, for example, Kass (1989) found that maize yield was higher when planted farther away from *Gliricidia* hedgerows. Results of economic analyses for the same systems indicated that the alley cropping system was not as profitable for N, with a lower market cost for fertilizer, as for K and P, with relatively higher market fertilizer prices.

The adoption of alley cropping systems appears widespread in low income areas of eastern Indonesia, southern Philippines, and Sri Lanka (Kang and Wilson 1987). In Nigeria, researchers from the International Institute of Tropical Agriculture (IITA) have found that although labor for pruning hedges is a major constraint, alley cropping with *Calliandra* spp. has given good results (Plucknett 1990). However the application of alley cropping techniques has its limitations: Sometimes crop monocultures are preferred for practical reasons, and sometimes the value of prunings from the hedgerows is higher than the value of crops. Planting the hedgerow trees at high density favors biomass production by trees, but may result in lower crop yields because of competition with trees for light and nutrient resources. On highly weathered soils, successful alley cropping may require the use of external inputs (e.g., liming to increase soil pH) to maintain levels of soil fertility adequate for the desired crop yields (Evensen et al. 1995).

### 7.5.1 THE NEED FOR INITIAL ECONOMIC INCENTIVES TO FACILITATE SYSTEM ESTABLISHMENT

When nutrient stocks are low, it may take a number of years for an agroforestry system to build up nutrients and soil carbon to the point where the system is profitable compared with monocultures. For example, three years after the establishment of the *Albizia*-sorghum alley cropping system previously discussed, production of the grain crop was still less than the control, despite the slow but continuous rise in available soil P in the agroforestry system. Because of the low initial productivity of agroforestry systems established on degraded soil, sometimes it is not economically feasible for a farmer to begin agroforestry systems.

Since the adoption of agroforestry systems involves the planting of selected trees, some initial capital will be needed to cover the first two to three years of establishment costs. This requirement can be a problem if farmers have no access to capital. In the Atlantic lowlands of Costa Rica, enriched fallow systems are more profitable than conventional agriculture or cattle; however, some assistance may be required to help farmers make the initial expenditures needed to plant the trees (Montagnini and Mendelsohn 1997). Small subsistence farmers often have no access to loans and so cannot afford to make even profitable investments. In Costa Rica, programs such as the Forestry Development Fund (FDF) provide loans for planting trees to small farmers. The farmers repay the loan by giving 30% of the income from harvesting the trees at maturity. Such programs could make sustainable development a reality by allowing small farmers to make sound long-term investments in their land.

## 7.6 CONCLUSIONS

One of the most important reasons for adoption of agroforestry systems in regions where commercial fertilizers are expensive or unavailable is the ability of such systems to recover, recycle, or efficiently utilize nutrients. This ability is often linked to mechanisms associated with woody or perennial species. While agroforestry systems can be profitable if established immediately after forest clearing, they often require a number of years to become profitable when established on degraded lands. For this reason, capital-limited farmers on poor soils may require subsidies to encourage establishment of agroforestry systems.

## 7.7 ACKNOWLEDGMENTS

We thank V. Derr and N. Muñiz-Miret for reviewing the manuscript.

## 7.8 REFERENCES

Adejuwon, J. O. and F. A. Adesina. Organic matter and nutrient status of soils under cultivated fallows: an example of *Gliricidia sepium* fallows from South Western Nigeria. *Agroforestry Syst.* 10 (1990):23–32.

Ae, N., J. Arihara, K. Okada, T. Yoshihara, and C. Johansen. Phosphorus uptake by pigeon pea and its role in cropping systems of the Indian subcontinent. *Science* 248 (1990):477–480.

Allen, B. J. Dynamics of fallow successions and introduction of robusta coffee in shifting cultivation in the lowlands of Papua, New Guinea. *Agroforestry Syst.* 3 (1985):227–238.

Anderson, A. B. Extraction and forest management by rural inhabitants in the Amazon estuary, in: *Alternatives to Deforestation: Steps Toward Sustainable Use of the Amazon Rain Forest,* A. B. Anderson, ed. (New York: Columbia University Press, 1990), 65–85.

Andreiesse, J. P. and R. M. Schelhaas. A monitoring study of nutrient cycles in soils used for shifting cultivation under various climatic conditions in Tropical Asia, III. The effects of land clearing through burning on the fertility level. *Agric. Ecosystems and Environ.* 19 (1987):311–332.

Aranguren, J., G. Escalante, and H. Herrera. Nitrogen cycle of tropical perennial crops under shade trees II. Cacao. *Plant and Soil* 67 (1982):259–269.

Balée, W. People of the fallow: A historical ecology of foraging in lowland South America, in: *Conservation of Neotropical Forests,* K. H. Redford and C. Padoch, eds. (New York: Columbia University Press, 1992), 35–57.

Bandy, D. E., D. P. Garrity, and P. A. Sanchez, The worldwide problem of slash-and-burn agriculture. *Agroforestry Today* 5 (1993):2–9.

Beer, J. Litter production and nutrient cycling in coffee (*Coffea arabica*) or cacao (*Theobroma cacao*) plantations with shade trees. *Agroforestry Syst.* 7 (1988):33–45.

Belsky, A. J., S. M. Mwonga, R. G. Amundson, J. M. Duxbury, and A. R. Ali. Comparative effects of isolated trees on their undercanopy environments in high- and low-rainfall savannas. *J. of Appl. Ecol.* 30 (1993):143–155.

Binkley, D., K. A. Dunkin, D. DeBell and M. G. Ryan. Production and nutrient cycling in mixed plantations of *Eucalyptus* and *Albizia* in Hawaii. *For. Sci.* 38 (1992):393–408.

Campbell B. M., P. Frost, J. A. King, M. Mawanza, and L. Mhlanga. The influence of trees on soil fertility on two contrasting semiarid soil types at Matopos, Zimbabwe. *Agroforestry Syst.* 28 (1994):159–172.

Chapin, F. S. The mineral nutrition of wild plants. *Annu. Rev. of Ecol. and Syst.* 11 (1980):233–260.

Chapin, F. S. Patterns of nutrient absorption and use by plants from natural and manmodified environments, in: *Disturbance and Ecosystems. Ecological Studies* vol. 44, H. A. Mooney and M. Godron, eds. (Berlin: Springer-Verlag, 1983), 175–187.

Cobbina, J. Growth and herbage productivity of *gliricidia* in *Panicum maximum* pasture as influenced by seed preparation, planting, and weed control techniques. *Agroforestry Syst.* 28 (1994/1995):193–201.

Cooperband, L. R. and T. J. Logan. Baseline soil characteristics of a humid tropical silvopastoral system and changes in selected soil properties. *Turrialba* 43 (1993):22–36.

Del Amo, R. S. and P. J. Ramos. Use and management of secondary vegetation in a humid-tropical area. *Agroforestry Syst.* 21 (1993):27–42.

Denevan, W. D., J. M. Treacy, J. B. Alcorn, C. Padoch, J. Denslow, and S. F. Paitan. Indigenous forestry in the Peruvian Amazon: Bora Indian management of forest fallows. *Interciencia* 9 (1984):346–357.

Evensen, C. I., T. S. Dierolf, and R. S. Yost. Decreasing rice and cowpea yields in alley cropping on a highly weathered Oxisol in West Sumatra, Indonesia. *Agroforestry Syst.* 31 (1995):1–19.

Fassbender, H. W., J. W. Beer, J. Heuveldop, A. Imbach, G. Enriquez, and A. Bonnemann. Ten-year balances of organic matter and nutrients in agroforestry systems at CATIE, Costa Rica. *For. Ecol. and Manage.* 45 (1991):173–183.

Fernandes, E. C. M., A. O'Kting'ati and J. Maghembe. The Chagga homegardens: a multi-storied agroforestry cropping system on Mt. Kilimanjaro (Northern Tanzania), in: *Agroforestry Systems in the Tropics,* P. K. R. Nair, ed. (Dordrecht, The Netherlands: Kluwer Academic Publishers, 1989), 309–332.

Fernández, R., F. Montagnini, and H. Hamilton, The influence of native tree species on soil chemistry in a subtropical humid forest region of Argentina. *J. of Trop. For. Sci.* 10 (1997):188–196.

Gill, A. S., K. S. Gangwar, and B. S. Sinsinwar. Productivity of perennial grasses in association with *Acacia albida* under different cutting schedules in dryland conditions. *Indian J. of Dryland Agric. Res. and Dev.* 5 (1990):68–71.

Glover, N. and Adams, N. eds. 1990. *Tree Improvement of Multipurpose Species. Multipurpose Tree Species Network Technical Series, vol. 2.* Winrock International Institute for Agricultural Development. Arlington, VA.

Gómez-Pompa, A. and A. Kaus. Traditional management of tropical forests in Mexico, in: *Alternatives to Deforestation: Steps Toward Sustainable Use of the Amazon Rain Forest,* A. B. Anderson, ed. (New York: Columbia University Press, 1990), 45–64.

Grewal, S., J. M. Juneja, K. Singh, and S. Singh. A comparison of two agroforestry systems for soil, water, and nutrient conservation on degraded land. *Soil Technol.* 7 (1994):145–153.

Grubb, P. J. The role of mineral nutrients in the tropics: A plant ecologist view, in: *Mineral Nutrients in Tropical Forest and Savannah Ecosystems,* J. Proctor, ed. (Boston: Blackwell Scientific Publications, 1989), 417–440.

Haggar, J. P. Trees in alley cropping: Competitors or soil improvers? *Outlook on Agric.* 23 (1994):27–32.

Haggar, J. P., E. V. J. Tanner, J. W. Beer, and D. C. L. Kass. Nitrogen dynamics of tropical agroforestry and annual cropping systems. *Soil Biol. and Biochem.* 25 (1993):1363–1378.

Han, X. The mobilization of aluminum- and iron-bound phosphorus by organic matter decomposition in highly weathered soils. Ph.D. dissertation, University of Georgia, Athens, GA, 1989.

Isichei, A. O. and J. I. Muoghalu. The effects of tree canopy cover on soil fertility in a Nigerian savanna. *J. of Trop. Ecol.* 8 (1992):329–338.

Jordan, C. F. *An Amazonian Rain Forest.* Carnforth, U.K.: UNESCO and The Parthenon Group. 1989.

Jordan, C. F. *Conservation: Replacing Quality with Quantity as a Goal for Global Management.* New York: John Wiley and Sons. 1995.

Kang, B. T. and G. F. Wilson. The development of alley cropping as a promising agroforestry technology, in: *Agroforestry. A Decade of Development,* H. A. Steppler and P. K. R. Nair, eds. (Nairobi, Kenya: ICRAF, 1987), 227–244.

Kang, B. T., Reynolds, L. and Atta-Krah, A. N. Alley farming. *Adv. in Agron.* 43 (1990):315–359.

Kass, D. L. Resultados de seis años de investigación de cultivos en callejones (alley cropping) en 'La Montaña,' Turrialba, Costa Rica. *El Chasqui (Costa Rica)* 19 (1989):5–24.

Kass, D. C. L., C. Foletti, L. T. Szott, R. Landaverde, and R. Nolasco, R. Traditional fallow systems of the Americas. *Agroforestry Syst.* 23 (1993):207–218.

Kellman, M. Mineral nutrient dynamics during savanna-forest transformation in Central America, in: *Mineral Nutrients in Tropical Forest and Savannah Ecosystems,* J. Proctor, ed. (Boston: Blackwell Scientific Publications, 1989), 137–151.

Lal, R. Agroforestry systems and soil surface management of a tropical alfisol: III: Changes in soil chemical properties. *Agroforestry Syst.* 8 (1989): 113–132.

Lantican, C. B. and Taylor, D. A., eds. *Compendium of National Research on Multipurpose Tree Species 1976–1990.* Arlington, VA: Winrock International Institute for Agricultural Development, 1991.

Le Houérou, H. N. Indigenous shrubs and trees in the silvopastoral systems of Africa, *Agroforestry: A Decade of Development,* H. A. Steppler and P. K. R. Nair, eds. (Nairobi, Kenya: ICRAF, 1987), 141–156.

Lescure, J., L. Emperaire, F. Pinton and O. Renault-Lescure. Nontimber forest products and extractive activities in the middle Rio Negro Region, Brazil, *Sustainable Harvest and*

*Marketing of Rain Forest Products,* M. Plotkin and L. Famolare, eds. (Washington, DC: Island Press, 1992), 151–172.

Matta-Machado, R. and Jordan, C. F. Nutrient dynamics during the first three years of an alley cropping agroecosystem in southeastern USA. *Agroforestry Syst.* 30 (1995):351–362.

Medina, E. Physiological ecology of trees and application to forest management. in: *Tropical Forests: Management and Ecology. Ecological Studies* 112, Lugo, A. E., and Lowe, C., eds. (New York: Springer-Verlag, 1995), 289–307.

Mesquita, R. 1995. The effect of different proportions of canopy opening on the carbon cycle of Central Amazonian secondary forest. Ph.D. dissertation, University of Georgia, Athens, GA.

Michon, G., F. Mary, and J. Bompard. Multi-storied agroforestry garden system in West Sumatra, Indonesia. in: *Agroforestry Systems in the Tropics* Nair, P. K. R., ed. (Dordrecht, The Netherlands: Kluwer Academic Publishers, 1989), 243–268.

Montagnini, F. *Sistemas Agroforestales. Principios y Aplicaciones en los Trópicos,* 2da. ed. (San José, Costa Rica: OTS/CATIE, 1992), 622.

Montagnini, F., K. Ramstad, and F. Sancho, F. Litterfall, litter decomposition, and the use of mulch of four indigenous trees species in the Atlantic lowlands of Costa Rica. *Agroforestry Syst.* 23 (1993):39–61.

Montagnini, F., A. Fanzeres, and S. G. da Vinha. Studies on restoration ecology in the Atlantic forest region of Bahia, Brazil. *Interciencia* 19 (1994):323–330.

Montagnini, F. 1995. Recuperación de áreas degradadas con la utilización de árboles nativos: Experiencias en tres regiones de Latinoamérica. III Congreso Latinoamericano de Ecología, Mérida, Venezuela. October 19–21, 1995.

Montagnini, F. and Mendelsohn, R. Managing forest fallows: Improving the economics of swidden agriculture. *Ambio* 26 (1997):118–123.

Nair, P. K. R. The role of trees in soil productivity and protection, in: *Agroforestry Systems in the Tropics.* P. K. R. Nair, ed. (Dordrecht, The Netherlands: Kluwer Academic Publishers/ International Council for Research in Agroforestry, Netherlands, 1989), 567–589.

Nair, P. K. R. The prospects and promise of agroforestry in the tropics, a review of technical and socioeconomic information with special emphasis to Africa. Report to the World Bank, Washington, DC. (1990) 121.

National Academy of Sciences. *Tropical legumes: Resources for the Future.* Board of Science and Technology for International Development. Washington, DC: National Research Council, 1979.

National Academy of Sciences. *Firewood Crops. Shrubs and Tree Species for Energy Production.* Board of Science and Technology for International Development. Washington, DC: National Research Council, 1980.

Nations, J. D. Xateros, chicleros, and pimenteros: Harvesting renewable tropical forest resources in the Guatemalan Petén, in: *Conservation of Neotropical Forests,* K. H. Redford and C. Padoch, eds. (New York: Columbia University Press, 1992), 208–219.

Onim, J. F. M., M. Mathuva, K. Otieno, and H. A. Fitzhugh. Soil fertility changes and response of maize and beans to green manures of *Leucaena, Sesbania,* and *Pigeonpea. Agroforestry Syst.* 12 (1990):197–215.

Padoch, C., J. Chota Inuma, W. De Jong, and J. Unruh. Amazonian agroforestry: A market-oriented system in Peru. *Agroforestry Syst.* 3 (1985):47–58.

Padoch, C. and W. De Jong. Traditional agroforestry practices of native and ribereno farmers in the lowland Peruvian Amazon, in: *Agroforestry, Realities, Possibilities, and Potentials,* H. Gholz, ed. (Dordrecht, The Netherlands: Martinus Nijhoff, 1987), 179–194.

Palm, C. A. Contribution of agroforestry trees to nutrient requirements of intercropped plants. *Agroforestry Syst.* 30 (1995):105–124.

Plucknett, D. L. International goals and the role of the international research centers, in: *Sustainable Agricultural Systems,* C. A. Edwards, R. Lal, P. Madden, G. Miller, and G. House, eds. (Delray Beach, FL: St. Lucie Press, 1990), 32–49.

Posey, D. A. The keepers of the forest. *Garden* 6 (1982):18–24.

Raintree, J. B. and K. Warner. Agroforestry pathways for intensification of shifting agriculture. *Agroforestry Syst.* 4 (1986):39–54.

Ramakrishnan, P. S. *Shifting Agriculture and Sustainable Development. Man and the Biosphere Series,* vol. 10. (New Jersey: Parthenon, 1992), 424.

Regnier, E. and R. Janke. Evolving strategies for managing weeds, in: *Sustainable Agricultural Systems,* C. A. Edwards, R. Lal, P. Madden, G. Miller, and G. House, eds. (Delray Beach, FL.: St. Lucie Press, 1990), 174–202.

Salazar, A., L. T. Szott, and C. A. Palm. Crop-tree interactions in alley cropping systems on alluvial soils in the upper Amazon basin. *Agroforestry Syst.* 22 (1993):67–82.

Sanchez, P. A., C. A. Palm, C. B. Davey, L. T. Szott, and E. C. Russell. Tree crops as soil improvers in the humid tropics? in: *Attributes of Trees as Crop Plants* M. G. R. Cannell and J. E. Jackson, eds. (Huntingdon, England: Institute of Terrestrial Ecology, National Environmental Research Council, Abbots Ripton, 1985), 327–350.

Sanchez, P. A. Science in agroforestry. *Agroforestry Syst.* 30 (1995):5–55.

Singh, G. B. Agroforestry in the Indian subcontinent: past, present, and future, 138. in: *Agroforestry: A Decade of Development,* H. A. Steppler and P. K. R. Nair, eds. (Nairobi; Kenya: ICRAF, 1987), 117–138.

Sips, P. *Management of Tropical Secondary Rain Forests in Latin America. Today's Challenge, Tomorrow's Accomplished Fact!?* (Wageningen, The Netherlands: National Reference Center for Nature, Forests and Landscape, Ministry of Agriculture, Nature Management and Fisheries, 1993), 72.

Soemarwoto, O. Homegardens: A traditional agroforestry system with a promising future, in: *Agroforestry: A Decade of Development,* H. A. Steppler and P. K. R. Nair, eds. (Nairobi, Kenya: ICRAF, 1987), 157–171.

Subler, S. and C. Uhl. Japanese agroforestry in Amazonia: a case study in Tomé-Açu, Brazil, in: *Alternatives to Deforestation: Steps Toward Sustainable Use of the Amazon Rain Forest,* A. B. Anderson, ed. (New York: Columbia University Press, 1990), 152–166.

Szott, L. T., C. A. Palm, and P. A. Sanchez. Agroforestry in acid soils of the humid tropics. *Adv. in Agron.* 45 (1991):275–300.

Unruh, J. D. Iterative increases of economic tree species in managed swidden-fallows of the Amazon. *Agroforestry Syst.* 11 (1990):175–197.

Vandenbeldt, R. J. Agroforestry in the semiarid tropics, in: *Agroforestry: Classification and Management,* K. MacDicken and N. Vergara, eds. (New York: John Wiley and Sons, 1990), 150–194.

Van Wambeke, A. *Soils of the Tropics. Properties and Appraisal.* New York: McGraw-Hill, 1992.

Vergara, N. T. Agroforestry: A suitable land use for fragile ecosystems in the humid tropics, in: *Agroforestry: Realities, Possibilities, and Potentials,* H. Gholz, ed. (Dordrecht, The Netherlands: Martinus Nijhoff, 1987), 7–20.

Vitousek, P. M. Litterfall, nutrient cycling, and nutrient limitation in tropical forests. *Ecology* 65 (1984):285–298.

Von Maydell, H. J. Agroforestry in the dry zones of Africa: past, present, and future, in: *Agroforestry: A Decade of Development,* H. A. Steppler and P. K. R. Nair, eds. (Nairobi, Kenya: ICRAF, 1987), 89–116.

Watters, R. F. *Shifting Cultivation in Latin America. FAO Forestry Development Paper No. 17.* (Rome: Food and Agriculture Organization of the United Nations, 1971), 305.

Wood, P. J. Principles of species selection for agroforestry, in: *Agroforestry: Classification and Management,* K. G. MacDicken and N. T. Vergara, eds. (New York: John Wiley and Sons, 1990), 290–309.

Yamoha, C. F., A. A. Abgoola, and G. F. Wilson. Nutrient contribution and maize performance in alley cropping systems. *Agroforestry Syst.* 4 (1986):247–254.

Young, A., *Agroforestry for Soil Conservation.* (Wallingford, U.K.: C.A.B. International and International Council for Research in Agroforestry, 1989), 276.

# 8 Nitrogen Fixation of Leguminous Trees in Traditional and Modern Agroforestry Systems

*James A. Bryan*[1]

**TABLE OF CONTENTS**



## 8.1 INTRODUCTION

Farmer use of biological nitrogen fixation has a long history in agricultural systems throughout much of the world, since early farmers in the Americas, in Africa, in Europe, in Asia, and in the Pacific Islands recognized the benefits of legumes and other $N_2$-fixers in maintaining or restoring their soil's fertility and incorporated $N_2$-fixing plants into their agricultural systems (Fred et al. 1932, Nutman 1965, Cohen 1977, Liebman 1987, Altieri 1995). Many traditional agricultural systems used trees

[1] Program Director, Tropical Resources Institute, Yale University, School of Forestry and Environmental Studies, New Haven, CT 06511, USA. Tel.: (203) 432-3660; Fax: (203) 435-5942; e-mail: james.bryan@yale.edu

0-8493-2206-5/00/$0.00+$.50

as part of their integrated programs (Bene et al. 1977, Soemarwoto et al. 1985, Budowski 1987, King 1987). In the tropics, especially, $N_2$-fixing trees have been used in woodlots, orchards, and other forest patches; household and compound gardens; roadsides, trailsides, living fences, boundary markers, and other edges; in various fallows; and other integrated tree-farming systems. In both tropical and temperate regions, $N_2$-fixing annuals may be used as nitrogen-providing cover crops; the

$N_2$-fixing components in agroforestry need not always be in the trees, although the restorative effects of herbaceous nitrogen-fixers may not be as long-lasting as those of trees and shrubs (Kang et al. 1990).

A wide variety of leguminous trees are still used in traditional perennial production systems, in a continuum of management systems ranging from the exploitation of naturally occurring trees and shrubs, to protection of favored trees from clearing for pasture or agriculture, to more or less intentional management for regrowth, to planting and tending of trees as diverse and important components of agricultural production systems (Wilken 1977, Le Houérou 1980, Alcorn 1981, Le Houérou 1987, Singh 1987, Soemarwoto 1987, von Maydell 1987, Michon and de Foresta 1999). $N_2$-fixing legume trees form the heart of modern agroforestry, as well, both in newly developing systems such as alley cropping and in modern adaptations of traditional systems, such as enhanced fallows.

$N_2$-fixing trees have historically played important roles in much of world agriculture, and have been the focus of a large proportion of modern agroforestry research. In many agriculture production systems, they provide needed soil nitrogen and increase protein production. Serious experimental difficulties in quantifying $N_2$-fixation are yet to be surmounted, and quantifying the benefits of $N_2$-fixation, as distinct from other effects of polyculture with trees, remains challenging for many of the agroforestry systems in which $N_2$-fixing trees play an integral part. Nevertheless, due to their $N_2$-fixation, their rapid growth, and their products, leguminous trees remain valuable in both traditional and modern agroforestry systems.

## 8.2 WOODLOTS, GROVES, FOREST PATCHES, AND BLOCK PLANTINGS

Woodlots, groves, and forest patches have been kept during clearing for agriculture or grown back from parcels of agricultural land, by planting or abandonment, in farming systems worldwide. Whether because of their ubiquity, their utility, their profitability or farmers' recognition of their benefit to soil fertility for nearby crop production, $N_2$-fixing legume trees are components of these forest patches in many traditional patch agroforestry systems, and are grown by farmers in many regions. In many temporate regions and at higher elevations in the tropics, $N_2$-fixing alders (*Alnus* spp.) also play a prominent part in restoration of fields to forest patches.

Because of widespread farmer preference for trees in patches outside their fields rather than hedgerows in their fields, the question must also be raised to what extent mulch from blocks of trees located near fields can meet the biological benefits of alley cropping without the tree-crop competition, such as maintenence of soil organic matter and soil organic matter nitrogen. Fast-growing $N_2$-fixing legumes such as *Leucaena leucocephala* are sometimes grown in blocks near cornfields in southern Mexico; based on comparison with blocks of *L. leucocephala* grown in trials in several regions, these blocks could provide up to 500–600 kg N $ha^{-1}$ $yr^{-1}$, including both fixed N and N taken up from the soil (reviewed by Giller and Wilson 1991). Varying amounts of fixed nitrogen make their way from forest patches into cropland as mulch or manure.

While the physical separation of trees from crops has the advantage of avoiding tree-crop competion, the nitrogen contributions of tree root turnover in intercropping is also absent from the cropland, and may result in decreased benefit to crops. For example, Böhringer and Leihner (1997) found that *Gliricidia sepium* and *Flemingia macrophylla* grown in blocks were less effective than alley hedges at recycling soil organic matter and nutrients in poor soils in Benin. Given the widespread presence of $N_2$-fixing trees and shrubs in farm plots, further information on $N_2$-fixation in such patches would be useful. However, farm forest patches may be either mature residual forest or regeneration, or intermediate combinations, and both the extent of $N_2$-fixation and the experimental difficulties in measuring it increase with forest maturity. In many forests, the proportion of $N_2$-fixing trees decreases with forest age, and it is generally assumed, with little experimental evidence, that the rate of symbiotic $N_2$-fixation dwindles with forest maturity, even where a mature legume component remains.

The $N_2$-fixing trees in many woodlots are an important feed source for livestock. They can also provide an important habitat for birds and other wildlife (Greenberg 1996).

## 8.3 NITROGEN FIXATION IN HOUSEHOLD ORCHARD GARDENS

Homegardens, household gardens, dooryard gardens, kitchen gardens or household orchard gardens, mixtures of trees, shrubs, and annuals grown around or near homes throughout the tropics, form part of the residence and agricultural production system for rural families. Since the beginnings of agriculture, the trees grown in household gardens have provided shade, fuelwood, fodder, fruit, flowers, honey, medicinals, poles, timber, many other products, and protection for the gardeners, their animals, and their soils (Anderson 1952, Michon 1983, Soemarwoto et al. 1985, Soemarwoto 1987, Padoch and de Jong 1991).

Throughout the tropics, people grow $N_2$-fixing leguminous trees in their yards. Whether or not intended by the homegardeners, the leguminous trees in many gardens are a major contributor to soil nitrogen and, sometimes through animal protein, human dietary protein (Cleveland and Soleri 1987, Okafor and Fernandes 1987, Christanty 1990). To some extent because of their $N_2$-fixing trees, household gardens generally provide stable and continuous production (Perera and Rajapakse 1991; Soemarwoto 1987; Jose and Shanmugaratnam 1993; Jensen 1993b), even in regions where cleared fields are productive for only a single season (Covich and Nickerson

1966). Household gardens sometimes are used in restoration of areas degraded by other agricultural uses (Kimber 1966, Boonkird et al. 1984, Everett 1995); $N_2$-fixing trees are considered useful resources in soil restoration where soil nitrogen is low or where the establishing trees must grow rapidly to crowd out competing weeds.

Many observers have commented that homegardens maintain high organic matter content and high soil fertility; however, the extent of $N_2$-fixation in household gardens has apparently not been measured, or generally even roughly estimated, in the many homegarden ecosystems and social systems in which it is probably important for soil fertility and dietary protein. (Torquebiau 1992). Though each has major methodological limitations, the available methods to estimate $N_2$-fixation in homegarden are selection of most likely $N_2$-fixing sources, with measurement of N fixed indirectly by acetylene reduction or $^{15}N$; interpolation from other, more controlled environments, like hedgerows; and comparisons of total N in homegarden. Interpolation of $N_2$-fixed in homegardens from that measured in other environments like hedgerows would admit the error introduced by comparison of the fast-growing juvenile trees of the hedgerows compared to the older trees (Jose and Shanmugaratnam 1993), often in shaded, forest-derived environments of the tropical homegarden. While it would include the nitrogen brought in by animals and humans and thus probably exceed the nitrogen fixed, the measure of nitrogen most relevant to human well-being would probably be the comparison of total nitrogen in homegardens with that in other areas.

In the only study found considering $N_2$-fixation in a homegarden, Jensen (1993) presents a detailed nutrient budget for a west Javanese homegarden, estimating (by acetylene reduction) 18 kg N nitrogen $ha^{-1}$ $yr^{-1}$ nitrogen fixed in a homegarden with several legumes (the trees *Parkia speciosa* and *Pithecellobium lobatum* and the climber *Lablab niger*) (Jensen 1993a). If the legume nodules were not assayed in this study (Jensen 1993b), this figure could represent an underestimate of $N_2$-fixed. Jensen also found that the homegarden was fertilized by litterfall 77 kg N $ha^{-1}$ $yr^{-1}$, pruning and weeding 104 kg N $ha^{-1}$ $yr^{-1}$, chicken manure 42 kg N $ha^{-1}$ $yr^{-1}$, a total of 223 kg N $ha^{-1}$ $yr^{-1}$, and that agricultural production included 70 kg N $ha^{-1}$ $yr^{-1}$. It can be speculated that the actual amount of $N_2$-fixed in the homegarden might include a greater proportion of the 293 kg N $ha^{-1}$ $yr^{-1}$ cycled through and from the multistoried garden.

Several tree species with commercial products, such as coffee, cacao, and citrus, have been introduced to homegardens worldwide; a few nodulating legume trees, such as *L. leucocephala* and *Calliandra calothyrus,* have been adopted throughout the tropics. Regional favorites such as the *Parkia* species of Africa, Southeast Asia, and Indonesia and the *Inga* species of the Americas have tended to remain local. Many of the legume tree species most often grown in homegardens worldwide are nonnodulating species such as *Tamarindus indica, Cassia* and *Senna* species, and *Delonix regia.* Though generally assumed not to fix nitrogen, the nonnodulating legumes are high in protein content, appear to enrich rather than deplete their soils in nitrogen, and in some cases have been found to facilitate the growth of companion plants. Recent studies indicate that the nonnodulating legumes may also fix nitrogen, at much lower rates than the nodulating legumes (Bryan 1995, Bryan et al. 1996).

## 8.4 NITROGEN-FIXING SHADE TREES FOR TREE PLANTATIONS

Several important tree crops benefit from partial shelter from the sun: coffee, cacao, and tea, for example, are often grown in the shade of other trees, in many cases $N_2$-fixing legumes (reviewed by Budowski et al. 1984, Kass et al. 1997, Beer et al. 1998). An immense range of shade may be used in plantations, even for the same crop tree, such as coffee, from the mono-specific *Erythrina poeppigiana* shade trees in many Costa Rican coffee plantations, regularly pollarded and of uniform shape and size, to collections of diverse species and sizes of *Inga,* to mixed shade of $N_2$-fixing trees and valuable timber, citrus, or other fruit trees and even to closed, forest-like mixtures of large trees with coffee scattered in the understory. Recently, many growers have planted high-production, sun-tolerant coffee varieties, provided high inputs of fertilizer and other agrochemicals, and removed their shade trees. Although open-grown, high-input coffee yields are generally greater, especially on the best sites, varying coffee prices on the world market sometimes make the high-input production unprofitable (Galloway and Beer 1997), and the use of $N_2$-fixing shade trees in coffee plantations is again increasingly being found to have many benefits, on both poor and fertile sites. In some markets, shade-grown coffee brings premium prices, for its higher quality, its ecological benefits, or both.

$N_2$-fixing trees have the same positive effects on tree plantation soils that they have in conjunction with annual crops (carefully reviewed by Beer et al. 1998): erosion prevention, moderation of temperature, and increased soil organic matter, which leads to increased soil porosity, aeration, water infiltration, and water retention, decreased runoff, increased CEC and generally improved soil nutrient status and generally greater tolerance of acid soil conditions. $N_2$-fixing trees may provide part or all of the nitrogen fertilizer needed for the crop tree. The amount of nitrogen fixed by *I. jinicuil* in a coffee plantation in Mexico was estimated by acetylene reduction of removed nodules at 47 kg N $ha^{-1}$ $yr^{-1}$ (Roskoski 1981, Roskoski 1982). Acetylene reduction activity of nodules of *E. poeppigiana* shade trees in a Costa Rican coffee plantation appeared to be in the same range (Lindblad and Russo 1986). Leguminous shade trees such as *E. poeppigiana* may contribute some 6,900–20,000 kg organic material $ha^{-1}$ $yr^{-1}$ in coffee plantations (Beer 1988), with up to 330 kg N $ha^{-1}$ $yr^{-1}$ (Russo and Budowski 1986).

Shaded coffee plantations, in many respects similar to forest habitats, may support a diversity of both migratory and resident birds similar to that in remaining forests (Greenberg 1996); open grown coffee generally supports less diversity of bird species. Taxonomically and structurally diverse shade systems with high and unpruned trees may be needed, however, to provide suitable habitats for bird species diversity (Greenberg et al. 1997), so inclusion of timber trees for shade may provide better bird habitats than $N_2$-fixing shade trees alone. Some $N_2$-fixing trees, such as *Mimosa scabrella,* serve the combined purpose of shade, $N_2$-fixation, timber production, and a high canopy (Musálem 1995). Other forms of biodiversity, such as insects, are also supported by shade trees (Perfecto et al. 1996). Shade from $N_2$-fixing trees such as *Inga,* alone or in combination with other non-$N_2$-fixing trees, reduces weed germination and growth in coffee plantations and thereby reduces herbicide use (Nestel and Altieri 1992). Ground water pollution may be decreased (Beer et al. 1998).

Many trees of the large *Inga* genus have long been favored as shade trees for coffee in many regions of the Americas (Cook 1901, León 1966, Pennington 1997). The most important species for coffee shade have included *I. vera* in the Caribbean Islands, Mexico, and Central America (Marrero 1954, Pennington 1997), *I. jinicuil* and *I. paterno* in Mexico, and *I. edulis, I punctata,* and *I. oerstediana* in both South America and Central America. Several *Inga* species may be grown in the same plantation. In addition to their environmental benefits, they provide fuelwood for coffee workers and many have sweet, edible pulp around the seeds, which is highly appreciated in many regions of South America, Central America, and Mexico (León 1966, Bryan 1995, Lawrence 1995, Pennington 1997).

The $N_2$-fixing trees *Gliricidia sepium* and *Inga feuilli* have been used for cacao shade since pre-Columbian times (León 1966, Pennington 1997). Long-term experiments in Costa Rica (Fassbender 1998) indicate that both the $N_2$-fixing *Erythrina poepigiana* and the valuable non-$N_2$-fixing timber tree *Cordia alliodora* produce adequate litter to increase soil organic matter and, with annual input of fertilizer (average 87.5 kg N, 18.2 kg P, and 11.5 kg K $ha^{-1}$ $yr^{-1}$), maintain or improve most soil nutrient levels (except exchangeable K, which decreased markedly). Production of cacao was equal with the $N_2$-fixing and non-$N_2$-fixing shade trees; but it can be assumed that the $N_2$-fixation would be more important without fertilizer inputs.

Two variations of the use of $N_2$-fixation in tree plantations should also be mentioned here. In addition to their use for shade, fast-growing $N_2$-fixing trees have also been grown in the understory of other crop trees, with benefits both to soils and plantation trees. In the understory of coconuts in degraded soils in Sri Lanka, for example, the $N_2$-fixing trees *Acacia auriculiformis, C. calothyrsus, G. sepium* and *L. leucocephala* increased soil OM in the upper soil horizons, lowering bulk density, improving soil aeration, and increasing water availability, and thereby improved coconut root growth in degraded, marginal soil (Vidhana Arachchi and Liyanage 1998).

Herbaceous legumes are used in tree plantations as cover crops, for example in rubber plantations, where they protect the soil from otherwise serious erosion, provide nitrogen inputs, and, in many cases, supply protein-rich fodder to livestock.

## 8.5 NITROGEN-FIXING TREES AND SHRUBS FOR PASTURE IMPROVEMENT AND FODDER

Nitrogen-fixing trees and shrubs provide livestock feed in woodlots, homegardens, and many other traditional farming systems: $N_2$-fixing trees are protected when land is cleared for pastures or crops; grown on borders and boundaries, in fencerows and by paths, roads and railways; kept on the banks of canals, rice paddies, streams, and river banks; and planted in various designs as windbreaks, all of which may be integral sources of feed for livestock. Leguminous trees and shrubs are both lopped for fodder and browsed directly. Woody legume hedgerows are often suitable for large-scale mechanical farming: the alleys, with herbaceous pasture, can be mechanically mowed to control invasion of unpalatable weeds. In modern agroforestry systems, leguminous trees are regularly chosen for the hedgerows in alley farming, the contour hedgerows on steep slopes, and the fast-growing pioneer tree species for accelerating fallow regrowth and pasture reclamation, in which, among their other protective and productive functions, they provide protein for livestock (Benavides 1994, Gutteridge and Shelton 1994, Paterson et al. 1996). To an extent not yet well quantifiable, the $N_2$-fixation of trees and shrubs is a resource for provision of protein to livestock, directly as fodder, as a component of pasture forage, and in maintaining pasture productivity in both traditional and modern production systems.

With worldwide human population increases coupled with growing appetites for meat and dairy products, conversion of forest and marginal land for livestock production has intensified. At the same time that animal products are required, pasture productivity on many newly converted lands is often shortlived, and productivity of existing pastures, in many cases, is decreasing at alarming rates; as pasture productivity declines, weed invasion intensifies (Heckadon-Moreno 1985, Le Houérou 1987, Serrão and Toledo 1990, Gutteridge and Shelton 1994, Boddey et al. 1997). Nitrogen is required both for feed growth and for plant protein for livestock nutrition. Therefore, livestock production requires N inputs through fertilizer, biological $N_2$-fixation, or both.

Most modern, high-production pastures are composed of either heavily fertilized grasses selected for both rapid growth and high protein content or mixtures of grasses and $N_2$-fixing herbaceous legumes. Without adequate N fertilizer, or if the $N_2$-fixing legume component declines, most pastures decline in production, change composition to low-nutrient-demanding and less nutritious grass species and weeds, and eventually to scrub and forest (Haynes and Williams 1993). The advantages of woody $N_2$-fixing plants over herbaceous legumes are their generally greater longevity and hardiness and therefore greater persistence in pastures, their provision of fodder in seasons when other feed is in short supply, their potential improvement of livestock environment through shading, their protection of soils from wind and water erosion by their perennial root and shoot growth, their long-lasting soil organic matter fractions, and their potential to grow on adverse sites and rehabilitate them (Le Houérou 1987, Gutteridge and Shelton 1994). An extreme example of adverse site adaptation is the *Prosopis* genus, which permits livestock raising in some of the driest deserts of the world (Zelada G. 1986). $N_2$-fixing trees and shrubs, even while being cleared or degraded in many regions, are becoming increasingly important resources for pasture management.

Protein is often the factor limiting growth of livestock, both in highly productive and in the lower production systems common in both dry and wet regions in the tropics. In regions of seasonal drought, protein is especially limiting during the latter part of dry seasons when fodder is scarce (Le Houérou 1987, Lesseps and Chipanda 1998); in pastures converted from rain forest, feed may be chronically too low in protein for effective ruminant digestion. The foliage, fruits, seeds, twigs, and green bark of most $N_2$-fixing—and indeed of nonnodulating—leguminous trees and shrubs also have high protein contents.

Even in high-production livestock raising systems, animal production is so closely tied to protein intake that growth or milk production may generally be stimulated by feeding protein supplements (Norton 1994c). The central questions about the continued or potentially increased use of $N_2$-fixing trees in livestock production systems are: in what circumstances, by what methods, and to what extent $N_2$-fixing trees and shrubs can (1) improve the quality and quantity of pasture production and add needed fodder sources; (2) increase livestock productivity; and (3) slow or reverse declines of pasture productivity. The inclusion of $N_2$-fixing trees and shrubs into pastures, broadly defined, changes the pastures, feeds, and animal growth in numerous ways. Here one would want to consider the effect of $N_2$-fixing trees and shrubs on the growth of associated grasses, the feed-production of the woody legumes themselves, and their effects on animal growth, and, to the extent possible, predict likely long-term effects of the $N_2$-fixing trees and shrubs on pastures.

The inclusion of herbaceous legumes in grass pasture mixes has been well-documented to increase grass production and grass protein content, and the proportion of herbaceous legumes required to optimize pasture production has been evaluated for numerous temperate and some tropical grass-legume combinations (Thomas 1995); however, the effects of $N_2$-fixing trees and shrubs on pasture grass production appear to have been much less studied (Kang et al. 1990). The addition of $N_2$-fixing trees and shrubs to grass pastures could theoretically increase the biomass production of pasture grasses, the protein content of the grasses, or both, especially if the woody component is managed to control wind or water erosion and improve soil physical and chemical qualities. Clearly, however, the trees' and shrubs' competition for sun, water, or nutrients can undermine the potential benefits to pasture production.

In field trials, total biomass production tends to be greater in woody legume-grass mixtures than in pure grass pastures, production of grass often somewhat less, protein content of the mixture higher than that of grass alone, and protein content of grass in association with the woody legume higher than in grass grown in monoculture. For examples: in an experiment in the highlands of Rwanda, contour hedges of *Calliandra calothyrsus* and *Sesbania sesban* increased the protein content of the grasses *Pennisetum purpureum* and *Setaria splendida,* without affecting overall grass production (Niang et al. 1998); *Erythrina poeppigiana,* grown with King grass (*P. purpureum*) in the humid tropics of Costa Rica, either did not affect or slightly reduced the total growth of King grass, but increased the grass's crude protein content; total biomass production of the combination of *P. purpureum* and *E. poeppigiana,* was considerably greater than that of King grass alone (Benavides et al. 1994).

Nutritional benefits of many hedgerow fodder components have been well-demonstrated. Fodder production by the woody legume component of pastures varies widely, but has been a major source of livestock feed, especially protein, in most livestock raising systems worldwide. Trees and shrubs provide all or most livestock feed in some regions (Le Houérou 1980, Le Houérou 1987); elsewhere, as a component of marginal quality grass pastures, $N_2$-fixing leguminous trees and shrubs provide protein-rich fodder that can elevate the N content of total feed above the critical threshold required for effective rumen function (Norton 1994b, Thomas 1995); in many settings, fodder from fast-growing woody legumes can increase livestock production by providing protein supplements to livestock on both high and low quality feed (Benavides 1994, Jones 1994b, Norton 1994c). Pastures, occupying vast and increasing expanses, seem ideally suited for recently developing alley farming practices; the effects of woody hedgerows on yields and quality of pasture grasses deserve additional research attention.

Livestock growth is enhanced by the inclusion of legumes, both herbaceous and woody, in diverse production systems. Much of the research attention regarding the effects of pasture legumes on livestock growth has been focused on herbaceous legume cover crops, which sometimes increase livestock growth dramatically. As a recent example, in the Cerrado of Bahia, Brazil, cattle growth was doubled to tripled by the addition of fodder legumes (*Arachis pintoi, Calopogonium muconoides, Desmodium ovalifolium,* and *Stylosanthes* sp. to grass pastures (Boddey et al. 1997).

Woody legumes can also increase animal productivity substantially. *L. leucocephala,* one of the most important fodder trees, which may produce up to 30 t $ha^{-1}$ $yr^{-1}$ edible dry matter when moisture is not limiting; 18 t $ha^{-1}$ $yr^{-1}$ in the seasonally moist climate of Ibadan, Nigeria, with approximately 1280 mm of rainfall (Van der Meersch et al. 1993); and 3–11 t $ha^{-1}$ $yr^{-1}$ in semiarid regions (Arora et al. 1986, reviewed by Shelton and Brewbaker 1994). In a review of 13 experiments comparing cattle growth on grass pastures and on grass pastures with *L. leucocephala,* cattle gained at least 70% more weight in over half the trials, less in only one (Jones 1994b). In fertile pastures in southeast Queensland, cattle grazing *Leucaena*/grass pastures gained 310–430 kg $ha^{-1}$, approximately twice the gains of cattle on grass-based pastures in the same environment. In other Australian grass/*Leucaena* pastures under ideal conditions, cattle liveweight gains of 273–300 kg $head^{-1}$ $yr^{-1}$ or 1422 kg $ha^{-1}$ $yr^{-1}$ have been recorded, higher growth rates than could be expected in the same environment on pastures of herbaceous legumes and grasses (Jones 1994b, Shelton and Brewbaker 1994). Milk production has also been found to increase an average of 14% for dairy cows with *L. leucocephala* supplements to grass pastures (Jones 1994b).

Several other leguminous trees and shrubs have also been widely evaluated for their fodder biomass and protein production in block plantings, hedgerows, or as living fences. Numerous reviews are now available of production, fodder values, and/or effects on grass pastures of other $N_2$-fixing woody trees and shrubs such as *Acacia* spp. (Gutteridge 1994a); *Calliandra calothyrsus* (Palmer et al. 1994, Evans 1996); *Erythrina* spp. (Kass 1994); *Faidherbia albida;* (Anon 1989, Lesseps and Chipanda 1998) *Gliricidia sepium* (Simons and Stewart 1994); other species of *Leucaena* (Hughes 1998, Stewart and Dunsdon 1998); *Paraserianthes falcataria* (Norton 1994c); *Prosopis* (East and Felker 1993); *Sesbania grandifolia* and *S. sesban* (Topark-Ngarm and Gutteridge 1990, Gutteridge 1994b); and numerous other species (Norton 1994b). Many but not all such nodulating woody legumes have a leaf crude protein content of 20–30%, similar to that of tropical herbaceous pasture legumes, (Norton 1994b), as well as to that of high protein temperate legume forage species such as alfalfa (*Medicago sativa*), and clover (*Trifolium* spp.) and of other high quality herbaceous legumes (Gutteridge 1994b; Kass 1994; Palmer, Macqueen and Gutteridge 1994; Shelton and Brewbaker 1994; Simons and Stewart 1994). Woody legume fruits, seeds, flowers, twigs, and other plant parts also have high protein contents.

While the addition of woody legumes to grass pasture generally raises the total biomass production and total feed protein content significantly (Benavides et al. 1994), in nutrient-poor pastures, the addition of woody legumes is particularly important: The additional protein provided by the woody legume may bring the total feed content over the 8% protein or 1.3% N threshold required for minimum ammonia level for effective rumen function (Norton 1994b, Thomas 1995). When added to diets of low quality forage grasses, woody legume fodder may increase the rate of weight gain or convert weight maintenance or loss to weight gain in cattle, sheep, and goats (Jones 1994b, Norton 1994c). In addition to their high protein values, part of the advantage of including woody legume fodder in ruminant diet is that the tannins in some tree leaves decrease their rate of digention in the rumen and therefore increase the amount of protein bypassing the rumen for digestion in the small intestine (Norton 1994b).

It should also be mentioned here that some woody legumes produce antinutritional factors (Jones 1994a, Norton 1994a). Since the production of antinutritional factors is often N-dependent, N-fixation may facilitate the production of nonprotein amino acids, for example, along with the production of nutritional protein. Tree legume species providing excellent palatability, digestibility and livestock growth rates in some regions may be unpalatable or even toxic elsewhere, depending on plant provenance, season, growth stage, freshness of fodder, management, percent of plant's fodder in diet, adaptation of animals and rumen bacteria populations to the antinutritional factor, and other conditions (Jones 1994a, Norton 1994a, Simons and Stewart 1994).

$N_2$-fixing trees and shrubs provide a multitude of protective benefits in pastures: erosion control, shade, moderation of temperature extremes, and increase of durable organic matter, which helps maintain soil structure and provides gradual release of nitrogen and other minerals. While it is generally recognized that soil organic matter is a major component in soil productivity, the comparative value of herbaceous and ligneous soil organic matter in pastures has apparently not been well-determined (reviewed by Staver 1989). With their rapid growth on clear and vulnerable sites, woody $N_2$-fixers are often important protectors against water and wind erosion. Woody legumes may also improve pasture microclimate, increase soil fertility in other minerals, and decrease the percent of pasture occupied by nonpalatable plants. $N_2$-fixing trees and shrubs are generally more persistent in pastures than forage legumes that are also planted for their nitrogen and protein, but which, especially if they are palatable, may disappear rapidly under grazing. Some woody legumes in grass mixtures, by contrast, may be grazed for many decades (Shelton and Brewbaker 1994).

## 8.6 IMPROVED FALLOWS

Consideration of improved fallows usually focuses on slash-and-burn agriculture in the tropics. In fact, soils depleted by agriculture are abandoned to fallow and forest regrowth worldwide. For example, many of the present forests of the eastern U.S. are fallow restoration of degraded farmlands. In well-managed traditional slash-and-burn or swidden-fallow agricultural systems with adequate available land, forest regrowth during a long fallow period causes a return of the soil fertility depleted during cropping. During the 10- to 30-year or even much longer fallow period, weeds are shaded out, pest cycles are broken, and soil organic matter is restored, with improvements in soil structure and available minerals, including nitrogen (Nye and Greenland 1960, Zinke et al. 1978). Now, with growing populations in much of the world, adequate land is often not available for fallow periods long enough to permit restoration of soil fertility. Nitrogen being a major nutrient required for plant growth and a mineral depleted in many agricultural soils, a serious consideration of agroforestry research has been under what circumstances agroforestry can accelerate the restoration or augmentation of available N without excessive loss of crop space, crop growth, or labor (Staver 1989, Fernandes and de Souza Matos 1995, Nair et al. 1999).

Biological $N_2$-fixation has been used to enhance fallow regrowth and the restoration of fertility in farming systems, both traditional and modern, throughout the world. Many traditional agroforestry systems make use of leguminous trees in fallows, whether for their effects on the soil, for the products they will produce, or because fast-growing legumes are natural pioneers with an abundant seed source in nearby forests. Traditionally, various species of $N_2$-fixing trees have been protected when forests are cleared for farming, providing advanced regeneration; nitrogen-fixing trees have been kept on field boundaries or in fields as a seed source for regeneration; and tree seedlings have been transplanted into fields during the production of annual crops. *Inga* species, for example, long favorites of forest dwellers and farmers in many regions of tropical America, have traditionally been planted or kept when land was cleared for gardens and planted into garden areas (Gordon 1982), and are also important to modern agroforestry.

In modern agroforestry systems, seedlings of fast-growing legumes such as *Sesbania sesban, Gliricidia sepium,* and *L. leucocephala* have been established during the cropping cycle, in an effort to protect the soil during the transition to fallow, speed the recovery of nitrogen in the field, provide useful crops during fallows, and provide a favorable environment for the invasion of other tree species. Biologically fixed nitrogen is especially important where industrial fertilizers are unavailable, such as in sub-Saharan Africa (Kang and Mulongoy 1992).

Several trials have demonstrated the capacity of trees, particularly $N_2$-fixing legumes, to increase fallow benefits in nitrogen supplied in organic matter, soil nitrogen, and growth of subsequent crops (reviewed by Rao et al. 1998). An improved fallow system in Zaire with *L. leucocephala,* after 11 months growth, had more than twice the plant biomass as natural fallow slightly older, in the same area; the nitrogen content of the 11-month *L. leucocephala* plant material (317 kg $ha^{-1}$) greatly exceeded that of six-year secondary forest regrowth (247 kg $ha^{-1}$) (Bebwa and Lejoly 1993). Inputs of nitrogen from $N_2$-fixing trees typically increase both soil nitrogen status and growth of subsequent crops.

In addition to $N_2$-fixation, the roots of perennials in fallows take up nitrate that might otherwise be leached (Mekonnen et al. 1997); it appears that one of the advantages of fast-growing $N_2$-fixing trees is that their root growth is also rapid, permitting the uptake of minerals leached from surface soil layers (Buresh and Tian 1998).

## 8.7 HEDGEROWS OF LEGUMINOUS TREES IN ALLEY CROPPING AND ALLEY FARMING

Agroforestry research is part of a complex of international efforts to increase food production in the face of worldwide shortening of fallows and accompanying declines in the fertility of cultivated land. Widespread traditional practices of growing perennials with or in alteration with annual crops had been observed to protect soil from erosion and prevent long-term loss of fertility. Much of the recent research effort has been an attempt to adapt those traditional practices, such as bush fallows, to the relatively new system of alley cropping, intended to permit productive cropping to proceed uninterrupted by prolonged fallow periods (Kang and Wilson 1987, Kang et al. 1990, Kang 1993).

Alley cropping is the growing of crops such as grains in the open spaces between rows of trees or shrubs that are pruned during the growing season to an extent depending on the needs of the crops and the spacing of the hedgerows. Even with low (25 cm) coppicing during crop growth as often as four times a year, some benefits of fallow occur concurrently with cropping, due to the maintenance of the extensive living woody root system, which continues to hold the soil, contributes to soil organic matter and permits rapid regrowth of stem and leaf material between crops (Kang and Mulongoy 1992). Most alley cropping trials in the past two decades have produced fast-growing nodulating leguminous trees in the hedgerows, to add nitrogen-rich organic matter directly to the alley crops, while perhaps providing numerous other products including high protein food or fodder. On the steep slopes that have come under increasing farming pressure, the hedgerows are planted on the contours, providing rapid soil cover, eventually forming terraces, arresting erosion and fertilizing the soil, as well as providing a great variety of other products (Kang et al. 1990). The agronomic, fodder-enhancing, and social benefits of alternating various woody species from hedgerow to hedgerow or mixing species within hedgerows is increasingly recognized by researchers (Van Noordwijk et al. 1995).

In contrast to more traditional systems such as homegardens, species for hedgerows are generally chosen from a small group of internationally recognized high producers, which have now been subjected to extensive trials for biomass and nutrient production and effects of companion plants. Most of the species used for hedgerows are nodulating legumes such as *Caliandra calothyrus, L. leucocephala, Inga edulis* and

*Paraserianthes (Albizia) falcataria* (Mimosoideae), and *Cajanus cajan, Flemingia macrophylla, Gliricidia sepium, Sesbania sesban, S. rostrata,* and *S. grandiflora* (Papilionoideae). The rapid growth of such trees and shrubs provides for rapid canopy closure between crops, protection of the soil from splash and wind erosion, and an abundance of nitrogen-rich foliage and stems, which, when clipped, mulch the crop, enhance soil organic matter, and fertilize the crop, and, along with their root turnover, improve soil structure and permit long-term supply of nitrogen and other minerals.

Early experiments with alley cropping demonstrated that hedgerows of fast-growing leguminous shrubs or trees grown in fertile soils of the moist tropics have the potential to produce clippings with 200–250 kg N $ha^{-1}$ $yr^{-1}$ (Kang et al. 1981). Numerous experiments over the past 20 years have demonstrated that the prunings of legume hedgerows, even in infertile, acidic or semiarid soils, can generally provide the roughly 80 kg N required for a low-yield maize crop (reviewed by Palm 1995). In a recent review of nitrogen yields in alley cropping in the forest-savanna transition zone in southern Nigeria, *Caliandra calothyrus, Flemingia macrophylla, Gliricidia sepium,* and *L. leucocephala* were found to yield 218, 149, 169, and 247 kg N $ha^{-1}$ $yr^{-1}$, respectively; root turnover of leguminous hedgerows also provides significant amounts of N in alley cropping systems, though largely unmeasured (Kang and Mulongoy 1992). It has been estimated that approximately 45% of hedgerow nitrogen may derive from N-fixation (Kang and Mulongoy 1992).

Increases in crop yield have been demonstrated experimentally by the addition of wood legume prunings to maize, upland rice, cassava, cowpeas, and other crops, in many conditions and many regions. In practice, the effects of alley cropping have proven variable, however, depending on climate, soils, width of hedgerows, frequency, timing, and extent of pruning, farmer use of woody material, and various other factors. In infertile and acidic soils, competition for nutrients often decreases rather than increases crop yields (Szott et al. 1991, Fernandes et al. 1993, Garrity et al. 1995, Palm et al. 1995, Van Noordwijk et al. 1995). In semiarid regions, loss of growth to competition for moisture often offsets benefits to crop production from added nitrogen, even when hedgerows are spaced much more widely than the 4 m spacing frequently used in the humid tropics (VanDenBeldt 1990, Jama Adan 1993).

Common sense as well as practice lends support to widespread farmer skepticism about alley cropping. In addition to their concerns for weediness of hedge trees and increased labor, the widely proposed systems of hedgerows with alley cropping have a fundamental flaw in use for $N_2$-fixation: The effective provision of N through turnover of tree roots in close proximity to crops is in direct competition for moisture in water-limited systems, for other nutrients in nutrient-limited soils, and for sunlight in sun-limited systems; in many circumstances $N_2$-fixation fails to provide the anticipated nitrogen fertilizer in a timely fashion; and, especially in the marginal fields that most require inputs of biological $N_2$-fixation for long-term crop production, the presence of $N_2$-fixing trees and shrubs may decrease rather than increase crop yields. From numerous studies in which inclusion of hedgerows has decreased crop yields, it can now be generalized that legume hedgerows will probably not benefit the alley crop in conditions of marginal rainfall, inadequate pruning, highly acidic soils, high soil fertility, or poorly adapted legumes. Nevertheless, as Vandermeer (1998) points out, in some instances the facultative effects of leguminous tree hedges exceed any competitive reductions in the alley crops; and alley cropping systems need to be optimized before the system can be rejected theoretically. Both theoretical benefits and numerous and diverse experimental successes of this artificial system justify continued development, hopefully in farmers' repeated and improved trials, before the alley cropping concept itself can be rejected. Hedgerows for pastures have received less research attention than alley cropping of annual crops, despite the vast problem of forest-clearing for pasture followed by low production and rapid loss of soil fertility of many tropical pasture lands. Hedgerow technologies appear excellently suited to pasture improvement, for several reasons directly related to livestock production: high protein fodder, widely needed to increase meat or milk production; provision of fodder during the dry season; convenience of large-scale pasture maintenance (such as mechanical mowing) between tree rows; control of weed growth in pastures; provision of shade for livestock; general protection of soil through maintenence of soil OM; access to lower level ground water; recycling of soil minerals from depths; and prevention of wind and water erosion. Reynolds and Jabbar (1995) found that supplimentation of grass diets with *Leucaena* forage in alley crops increased growth, survival of offspring, milk yield and intake of other feeds, in sheep, goats and cattle, with effects greater or less depending on season and livestock physiological condition. Regular pruning of the tree component was necessary to prevent shading and decline or death of grasses growing in the alleys. In considering plans to increase the nitrogen inputs and protein production through the use of $N_2$-fixing trees and shrubs, the palatability of plant foliage, and antinutritional qualities must be assessed (Tuah 1995). The cycle whereby atmospheric nitrogen becomes plant protein and then animal protein and manure remains fundamentally essential to human nutrition worldwide and to maintaining the fertility of agricultural soils, and applications of $N_2$-fixation to alley farming for animal production represent an important resource.

While alley cropping has not met some of its early expectations for combining crop improvement, soil protection, and farmer acceptance in the widespread geographic, edaphic, and economic annual production systems that appear to need the protection and inputs of perennials, farmer modifications of alley cropping for steep slopes appear to provide a particularly useful and desired tool for farming in steep and fragile slopes (Banda et al. 1994, Garrity and Mercado 1994). Fast-growing *Calliandra calothyrsus, Gliricidia sepium,* and *Leucaena* spp. are most often recommended for these purposes; in some cases, farmers replace the proposed $N_2$-fixing tree component with plants more useful to their immediate needs (Bannister and Josiah 1993), which may, in fact, successfully meet the overarching need of minimizing soil erosion.

## 8.8 BOUNDARY TREES: TRAIL AND ROADSIDE PLANTINGS, LIVING FENCES AND BOUNDARY MARKERS

Several other forms of linear plantings that make use of $N_2$-fixing trees have been part of traditional farming systems and remain widely used, in some cases where most other wood vegetation has been removed: trail- and roadside trees, boundary markers, and living fences.

Leguminous fruit trees such as *Inga* spp. have probably been planted on the edge of forest trails, either intentionally or by the discarding of seeds (Anderson 1952), since the early days of human prehistory. Roadsides remain favorite sites for trees in rural communities, even in regions otherwise deforested. Roadside trees provide fodder and fuel in addition to shade and other products. To the extent that the nitrogen fixed by roadside trees makes its way, even slowly and indirectly, to agricultural crops, and to the extent that nitrogen-fixing trees protect fields, for example by reducing wind speed or preventing off-site erosion, these traditional and popular systems should be considered successful forms of agroforestry. Where agroforestry systems have met farmer rejection instead of expect adaptation, the potential increase in livestock or human dietary protein through improvement of roadside plantings may be an adaptable $N_2$-fixing agroforestry technology.

Combining features directly useful to farmers with the direct and indirect benefits of $N_2$-fixing trees, and providing such conveniences as shade near but not interfering with crops, fencerow plantations have been adapted in many regions where alley cropping has not found favor with farmers.

## 8.9 CONCLUSION

$N_2$-fixing trees have been grown in traditional agroforestry systems in much of the world: household gardens, trail- and roadside plantings, and trees left in fields or on field borders. In some cases because of the multiple products and rapid growth, in other cases because of their perceived benefits to soil fertility and crop growth, $N_2$-fixing trees were an important component of preindustrial and precolonial agriculture. $N_2$-fixing trees have also been the major component in most of modern agroforestry.

However, the effects of $N_2$-fixation on soil fertility and crop yield remain difficult to quantify, especially in the more complex agroforestry systems. A careful study such as that of Hocking and Islam (1998) can quantify the negative yield effects of tree competition on nearby plants without addressing the probable positive effects of the same trees on crops beyond the root and shade zone. The long-term effects of the presence or absence of $N_2$-fixing trees on the maintenance or decline of ligneous organic matter remain an issue of concern and speculation.

In some cases, governments or international environmental groups may be eager to restore the tree cover of degraded areas, even if that is not the first priority of farmers. The advocacy of multipurpose $N_2$-fixing trees, often an attempt to reconcile different priorities and meet the perceived needs of multiple groups, sometimes leads to overdependence, overuse, or rejection of this valuable and diverse plant-microbe resource. Paradoxically, some of the trees now advocated by planners were recently rejected as old-fashioned by similar groups. As a hopeful note for restoration of degraded areas, another paradox concerning $N_2$-fixing trees can also be noted: A simplified, few-crop system like alley cropping may also provide a means for restoring greater diversity of tree species, as well as the biological diversity the augmented nitrogen and protein may support (Akinnifesi et al. 1995).

## 8.10 REFERENCES

Akinnifesi, F. K., H. J. W. Mutsaers, and H. Tijani-Eniola. Pioneer woody species in *Leucaena leucocepha* and *Senna siamea* hedgerows in a forest ecosystem in southern Nigeria, in: *Alley Farming Research and Development: Proceedings of the International Conference on Alley Farming.* 14–18 September 1992, B. T. Kang, A. O. Osiname, and A. Larbi eds. (Ibadan, Nigeria: IITA, 1995), 236–243.

Alcorn, J. B. Huastec noncrop resource management: Implications for prehistoric rain forest management. *Human Ecol.* 9(1981): 395–417.

Altieri, M. A. *Agroecology: The Science of Sustainable Agriculture.* Boulder, CO: Westview Press and Intermediate Technology Publications, 1995.

Anderson, E. *Plants, Man, and Life.* Berkeley: University of California Press. 1952.

Anon. Faidherbia albida *(Del.) A. Chev. (Synonym:* Acacia albida *Del.): A Monograph.* Nogent sur Marne Cedex, France: Centre Technique Forestier Tropical, 1989.

Arora, S. K., R. S. Paroda, U. N. Joshi, M. L. Saini, and J. V. Singh, Forage yield and quality of some *Leucaena* cultivars in the semiarid region of Hisar, India, *For. Ecol. and Manage.* 16(1986): 355–366.

Banda, A. Z., J. A. Maghembe, D. N. Ngugi, and V. A. Chome. Effect of intercropping and closely spaced *Leucaena* hedgerows on soil conservation and maize yield on a steep slope at Nicheu, Malawi. *Agroforestry Syst.* 27(1994): 17–22.

Bannister, M. E. and S. J. Josiah. Agroforestry training and extension: The experience from Haiti. *Agroforestry Syst.* 23(1993): 239–251.

Bebwa, B. and J. Lejoly. Soil organic matter dynamics and mineral nutrients content in a traditional fallow system in Zaire, in: *Soil Organic Matter Dynamics and Sustainability of Tropical Agriculture,* K. Mulongoy, and R. Merckx eds. (Chichester: John Wiley and Sons, 1993), 135–142.

Beer, J. Litter production and nutrient cycling in coffee (*Coffea arabica*) or cacao (*Theobroma cacao*) plantations with shade trees. *Agroforestry Syst.* 7(1988): 103–114.

Beer, J., R. Muschler, D. Kass, and E. Somarriba. Shade management in coffee and cacao plantations. *Agroforestry Syst.* 38(1998): 139–164.

Benavides, J. E., ed. *Arboles y arbustos forrajeros en América Central. Serie Técnica, Informe Técnico No 236.* Turrialba, Costa Rica: CATIE, 1994.

Benavides, J. E., R. A. Rodriguez, and R. Borel. Producción y calidad del forraje de king grass (*Pennisetum purpureum* x *P. typhoides*) y poró (*Erythrina poeppigiana*) en asociación, in: *Arboles y Arbustos Forrajeros en América Central. Serie Técnica, Informe Técnico No 236,* J. E. Benavides ed. (Turrialba, Costa Rica: CATIE, 1994), 441–452.

Bene, J. G., H. W. Beall, and A. Côte. *Trees, Food and People: Land Management in the Tropics.* Ottawa: IDRC, 1977.

Boddey, R. M., J. C. d. Sá, B. J. R. Alves, and S. Urquiaga, The contribution of biological nitrogen fixation for sustainable agricultural systems in the tropics. *Soil Biol. and Biochemistry* 29(1997): 787–799.

Böhringer, A. and D. E. Leihner. A comparison of alley cropping and block planting systems in subhumid Bénin. *Agroforestry Syst.* 35(1997): 117–130.

Boonkird, S. A., E. C. M. Fernandes, and P. K. R. Nair. Forest villages: An agroforestry approach to rehabilitating forest land degraded by shifting cultivation in Thailand. *Agroforestry Syst.* 2(1984): 87–102.

Bryan, J. A. *Leguminous Trees with Edible Beans, with Indications of a Rhizobial Symbiosis in Non-nodulating Legumes.* New Haven, CT: Yale University, 1995.

Bryan, J. A., G. P. Berlyn and J. C. Gordon. Toward a new concept of the evolution of nitrogen fixation in the leguminosae. *Plant and Soil* 186(1996): 151–159.

Budowski, G. The development of agroforestry in Central America, in: *Agroforestry: A decade of progress,* H. A. Steppler and P. K. R. Nair eds. (Nairobi, Kenya: ICRAF, 1987), 69–88. Kenya.

Budowski, G., D. C. L. Kass, and R. Russo. Leguminous trees for shade. *Pesquisa Agropecuária Brasileira* 19(1984): 205–222.

Buresh, R. J. and G. Tian. Soil improvement by trees in sub-Saharan Africa. *Agroforestry Syst.* 38(1998): 51–76.

Christanty, L. Homegardens in tropical asia, with special reference to Indonesia, in: *Tropical homegardens: Selected papers from an International Workshop, Institute of Ecology, Padjadjaran University, Bandung, Indonesia,* K. Landauer and M. Brazil eds. (Tokyo: United Nations University Press, 1990), 9–20.

Cleveland, D. A. and D. Soleri. Household gardens as a development strategy. *Human Organization* 46(1987): 259–270.

Cohen, M. N. *The Food Crisis in Prehistory: Overpopulation and the Origins of Agriculture.* New Haven: Yale University Press, 1977.

Cook, O. F. *Shade in Coffee Culture.* Washington, DC: U.S. Department of Agriculture, Division of Botany, 1901.

Covich, A. P. and N. H. Nickerson. Studies of cultivated plants in choco dwelling clearings, Darien, Panama. *Econ. Bot.* 20(1966): 285–301.

East, R. M. and P. Felker. Forage production and quality of four perennial grasses grown under and outside canopies of mature *Prosopis glandulosa* Torr. var. *glandulosa* (mesquite). *Agroforestry Syst.* 22(1993): 91–110.

Evans, D. O., ed. *International Workshop on the Genus Calliandra.* Morrilton, AR: Winrock International and the Taiwan Forestry Research Institute, 1996.

Everett, Y. Forest gardens of highland Sri Lanka—An indigenous system for reclaiming deforested land, in: *The Cultural Dimension of Development,* D. M. Warren, L. J. Slikkerveer, and D. Brokensha eds. (London: Intermediate Technology Publications Ltd. 1995), 174–184.

Fassbender, H. W. Long-term studies of soil fertility in cacao-shade trees-agroforestry systems: Results of 15 years of organic matter and nutrients research in Costa Rica, in: *Soils of Tropical Forest Ecosystems: Characteristics, Ecology, and Management,* A. Schulte and D. Ruhiyat eds. (Berlin: Springer-Verlag, 1998), 150–158.

Fernandes, E. C. M., C. B. Davey, and L. A. Nelson. Alley cropping on an acid soil in the upper Amazon: Mulch, fertilizer, and hedgerow pruning effects, in: *Technologies of Sustainable Agriculture in the Tropics, Special Publication* 56. (Madison, WI: American Society of Agronomy, 1993), 77–96.

Fernandes, E. C. M. and J. C. de Souza Matos. Agroforestry strategies for alleviating soil chemical constraints to food and fiber production in the Brazilian Amazon, *Chemistry of the Amazon: Biodiversity, Natural Products, and Environmental Issues,* P. R. Seidl, O. R. Gotlieb, and M. A. Coelho Kaplan eds. (Washington, DC: American Chemical Society, 1995), 34–50. Washington, D.C.

Fred, E. B., I. L. Baldwin, and E. McCoy. *Root Nodule Bacteria and Leguminous Plants.* Madison, WI: University of Wisconsin Press, 1932.

Galloway, G., and J. Beer. *Oportunidades para Fomentar la Silvicultura en Cafetales en América Central.* Turrialba, Costa Rica: CATIE, 1997.

Garrity, D. P., A. Mercado, and T. Solera. Species interference and soil changes in contour hedgerows planted on inclines in acidic soils in Southeast Asia, in: *Alley Farming Research and Development: Proceedings of the International Conference on Alley Farming, 14–18 September 1992,* B. T. Kang, A. O. Osiname, and A. Larbi eds. (Ibadan, Nigeria: IITA, 1995), 351–365.

Garrity, D. P. and A. R. Mercado. Nitrogen fixation capacity in the component species of contour hedgerows: How important? *Agroforestry Syst.* 27(1994): 241–258.

Giller, K. E. and K. J. Wilson. *Nitrogen Fixation in Tropical Cropping Systems.* Wallingford, U.K.: CAB International, 1991.

Gordon, B. L. *A Panama Forest and Shore: Natural History and Amerindian Culture in Bocas del Toro.* Pacific Grove, CA: Boxwood Press, 1982.

Greenberg, R. Managed forest patches and the diversity of birds in southern Mexico, in: *Forest Patches in Tropical Landscapes,* J. Schelhas and R. Greenberg eds. (Washington, DC: Island Press, 1996), 59–90.

Greenberg, R., P. Bichier, C. Angon Andrea, and R. Reitsma. Bird populations in shade and sun coffee plantations in Central Guatemala. *Conserv. Biol.* 11(1997): 448–459.

Gutteridge, R. C. Other species of multipurpose forage tree legumes, in: *Forage Tree Legumes in Tropical Agriculture,* R. C. Gutteridge and H. M. Shelton eds. (Wallingford, England: CAB International, 1994a), 97–100.

Gutteridge, R. C. The perennial *Sesbania* species, in: *Forage Tree Legumes in Tropical Agriculture,* R. C. Gutteridge and H. M. Shelton eds. (Wallingford, England: CAB International, 1994b), 49–64.

Gutteridge, R. C. and H. M. Shelton. The role of forage tree legumes in cropping and grazing systems, in: *Forage Tree Legumes in Tropical Agriculture,* R. C. Gutteridge and H. M. Shelton eds. (Wallingford, England: CAB International, 1994), 3–11.

Haynes, R. J. and Williams. Nutrient cycling and soil fertility in the grazed pasture ecosystem. *Adv. in Agron.* 49(1993): 119–199.

Heckadon-Moreno, S. La colonización campesina de bosques tropicales en Panamá, in: *The Botany and Natural History of Panama,* W. G. D'Arcy and M. Correa A. eds. (St. Louis: Missouri Botanical Garden 1985), 397–409.

Hocking, D. and K. Islam. Trees on farms in Bangladesh: 5. Growth of top- and root-pruned trees in wetland rice fields and yields of understory crops. Agroforestry Syst. 39(1998): 101–115.

Hughes, C. Leucaena: *A Genetic Resources Handbook.* Oxford: Oxford Forestry Institute, Department of Plant Sciences, University of Oxford, 1998.

Jama Adan, B. *Soil Fertility and Productivity Aspects of Alley Cropping* Leucaena leucocephala *and* Cassia siamea *Under Semiarid Conditions at Machakos, Kenya,* Gainesville, FL: University of Florida, (1993), 269.

Jensen, M. *Soil Conditions, Vegetation Structure and Biomass of a Javanese Homegarden. Agroforestry Syst.* 24(1993a): 171–186.

Jensen, M. Productivity and nutrient cycling of a Javanese homegarden. *Agroforestry Syst.* 24(1993b): 1897–201.

Jones, R. J. Management of anti-nutritive factors, with special reference to *Leucaena,* in: *Forage Tree Legumes in Tropical Agriculture,* R. C. Gutteridge and H. M. Shelton eds. (Wallingford, England: CAB International, 1994a), 216–231.

Jones, R. M. The role of *Leucaena* in improving the productivity of grazing cattle, in: *Forage Tree Legumes in Tropical Agriculture,* eds. R. C. Gutteridge and H. M. Shelton. (Wallingford, England: CAB International, 1994b), 232–244.

Jose, D. and N. Shanmugaratnam. Traditional homegardens of Kerala: A sustainable human ecosystem. *Agroforestry Syst.* 24(1993): 203–213.

Kang, B. T. Alley cropping: Past achievements and future directions. *Agroforestry Syst.* 23 (1993): 141–155.

Kang, B. T. and K. Mulongoy. Nitrogen contribution of woody legumes in alley cropping systems, in: *Biological Nitrogen Fixation and Sustainability of Tropical Agriculture,* K.

Mulongoy, M. Gueye and D. S. C. Spencer eds. (London: John Wiley and Sons, 1992), 367–375.

Kang, B. T., L. Reynolds, and A. N. Atta-Krah. Alley farming. *Adv. in Agron.* 43(1990): 315–359.

Kang, B. T. and G. F. Wilson. The development of alley cropping as a promising agroforestry technology, in: *Agroforestry: A Decade of Development.* H. A. Steppler and P. K. R. Nair eds. (Nairobi, Kenya: ICRAF, 1987), 227–243.

Kang, B. T., G. F. Wilson, and L. Sipkins. Alley cropping maize (*Zea mays*) and leucaena (*Leucaena leucocephala*) in southern Nigeria. *Plant and Soil* 63(1981): 165–179.

Kass, D. C. L., R. Sylvester-Bradley, and P. Nygren. The role of nitrogen fixation and nutrient supply in some agroforestry systems of the Americas. *Soil Biol. and Biochem.* 29(1997): 5–6.

Kass, D. L. *Erythrina* Species—pantropical multipurpose tree legumes, in: *Forage Tree Legumes in Tropical Agriculture,* R. C. Gutteridge, and H. M. Shelton eds.(Wallingford, England: CAB International, 1994), 84–96.

Kimber, C. T. Dooryard gardens of Martinique. *Yearbook, Assoc. of Pacific Coast Geographers* 28(1966): 97–118.

King, K. F. S. The history of agroforestry, in: *Agroforestry: A Decade of Development,* H. A. Steppler, and P. K. R. Nair eds. (Nairobi, Kenya: ICRAF, 1987), 3–11.

Lawrence, A. Farmer knowledge and use of *Inga* species. in: *Nitrogen fixing trees for acid soils: Proceedings of a Workshop, July 3–8, 1994,* D. O. Evans and L. T. Szott eds. (Turrialba, Costa Rica: CATIE, 1995), 123–134.

Le Houérou, H. N., ed. *Browse in Africa.* Addis Ababa: ILCA, 1980.

Le Houérou, H. N. Indigenous trees and shrubs in the silvopastoral systems of Africa, in: *Agroforestry: A Decade of Development,* H. A. Steppler and P. K. R. Nair eds. (Nairobi, Kenya: ICRAF, 1987), 139–156.

León, J. Central American and West Indian species of *Inga* (Leguminosae). *Annals of the Missouri Botanical Garden* 53(1966): 265–359.

Lesseps, R. and Chipanda, T. Pods of leguminous trees for cattle feed in Zambia, in: *Nitrogen Fixing Trees for Fodder Production: Proceedings of an International Workshop,* J. N. Daniel and J. M. Roshetko eds. (Morrilton, AR: Winrock International, 1998), 181–186.

Liebman, M. Polyculture cropping systems, in: *Agroecology: The Scientific Basis of Alternative Agriculture,* M. A. Altieri ed. (Boulder, CO: Westview Press, 1987), 115–125.

Lindblad, P. and R. Russo. $C_2H_2$ reduction by *Erythrina poepppigiana* in a Costa Rican coffee plantation. *Agroforestry Syst.* 4(1986): 33–37.

Marrero, J. Especies del género *Inga* usadas como sombra de café en Puerto Rico. *Caribbean Forester* 15(1954): 54–71.

Mekonnen, K., R. J. Buresh, and B. Jama. Root and inorganic nitrogen distributions in sesbania fallow, natural fallow, and maize fields. *Plant and Soil* 188(1997): 319–327.

Michon, G. Village forest gardens in west Java, in: *Plant Research and Agroforestry: Proceedings of a Consultative Meeting,* P. A. Huxley ed. (Nairobi, Kenya: ICRAF, 1983), 13–24.

Michon, G. and H. de Foresta. Agro-forests: Incorporating a forest vision in agroforestry, in: *Agroforestry in Sustainable Agricultural Systems,* L. E. Buck, J. P. Lassoie, and E. C. M. Fernandes eds. (Boca Raton: CRC Press, 1999), 381–406.

Musálem, M. A. La Bracatinga: Introducción, crecimiento, manejo y utilización en asociación con cafetal; Una combinación promisoria para los trópicos de México, in: *Nitrogen Fixing Trees for Acid Soils: Proceedings of a Workshop. July 3–8, 1994,* D. O. Evans and L. T. Szott eds. (Turrialba, Costa Rica: CATIE, 1995), 113–129.

Nair, P. K. R., R. J. Buresh, and D. N. Mugendi. Nutrient cycling in tropical agroforestry systems: Myths and science, in: *Agroforestry in Sustainable Agricultural Systems,* L. E. Buck, J. P. Lassoie, and E. C. M. Fernandes eds. (Boca Raton: CRC Press, 1999), 1–31.

Nestel, D. and M. A. Altieri. The weed community of Mexican coffee agroecosystems: Effect of management upon plant biomass and species composition. *Acta Ecologica* 13(1992): 715–726.

Niang, A., E. Styger, A. Gahamanyi, D. Hoekstra, and R. Coe. Fodder-quality improvement through contour planting of legume-shrub/grass mixtures in croplands of Rwanda highlands. *Agroforestry Syst.* 39(1998): 263–274.

Norton, B. W. Antinutritive and toxic factors in forage tree legumes, in: *Forage Tree Legumes in Tropical Agriculture,* R. C. Gutteridge, and H. M. Shelton eds. (Wallingford, England: CAB International, 1994a), 202–215.

Norton, B. W. The nutritive value of tree legumes, in: *Forage Tree Legumes in Tropical Agriculture,* R. C. Gutteridge and H. M. Shelton eds. (Wallingford, England: CAB International, 1994b), 177–191.

Norton, B. W. Tree legumes as dietary supplements for ruminants, in: *Forage Tree Legumes in Tropical Agriculture,* R. C. Gutteridge and H. M. Shelton eds. (Wallingford, England: CAB International, 1994c), 192–201.

Nutman, P. S. Symbiotic nitrogen fixation, in: W. V. Bartholomew and F. E. Clark eds. *Soil Nitrogen,* (Madison, WI: American Society of Agronomy, 1965), 10.

Nye, P. H. and D. J. Greenland. *The Soil Under Shifting Cultivation.* Bucks, England: Commonwealth Agricultural Bureaux, Farnham Royal, 1960.

Okafor, J. C. and E. C. M. Fernandes. Compound farms of southeastern Nigeria: A predominant agroforestry homegarden system with crops and small livestock. *Agroforestry Syst.* 5(1987): 153–168.

Padoch, C. and W. de Jong. The house gardens of Santa Rosa: Diversity and variability in an Amazonian agricultural system. *Econ. Botany* 45(1991): 166–175.

Palm, C. A. Contribution of agroforestry trees to nutrient requirements of intercropped plants. *Agroforestry Syst.* 30(1995): 104–124.

Palm, C. A., A. A. Salazar, L. T. Szott, and E. C. M. Fernandes. Long-term performance of alley cropping on an acid soil of the Amazon basin, in: *Proceedings: Second International Alley Farming Workshop,* B. T. Kang ed. (Ottawa: IDRC, 1995), 12–23.

Palmer, B., D. J. Macqueen, and R. C. Gutteridge. *Calliandra calothyrsus*—a multipurpose tree legume for humid locations, in: *Forage Tree Legumes in Tropical Agriculture,* R. C. Gutteridge and H. M. Shelton eds. (Wallingford, England: CAB International, 1994), 65–74.

Paterson, R. T., J. Daniel, G. R. Korwar, P. S. Pathak, M. D. S. Liyanage, M. G. Ejercito, R. Lesseps, S. Novaho, T. Kato, A. S. Gill, J. M. Roshetko, and A. N. F. Perera. Fodder production systems, in: *Nitrogen Fixing Trees for Fodder Production: A Field Manual,* J. M. Roshetko and R. C. Gutteridge eds. (Morrilton, AR: Winrock International, 1996), 25–43.

Pennington, T. D. *The Genus* Inga: *Botany.* Kew: Royal Botanical Garden, 1997.

Perfecto, I., R. A. Rice, R. Greenberg and M. E. Van der Voort. Shade coffee: A disappearing refuge for biodiversity. *BioScience* 6(1996): 598–608.

Rao, M. R., P. K. R. Nair, and C. K. Ong. Biophysical interactions in tropical agroforestry systems. *Agroforestry Syst.* 38(1998): 3–50.

Reynolds, L. and M. A. Jabbar. Livestock in alley farming systems: ILCA's experience, in: *Alley Farming Research and Development: Proceedings of the International Conference on Alley Farming, 14–18 September,* B. T. Kang, A. O. Osiname and A. Larbi. IITA eds. (Ibadan, Nigeria: 1995), 52–69.

Roskoski, J. P. Nodulation and $N_2$-fixation by *Inga jinicuil,* a woody legume in coffee plantations. I. Measurements of nodule biomass and field $C_2H_2$ reduction rates. *Plant and Soil* 59(1981): 201–206.

Roskoski, J. P. Nitrogen fixation in a Mexican coffee plantation. *Plant and Soil* 67(1982): 283–291.

Russo, R. and G. Budowski. Effect of pollarding frequency on biomass of *Erythrina poeppigiana* as a coffee shade tree. *Agroforestry Syst.* 4(1986): 145–162.

Serrão, E. A. and J. M. Toledo. The search for sustainability in Amazonian pastures, in: *Alternatives to Deforestation: Steps Toward Sustainable Use of the Amazon Rain Forest,* A. B. Anderson ed. (New York: Columbia University Press, 1990), 195–214.

Shelton, H. M. and J. L. Brewbaker. *Leucaena leucocephala*—the most widely used forage tree legume, in: *Forage Tree Legumes in Tropical Agriculture,* R. C. Gutteridge and H. M. Shelton eds. (Wallingford, England: CAB International, 1994), 15–29.

Simons, A. J. and J. L. Stewart. *Gliricidia sepium*—a multipurpose forage tree legume, in: *Forage Tree Legumes in Tropical Agriculture,* R. C. Gutteridge and H. M. Shelton eds. (Wallingford, England: CAB International, 1994), 30–48.

Singh, G. B. Agroforestry in the Indian subcontinent: Past, present, and future, in: *Agroforestry: A Decade of Development,* H. A. Steppler and P. K. R. Nair eds. (Nairobi, Kenya: ICRAF, 1987), 117–138.

Soemarwoto, O. Homegardens: a traditional agroforestry system with a promising future, in: *Agroforestry: A Decade of Development,* H. A. Steppler and P. K. R. Nair eds. (Nairobi, Kenya: ICRAF, 1987), 157–170.

Soemarwoto, O., I. Soemarwoto, E. Karyono, E. M. Soekartadiredja, and A. Ramlan. The Javanese homegarden as an integrated ecosystem. *Food Nutr. Bull.* 7(1985): 44–47.

Staver, C. Shortened bush fallow rotations with relay-cropped *Inga edulis* and *Desmodium ovalifolium* in west-central Amazonian Peru. *Agroforestry Syst.* 8(1989): 173–196.

Stewart, J. L. and A. J. Dunsdon. Preliminary evaluation of potential fodder quality in a range of *Leucaena* species. Agroforestry Syst. 40(1998): 177–198.

Szott, L. T., C. A. Palm, and P. A. Sanchez. Agroforestry in acid soils of the humid tropics. *Adv. in Agron.* 45(1991): 275–301.

Thomas, R. J. Role of legumes in providing N for sustainable tropical pasture systems. *Plant and Soil* 174(1995): 103–118.

Topark-Ngarm, A. and R. C. Gutteridge. Fodder productivity of perennial *Sesbania* species, in: *Perennial* Sesbania *Species in Agroforestry Systems,* B. Macklin and D. O. Evans eds. (Waimanalo, Hawaii: Nitrogen Fixing Tree Association: 1990), 79–88.

Torquebiau, E. Are tropical agroforestry homegardens sustainable? *Agric. Ecosystems and Environ.* 41(1992): 189–207.

Tuah, A. K. The role of multipurpose trees and shrubs in ruminant nutrition, in: *Alley Farming Research and Development: Proceedings of the International Conference on Alley Farming, 14–18 September 1992,* B. T. Kang, A. O. Osiname, and A. Larbi eds. (Ibadan, Nigeria: IITA, 1995), 179–186.

Van der Meersch, M. K., R. Merckx, and K. Mulongoy, K. Evolution of plant biomass and nutrient content in relation to soil fertility changes in two alley cropping systems, in: K. Mulongoy and R. Merckx eds. *Soil Organic Matter Dynamics and Sustainability of Tropical Agriculture.* (Chichester: John Wiley and Sons, 1993), 143–154.

Van Noordwijk, M., S. M. Sitompul, K. Hairiah, E. Listyarini, and M. Syekhfani. Nitrogen supply from rotational and spatially zoned inclusion of leguminosae for sustainable maize production on an acid soil in Indonesia, in: *Plant-Soil Interactions at Low pH: Principles and Management.* R. A. Date, N. J. Grundon, G. E. Rayment, M. W. Probert eds. (Dordrecht, The Netherlands: Kluwer Academic Press, 1995), 779–784.

VanDenBeldt, R. J. Agroforestry in the semiarid tropics, in: *Agroforestry: Classification and Management,* K. G. MacDicken and N. T. Vergara eds. (New York: John Wiley and Sons, 1990), 45–57.

Vidhana Arachchi, L. P. and M. d. S. Liyanage. Soil physical conditions and root growth in coconut plantations interplanted with nitrogen-fixing trees in Sri Lanka. *Agroforestry Syst.* 39(1998): 305–318.

von Maydell, H.-J. Agroforestry in the dry zones of Africa: Past, present, and future. in: *Agroforestry: A Decade of Development,* H. A. Steppler and P. K. R. Nair eds. (Nairobi, Kenya: ICRAF, 1987), 89–116.

Wilken, G. C. Integrating forest and small-scale farm systems in Middle America. *Agroecosystems* 3(1997): 291–302.

Zelada, G. L. The influence of the productivity of *Prosopis tamarugo* on livestock production in the Pampa del Tamarugal—a review. *For. Ecol. and Manage.* 16(1986): 15–31.

Zinke, P. J., S. Sabharsri, and P. Kunstadter. Soil fertility aspects of the Lua' forest fallow system of shifting cultivation, in: *Farmers in the Forest,* P. Kunstadter ed. (Honolulu: University Press of Hawaii, 1978), 21–34.

# 9 Species Interactions, Stand Structure, and Productivity in Agroforestry Systems

*Matthew J. Kelty*[1]

## TABLE OF CONTENTS



## 9.1 INTRODUCTION

The social and economic objectives of agroforestry include increasing agricultural revenues or subsistence yields, diversifying outputs, stabilizing yields over time, reducing the land area required for subsistence, and reducing economic and environmental costs of production. Achieving these social and economic objectives depends upon creating agroforestry systems with two ecosystem characteristics: 1) high levels of biological productivity and 2) ecological sustainability with limited artificial

[1] Department of Forestry and Wildlife Management, University of Massachusetts, Amherst, MA 01003; Tel: (413 545-1799; Fax (413) 545-4358; kelty@forwild.umass.edu.

0-8493-2206-5/00/$0.00+$.50

inputs of resources (Anderson and Sinclair 1993, Nair 1993, Sanchez 1995). The attainment of these ecological goals is sought by combining trees with shrubs, vines, or herbaceous plants to take advantage of ecological interactions among component species.

The goal of high productivity is assessed in experimental plantations by the measurement of biomass yield of harvestable products and ideally of total plant biomass as well. The goal of sustainability is to maintain the physical resources of a site over multiple harvesting cycles, and can be assessed by analysis of trends in long-term productivity on the site. When considering alternative systems, these ecological goals are expressed as two questions: Can productivity levels be increased over current agricultural and forestry practices, and can high levels of production be maintained for long periods on the same site? Many agroforestry research projects address only the first of these questions directly, because of the time commitment necessary to examine the second. Direct measurement of changes in nutrients, water, and soil structure are often used to make predictions of long-term sustainability from relatively short-term projects.

Agroforestry plantations may produce timber, fuelwood, and fodder from the tree components in addition to food crops, but the main interest frequently has been with the food crops, particularly in modern research and development efforts, because they have much higher value than many nonfood tree crops. Trees are often incorporated, primarily to improve upon existing agricultural practices. In this sense, agroforestry is a logical extension of agricultural intercropping (the combination of two or more nontree crop species). Thus, it is not surprising that the concepts of species interactions and the experimental approaches that are used in agroforestry are derived from agricultural intercropping experience. Although much has been gained by this transfer of ideas, limitations do exist. The addition of long-lived, large-stature woody plants adds greater structural complexity to stands, and calls for analysis of experimental plantings over an extended time period, closer to that used for silviculture than for agriculture. In addition, silvicultural planning is concerned with the development of complex, multistoried stand structure of the kind that occurs in many agroforestry plantations, so stand structural models created for forestry should be helpful in interpreting the relative performance of various plantation designs (see Ashton and Ducey, this volume).

This paper will: 1) review the concepts and evaluation methods of agricultural intercropping research as applied to agroforestry; 2) review a model of forest development that focuses on stand structure; 3) consider the effect of species interactions on production in some common agroforestry plantation designs, within the framework of the forest development model.

## 9.2 SPECIES INTERACTIONS

Vandermeer (1989) categorized all species interactions involved in agricultural intercropping as either "competitive" or "facilitative." This approach has been applied to species mixtures in agroforestry (Anderson and Sinclair 1993) and in silviculture (Kelty 1992), and the principles will be reviewed here briefly.

Competitive interactions among plants in monocultures are intense (given a sufficiently high plant density) because all plants are genetically similar, having similar root and crown structure, phenology, and resource requirements. Above a threshold, increasing plant density generally does not increase total biomass production per unit land area; stand-level production remains constant, but individual plant biomass decreases. To increase stand-level production above this level (without directly altering site conditions), it is necessary to add plants of different species.

The key to designing mixtures is to find combinations of species with niche differences that reduce the intensity of competitive interactions. If two species are similar in structure, phenology, and resource use (i.e., occupy similar niches), competition in a mixed stand will be similar to that in monoculture, and mixtures will likely have production rates equal to the mean of the monocultures (weighted by the relative density of each species in the mixture). However, in such mixtures, one species may be so much more successful in the niche that it eliminates the other ("the competitive exclusion principle"). Exclusion generally does not occur in the limited time period between harvests in managed systems, but one species may greatly dominate such a mixture.

Two terms are used to describe the relationship of plant species with substantial niche differences. Species with good "ecological combining ability" (Harper 1977) are those that differ sufficiently to coexist in a mixed stand. A subset of those species combinations may exhibit "complementarity" if the mixture takes up resources more completely and/or uses those resources more efficiently such that the productivity of the mixture exceeds that of the weighted mean of the monocultures of the component species (Ong, et al. 1996). The objectives of some agroforestry practices may be met with species combinations that exhibit ecological combining ability alone. These mixtures can be used to diversify crop outputs, reduce the impact of failure of any single crop species, and achieve facilitative interaction effects (described below), without increasing total system production. They might also reduce susceptibility to pathogens and pests that are dependent upon high densities of the host plant species, thus increasing yield by reducing losses rather than by increasing primary productivity. However, there is also considerable interest in identifying species combinations that have enhanced productivity through complementarity of resource use. This is not only to increase harvest yields, but to seek to maintain tight control over resources in order to limit nutrient loss.

The search for complementary sets of species is part of a fundamental question about the relationship of species diversity to ecosystem function: Does total resource capture and productivity increase incrementally as plant species are added to an ecosystem? If the answer is "yes," it indicates that complemenarity is common—each added species is able to use some fraction of resources more efficiently than the existing set of species. However, there is rather little evidence to answer this fundamental question. In some cases, mixtures have been found to have better nutrient retention than monocultures (Ewel, et al. 1991) and to maintain more stable production rates over multiyear periods with fluctuating resource levels (Tilman and Downing 1994), but the demonstration of complementarity in production has remained rather elusive. Some species are highly effective in monoculture in capturing resources and producing high biomass yields, approaching the productivity rates of highly diverse stands (Berish and Ewel 1988). In a review of intercropping studies of two-species mixtures, Trenbath (1974) found that few experiments showed evidence of complementarity; the yield of most mixtures was close to the mean productivity of the monocultures of component species. Studies of model ecosystems with more than two species (Naeem, et al. 1994, Hooper 1998) have indicated that production does not consistently increase with increasing number of species; ecosystem function appears to be dependent upon the functional attributes of the individual species, rather than on species number. These studies have also indicated that the effects of adding species on ecosystem function diminishes very rapidly, with the greatest effects occurring with the first few species added to a monoculture — if those species are from different functional groups.

One approach to identifying compatible species for agroforestry is to combine species of different life forms (canopy tree, subcanopy tree, shrub, vine, perennial herb, annual herb), because life form is an indicator of the different fundamental roles that plant species play in the ecosystem (Ewel and Bigelow 1996). However, other important functional attributes, such as nitrogen fixation, phenology, and canopy density, may vary as much within as among life forms. The key to reducing competitive interactions in agroforestry systems appears is to combine a small number of carefully chosen species that differ in life form or in other important functional characteristics.

Facilitative interactions are those in which one species directly benefits another growing in mixture, rather than simply reducing the intensity of competition. Four possible mechanisms of facilitation are: 1) increased efficiency of nutrient cycling, in which one species takes up nutrients unavailable to others and then releases those nutrients through litterfall, making them available to plants of all species; this is especially important when one species symbiotically fixes atmospheric nitrogen, thus producing litter rich in nitrogen; 2) improved soil structure leading to better water and nutrient retention; 3) improved moisture status of understory plants resulting from shading of overstory trees, or possibly from trees directly increasing moisture in upper soil horizons by "hydraulic lift" (Horton and Hart 1998; described further below); 4) reduction of loss to insect pests (Dix, et al. 1995), pathogens, or weeds. The use of nitrogen-fixing species has received the most attention of these in agroforestry research. The objective of designing facilitative interactions in plantations in practical economic terms is to achieve high levels of productivity without incurring the costs of fertilizers, pesticides, or irrigation.

The principles relating to competition and facilitation have been used to select species combinations in agricultural intercropping, mixed-species silvicultural plantations, and agroforestry plantations. Combinations of species may be chosen entirely to achieve high productivity through complementarity; others are chosen to achieve specific facilitative interactions. The latter is often the case in agroforestry, especially where the main goal is to add trees to an existing food crop system to achieve greater long-term sustainability. However, facilitation cannot be achieved without the potential for some degree of competition between tree and crop species. Thus, a balancing act is often required in designing species combinations, in which trees must be selected to carry out the facilitative function without negating the benefit to the main crop through interspecific competition.

This balancing act is demonstrated well by the cultivation of cacao (*Theobroma cacao*) and coffee (*Coffea* sp.) crops beneath shade trees (Willey 1975, Beer 1987). In these plantations, the tree roots take up nutrients from deeper soil layers beyond the reach of the roots of understory crops; litterfall from the trees returns organic matter and nutrients to the soil surface, improving conditions for the crops. However, competition in the form of light interception and nutrient uptake in the upper soil level by the overstory trees reduces crop production. If soil fertility levels are high, crop production may be lower in plantations with shade trees than in those without, because the facilitative effect of the trees on nutrient levels is unimportant. It is hypothesized that the balance may be tipped in the opposite direction if fertility levels are low—with tree-crop combinations having higher production than crop monocultures, but definitive empirical support for this result is lacking.

## 9.3 EVALUATION OF SPECIES INTERACTION EFFECTS ON PRODUCTIVITY

In both silviculture and agriculture, the critical measure used to evaluate practices involves the economically useful portion of biomass production. The same is true of agroforestry, even though the component species may have quite different patterns of allocation to the harvestable fraction (e.g., fuelwood makes up a larger fraction of tree biomass than do seeds or fruits of most food crop plants). However, assessment of total biomass production is useful in any system to shed light on the nature of the interactions among species. It is particularly important in agroforestry because the total mass of plants is the key to the sustainability function—retaining nutrients, shading soil, reducing soil movement, and resisting weed invasion.

A common approach to evaluating the effects of interactions in agricultural intercropping practices is to compare the yield of mixed plantings (polycultures) with that of the monocultures of each of the component species. These are used to calculate the "land equivalent ratio" (LER) (Vandermeer 1989, Nair 1993, Anderson and Sinclair 1993) as follows:

$$LER = RY_{(a)} + RY_{(b)}$$

where $RY$ (relative yield of species $a$ or $b$) is the ratio of the yield of each species in polyculture to that species in monoculture; yields are expressed per unit land area, and can be of total biomass or harvestable biomass. If total plant density is held constant for monoculture and polyculture plots, the set of plots forms a "replacement series" (Harper 1977), and the sum of the relative yields is called the "relative yield total" (RYT). This design is followed in many basic studies of species interactions, but it is not necessarily logical for applied studies, especially those involving species of greatly different individual plant size. Density of each plot should be optimum (producing highest yield) for each monoculture and each mixture (and of each proportional mix of species); this may be known for some monocultures but rarely for any mixtures. It would be ideal to incorporate a range of densities into the experimental design, but often this is not possible; mixture densities are usually based on knowledge of each species' performance in monoculture. The term *LER* is used in place of *RYT* where constant density across experimental plots is not specified (Nair 1993).

This approach can be extended from agricultural intercropping to agroforestry. A hypothetical example of a shrub or herbaceous crop grown beneath a canopy of a timber tree species can be used to demonstrate *LER* analysis. Consider a situation in which the overstory tree species is grown in polyculture at a density that is 0.2 of that of the monoculture timber plantation, and the understory crop is spaced at a density of 0.8 of that of the crop monoculture. If individual overstory trees grew at the same rate in polyculture as in monoculture, the $RY_{(trees)}$ would be 0.2. However, if the reduced competition in the polyculture allowed trees to double their size in monoculture; $RY_{(trees)}$ would be 0.4. If the balance of facilitation and competition resulted in a limited increase in production per crop plant, $RY_{(crop)}$ may reach 0.9. Thus, in this example, each of the component species has benefitted in production rates from the interactions; this inference can be drawn from the result that *RY* of each species is greater than its relative density in the polyculture. However, *RY* of each species is less than 1.0, indicating that neither species could overcome its lower density in the polyculture to reach the absolute yield levels found in monoculture. The *LER* in this hypothetical example is 1.3; indicating that it would require 1.3 ha to produce as much in monoculture as can be produced in 1.0 ha of polyculture.

In order for yield comparisons to be meaningful, it is necessary to measure monoculture and polycultures under similar site conditions and with similar management regimes. For example, if the polyculture is unfertilized, then monocultures must be unfertilized as well. It is often of interest to include other management regimes in an experiment—the main interest may be to compare an unfertilized polyculture with a fertilized monoculture. This may be the critical test for deciding upon adopting a polyculture design, and it may be tempting (because of time and land requirements for each treatment) to use only this comparison. However, such an experimental design will not allow the assessment of responses that result from the species interactions alone.

Even when plots are similar, one limitation of the *LER* approach is that these calculations cannot separate the effects of the different interactions. In the example described above, the crop yield in polyculture is decreased from competition effects and increased from facilitation effects, with both incorporated into $RY_{(crop)}$. Experimental designs that differentiate between these mechanisms are extremely helpful in understanding interactions and designing more productive combinations, but they require additional plots. One plot needed to separate these effects would be a polyculture in which overstory leaf litter is trapped and removed—to test competition effects without the facilitative effects on nutrient cycling. Clearly, such plots would be difficult to maintain. Experiments of this kind have been carried out in some studies of alley cropping, and will be discussed further below.

An additional complication exists when applying the *LER* approach to agroforestry. In agricultural intercropping, most interest is in increasing current yields, so

*LER* calculations over a small number of years can provide the important data. However, many agroforestry practices are designed for their effects on sustainability, and the most important questions can be answered only from data collected over a longer time span. Facilitation effects in systems such as the shade tree-crop plantation may not be seen in the early years after establishment, because of the high initial fertility of the soil. Differences among plots may occur only on the time scale in which significant fertility losses occur from leaching. Thus, the response of greatest interest is not the initial *LER* value, but the change in *LER* with time. This is similar to silvicultural approaches to growth and yield analysis, which model productivity changes over the course of stand development.

## 9.4 A DEVELOPMENTAL MODEL OF FOREST STAND STRUCTURE

Creating an agroforestry plantation design consists of defining a particular stand structure, specified by the ages, sizes, and spatial arrangement of plants, in addition to the choice of species. These components of stand structure are also important in silviculture, and a model of forest stand development that focuses on structure has been created as a tool for planning silvicultural treatments (Oliver and Larson 1996). The model defines four developmental stages with distinctive structures (Figure 9.1). The focus of the model is on the species of the main canopy, but there are also corresponding differences in the nature of interactions between canopy trees and smaller plants (shrubs, understory trees, vines, and herbs), so the model can be used for interpreting interactions in agroforestry stands as well. This model applies most directly to forests in climates where precipitation is sufficient to allow closed-canopy stands to form, but it can be applied with modifications to more arid climates.

The model begins with the "stand initiation" stage, which follows a major disturbance (e.g., fire, windstorm, timber harvesting) that removes all or most of the existing canopy over a large area. New trees become established from seeds or surviving root stocks, but because of their small size, they do not dominate resources at this early stage. Shade-intolerant shrubs and herbaceous plants make up much of the vegetation, and may dominate during stand initiation. As trees rise above other species and form a closed canopy, the stand enters the "stem exclusion" stage. Competition among the main canopy trees results in size differentiation and mortality of suppressed trees. Nearly complete capture of light, water, and nutrients by trees eliminates understory plants and prevents the establishment of new seedlings. As the stand ages further, a number of environmental changes may take place: Solar radiation changes in its pattern of distribution, as the patches of direct sun and deep umbral shade beneath a low canopy are replaced by a greater area in diffuse penumbral shade beneath a taller canopy; abrasion between crowns caused by wind sway may create small breaks in the canopy that increase the transmission of direct radiation; as the mass of the forest floor increases, decomposition of organic matter may increase nutrient availability and create an enriched carbon dioxide concentration in the air layer near the ground, thus improving photosynthetic efficiency of seedlings. The importance of these effects likely varies widely among forest types, but little direct

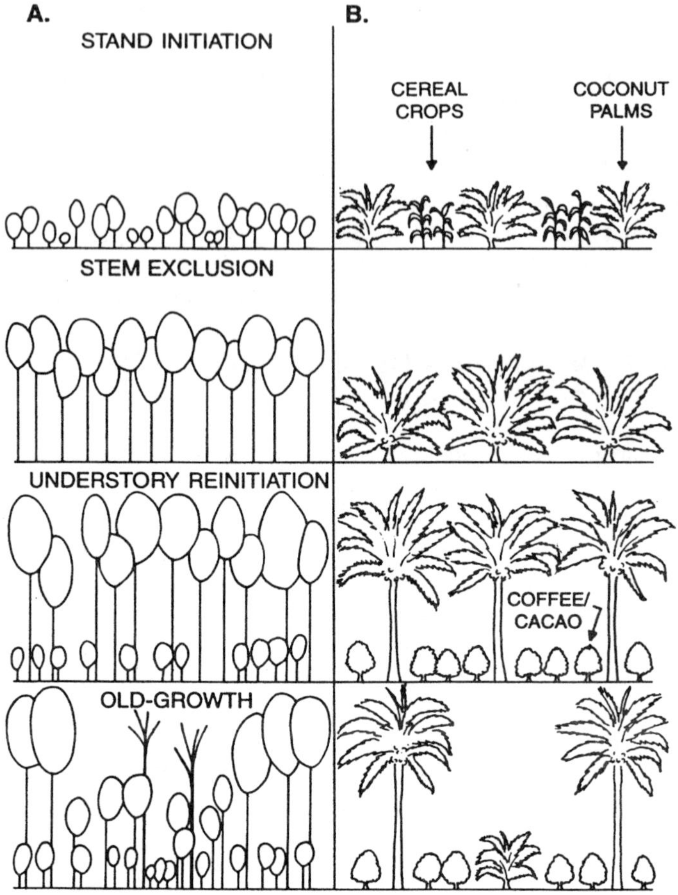


**FIGURE 9.1** Diagram of changes in stand structure with age since initial establishment. (A) Forest developmental stages in unmanaged stands, based upon Oliver and Larson (1996); only main canopy tree species are illustrated. (B) Hypothetical progression of agroforestry practices in a coconut palm plantation through the same developmental stages, based upon Nair (1993). See text for further description.

study has been made of these processes. As these changes occur, stands enter the "understory reinitiation" stage, with seedlings of shade-tolerant trees, shrubs, herbs, and vines becoming established. At greater stand ages, main-canopy trees approach their maximum lifespan and are subject to the effects of wind, disease, and insect attacks. As these trees die, the stand enters the "old-growth" stage, defined by the onset of canopy gap formation. Species of low shade tolerance may become established, if gaps are large enough. Release of established regeneration and establishment of new seedlings occurs in gaps, gradually producing an uneven-aged stand.

The parallel between this model of stand development and agroforestry practices can be seen in the description by Nair (1993) of the interplanting of various crop species in coconut (*Cocos nucifera*) plantations, summarized below and in Figure 9.1. During approximately the first eight years after establishing a coconut palm plantation (stand initiation), a variety of light-demanding cereals and grain legumes can be grown, which escape competition simply because coconut palms are slow to develop. During the next stage (stem exclusion) from age 8 to 25 years, the closed palm canopy precludes growing crops, but after age 25 (understory reinitiation) moderately shade tolerant species (coffee, cacao) can be grown in the understory beneath the palm canopy. Continuing beyond Nair's description, if the mature palms were eventually felled singly or in small groups, gaps would be created in which moderately intolerant species could be cultivated and young palms could be planted. This would create a uneven-aged plantation analogous to the old-growth structure of the forest development model.

A principal concept of the forest development model is that the first three structural stages are temporary, progressing to the old-growth stage in which the process of gap formation and closure continues for an indefinite period. This transitory nature of structure is recognized in the description of coconut plantation management and in other applications of agroforestry, but many agroforestry systems focus on creating a particular structure and then more or less arresting succession in that structure, in order to gain the benefits of the species interactions of that stage. It is useful, therefore, to consider each of the four stages in relation to the major agroforestry practices that incorporate the structure of that stage.

## 9.5 STAND INITIATION STAGE—ALLEY CROPPING AND TAUNGYA

Two well-known agroforestry practices incorporate the basic structure found in the stand initiation stage. "Taungya" is a rather close mimic of the developmental process that occurs in unmanaged forests that regenerate after natural disturbances, whereas "alley cropping" bears a closer resemblance to conventional row-cropping agricultural practices.

Taungya refers to the cultivation of food crops between trees seedlings which have been planted as a timber plantation. This practice had its modern origin in the nineteenth century for the establishment of teak (*Tectona grandis*) plantations in Southeast Asia, but was based upon traditional practices in Europe (Blanford 1958, cited in Nair 1993). It has since been used in plantation establishment for a large number of timber species throughout the tropical world. Crops such as dryland rice (*Oryza sativa*), maize (*Zea mays*), and soybeans (*Glycine max*) are grown among the trees for the first few years, ending when the trees begin to form a closed canopy. Weeding and pruning of the trees can be incorporated into the tending treatments of the food crops, thereby reducing silviculture plantation establishment costs.

One example of the use of taungya in Java was described by Wiersum (1982). In this system, seeds of teak and *Leucaena leucocephala* (as a nurse crop) are sown in alternate rows on lands managed by the forest service. Farm families have contracts, usually for a plot of land of about 0.5 ha, to plant crops between the rows of trees. Their responsibility includes planting tree seedlings to replace germination failures, thinning clumps of teak seedlings, and periodically cutting back *Leucaena* trees. The cropping period usually lasts for two years, but the plantation tending responsibilities continue for an additional three years after cropping ceases. Most of the 40,000 ha of teak being established each year in the 1970s in Java were a product of variations on this taungya system. Another example, on a much smaller scale of hundreds of hectares, is the practice of establishing bigleaf mahogany (*Swietenia macrophylla*) by the taungya system in Belize in the period of 1930–1965 (Mayhew and Newton 1998). Maize, plantains (*Musa* sp.), and cassava (*Manihot esculenta*) were planted, usually in mixture, with mahogany seeds sown about one month after the food crops. In this case, crops were grown for only one year in each plantation. Similar methods have been used throughout the Caribbean.

The success of taungya depends somewhat on the ecological combining ability of the tree and crop species, but the limitations on species choice are not severe. Competition from the trees is restricted by the desired low planting density, and the key is to grow food crops that can capture resources not being used by the trees in the period before their crowns and roots meet to form closed canopies. Weed species of herbs, shrubs, and trees commonly invade and grow vigorously in young plantations, using these resources. Taungya replaces those weed species with agriculturally productive plants. There is no attempt to stop the progress of the natural development from stand initiation to stem exclusion (which leads to the elimination of the agricultural plants). The cropping period is sometimes prolonged by increasing the spacing at which trees are planted, but the purpose is not to make the system permanent.

In contrast, alley cropping was devised nearly exclusively to improve agricultural crop production, and depends on holding a plot in the stand initiation structure permanently (i.e., without any planned shifts in structure as part of the system). This practice consists of establishing trees in hedgerows with crops such as maize, sorghum (*Sorghum* sp.), or rice grown in the space ("alleys") between. The tree crowns are pruned to limit their expansion over the crops, and the pruned branches are spread as green mulch in the alleys to serve as a source of nutrients and organic matter. Legumes are favored as hedgerow trees in order to produce nitrogen-rich mulch. The trees are incorporated in this system only for their facilitation effect, which must be greater than the competitive effect from the trees on the crop, in order for the system to meet its goal (Sanchez 1995).

In most cases of evaluating the success of alley cropping trials, only the food crop is of interest, because no yield is derived from the hedges (harvesting fodder or other products would reduce the facilitation effect by reducing the mulch available to the crop). Thus, *LER* (or *RYT*) is not of interest, but only *RY* of the food crop, which can be calculated by comparing plots with the food crop grown in alleys to plots with the food crop grown in monoculture, as described earlier. It is also possible to conduct further analyses that separate the facilitation effect from the competitive effect by including additional plots. Sanchez (1995) and Ong (1996) have presented a design for conducting such an experiment, as well as an equation to summarize the results. The "tree-crop interaction equation" in its simplest form is as follows:

$$I = F - C$$

where $I$ is the interaction (similar, but not identical, to the $RY$ of the crop species); $F$ (the fertility effect) is the difference in crop yield between monoculture plots with mulch applied (gathered from trees grown off the experimental plots) and monoculture plots without mulch applied, expressed as a proportion of the crop yield of the monoculture without mulch; and $C$ (the competitive effect) is the difference in crop yield between polyculture plots without mulch (hedgerow trees are pruned but not applied as mulch) and monoculture plots without mulch, again expressed as a proportion of the crop yield of the monoculture without mulch. There are other arrangements of plots that can be used to determine $F$ and $C$. $F$ could be determined as the difference between mulched and non-mulched polycultures instead of monocultures, and $C$ could be determined from the difference between polycultures and monocultures with both of them mulched instead of non-mulched. The key is that the additional plots give the fertility (facilitation) effect while holding competition constant, and the competition effect while holding the fertility (facilitation) effect constant. This approach could be applied to other agroforestry systems as well, but it would be very difficult in many of them. It is possible in alley cropping because the principal facilitation effect is controlled by management—it would not occur without pruning and mulching, so it can be manipulated independent of the competition effect of the hedgerows.

The interaction effect $I$ is essentially equivalent to calculating the $RY$ of the crop species, although it differs mathematically; the relationship is $I = RY - 1.0$. Thus, an $RY$ of 1.3 would equal an $I$ of 0.3, and either would be interpreted as showing that the crop species in polyculture had produced 30% more than in monoculture per unit of total plot area.

Sanchez (1995) applied this method of analysis to the results of a number of detailed alley cropping studies from India, Africa, Indonesia, and Central and South America. He noted that alley cropping often does not fare well in meeting the criteria of increasing crop yield, but with differences in climate and soil conditions affecting the results. In semiarid and subhumid climates, the competition effect is great because of competition for water between the hedgerow trees and the crop plants. In fertile soils in these climates, the fertility effect is low because water rather than nutrients are limiting to crop growth, so the fertility effect does not balance the high competition, resulting in a negative interaction effect ($RY < 1.0$ or $I < 0$). In less fertile soils in these dry climates, the fertility effect is greater and more than balances the competition effect ($RY > 1.0$ or $I > 0$), but the positive interaction is small, because water is still the main limiting factor. In the humid tropics, the fertility effect is large because of the greater production of mulch by hedgerow trees. However, the greater tree production means that shading and nutrient uptake by the trees are both greater, so the competition effect is greater than in dry climates. On low fertility soils in these humid climates, root competition appears to be the main problem, resulting generally in negative interaction effects. On high fertility soils, shading is the biggest problem, but this can be controlled more readily by pruning—and these are the sites that are most likely to show positive interaction effects (Sanchez 1995). However, high-rainfall, high-fertility sites are the ones with the least need for improvement of current agriculture with agroforestry.

Three observations can be made about these results from alley cropping trials: 1) The objective of achieving $RY > 1.0$ for the food crop is a much more difficult goal to hope to attain than the goal of achieving a combined $LER > 1.0$ which is used for most agricultural intercropping practices. For example, if maize is grown in alleys that comprise 80% of a plot with the remaining 20% consisting of hedgerow trees, the alley cropping criterion for success requires that the maize yield exceed that of an equal-sized plot of 100% maize. Even if there were no competition between the species, the fertility effect must compensate for the 20% of plot area not planted to maize, just to achieve $RY = 1.0$ for maize. Such a result was found in an alley cropping experiment on a high fertility site in Kenya (Jama, et al. 1995). In plots composed of 80% maize/20% senna (*Senna siamea*) hedgerows, $RY_{(maize)} = 1.0$, $RY_{(senna)} = 0.3$, and $LER = 1.3$. Thus, both species benefited from the interactions (both had $RY$ values greater than their proportional site occupancy at planting), but the system would not be considered successful if increasing maize yield to $RY > 1.0$ is the criterion for success.

2) Maintaining a semipermanent structural arrangement of shade-intolerant crop species between trees is an increasingly difficult task with increasing age from initial establishment. Size of tree crowns can be controlled by pruning, but roots will continue to expand, increasing below-ground competition. In fact, shoot pruning appears to restrict roots more to the topsoil, increasing competition with crop plants (Schroth 1995). It may be possible to select tree species with predominantly deep rooting, but competition will continue to be a serious limitation, simply because of the perennial nature of trees.

3) The evaluation of alley cropping trials soon after initial planting may be inappropriate in most cases. The importance of the fertility effect is likely to increase with time, as a result of nutrient removal in harvests and of nutrient leaching, especially in monocultures. If nutrient levels decline in fields of monoculture crops, the relative yield of the polyculture will increase even if the absolute yield remains constant (Figure 9.2). It is this pattern that should be of greatest interest, but detailed alley cropping trials have not been maintained for periods long enough to test this hypothesized result.

## 9.6 STEM EXCLUSION STAGE—TREE GARDENS

The defining characteristic of the stem exclusion stage is a dense tree canopy which prevents the vigorous development of understory vegetation; thus, it is not useful as a model for the design of agroforestry systems that incorporate herbaceous plants or shrubs as a major component. However, a mixture of timber, fuelwood, fruits, spices,

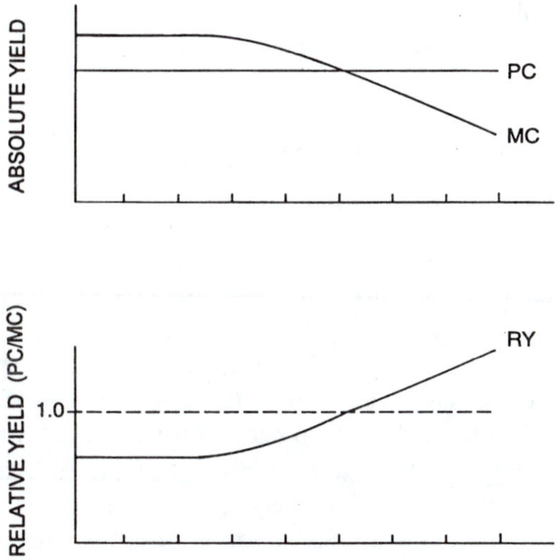


**FIGURE 9.2** A generalized view of results of an agroforestry experiment, showing the pattern of absolute yields and relative yield (*RY*) of the nontree food crop species over time (yield of the tree species is not of interest in this example). The yield of the crop in the polyculture (PC) may be lower than that of the crop monoculture (MC) initially, due to competition effects. If monoculture crop yields decrease with successive croppings, a stable crop yield in the polyculture will result in an increasing relative yield, as illustrated. This emphasizes the need for long-term analyses of experimental systems.

and medicines can be produced in multistoried stands of trees and vines which have the basic structure of the stem exclusion stage. Such stands, called "tree gardens" are a traditional form of management in Java on farms where annual food crops are grown in separate cultivated fields (Wiersum 1982). Tree gardens have also been developed as a newly created practice in Indonesia as an extension of taungya (Kartasubrata and Wiersum 1995). Generally, taungya has led quickly to complete dominance of teak or other timber trees in Indonesian plantations, as annual crop production is abandoned after the first two or three years of plantation growth. More recently, as a result of reform of land policies, taungya farmers have been allowed to plant fruit and other multipurpose trees on a portion of the plantation area and carry them through the full teak timber rotation, thus creating a combined subsistence/commercial tree garden.

In arid and semi-arid climates, forest stands may not form closed canopies, so the stem exclusion stage does not occur. A protracted stand initiation period grades into understory reinitiation as the canopy trees rise above non-tree vegetation. Intolerant understory species may persist or gradually be replaced by more tolerant ones. In humid climates, this process can be mimicked by using thinning operations to maintain a sparse canopy, first to lengthen the stand initiation period and later to hasten the establishment of understory species of moderate shade tolerance. This would make it possible to have continuous production of shrub and herb crops, but only by avoiding the limitations of the stem exclusion stage.

## 9.7 UNDERSTORY REINITIATION STAGE—SHADE TREE-CROP COMBINATIONS

The structure associated with the understory reinitiation stage is used widely for growing crop species of at least moderate shade tolerance such as coffee, cacao, tea (*Camellia sinensis*), cassava, and a variety of spices beneath an overstory of shade trees (Nair 1993, Herzog 1994). A number of benefits to understory crop production are hypothesized as occurring as a result of facilitative interactions with overstory trees. One of these is similar to the role of hedgerow trees in alley cropping—to limit the leaching of nutrients and maintain rapid rates of cycling within the tree-crop system; another is that the shade itself may have importance in some climates by reducing leaf temperature and evapotranspiration of crops, and by reducing photo-oxidation of chlorophyll in crop foliage. Additional possible benefits of shading include suppression of shade-intolerant weeds, and reduction of insect and disease effects. All of the above are facilitative effects, which would come with potentially even greater competitive effects than occur in alley cropping, because shade trees are scattered among the crop plants and are allowed to grow much taller than the crops. However, the overstory often produces an additional benefit in the form of harvestable products, usually timber in large commercial operations, but also fuelwood, fruits, and medicines, particularly in small farmholdings.

Although these systems are widely used, there is little documentation of the interactions involved. Some aspects of the balance between competition and facilitation in shade tree-crop combinations can be seen in the analysis of productivity and nutrient dynamics in two plantations in Costa Rica over a 10-year period (Beer et al. 1990, Fassbender et al. 1991). A plantation of the timber species *Cordia alliodora* and cacao produced litterfall rates similar to those of unmanaged lowland tropical forests, with half of the litter fall supplied by the *Cordia* trees. Soil organic matter increased over the 10-year period, suggesting the long-term sustainability of the mixture in terms of soil fertility, and the importance of the overstory trees in this process. The rate of nutrient cycling and soil organic matter accumulation was higher in adjacent cacao plantations in which the nitrogen-fixing legume *Erythrina poeppigiana* was used as the shade tree species; in this case, the tree crowns were pruned two or three times per year, thereby doubling the natural rate of litter input. Although soil and nutrient conditions were better in the *Erythrina* plantations, the cacao harvests were comparable from the two systems (possibly because greater interception of light by *Erythrina,* in spite of the pruning, offset the better nutrient conditions), and there was no timber yield derived from the *Erythrina* overstory. Other experience in the region (Perfecto et al. 1996) indicates that overstory trees significantly reduce the production of cacao (or coffee) compared to that from monoculture crop plantations with fertilization, but there were no monoculture plots on the Costa Rica study site to allow direct comparisons for investigating the nature of interactions further.

One project designed to study these interactions directly uses "model ecosystems" which mimic the shade tree-crop plantation structure in a set of plots growing on a site with high rainfall and soil fertility in Costa Rica (Haggar and Ewel 1997). Three tree species were each planted in monoculture and in polyculture with an understory of a palm (*Euterpe oleracea*) and a large perennial herb (*Heliconia imbricata*). In all plots, trees were planted at a uniform, high density such that closed overstory canopies formed within the first two years. At three years, the most productive tree species in monoculture (*Hyeronima alchorneoides*) developed a high leaf area index and a dense network of fine roots; competition for light and soil resources was intense enough that the understory species survived but added little to the total biomass production of the system. Thus, *Hyeronima* monocultures and polycultures had equal productivity. In contrast, the palm/herb understory flourished beneath the canopies of the two tree species with lower productivity (*Cordia alliodora* and *Cedrela odorata*), and these polycultures had higher productivity than the respective tree monocultures, indicating complementary resource use between the understory monocots and the tree species. Furthermore, biomass production of the *Cordia* and *Cedrela* polycultures was similar to that of the *Hyeronima* plantations. Thus, in this case, the tree species with highest productivity in monoculture was not the best choice for combining with understory plants to achieve high total production in a mixture. The two overstory tree species having high resource-use efficiency but relatively low resource capture, left unused resources available to understory plants. It is important to note, though, that the monoculture of the most productive tree species was able to achieve productivity equal to that of any of the polycultures, at least at this early stage of development.

Patterns of development past three years in these plantations indicate the importance of assessing interactions over extended periods. For example, the *Euterpe/Heliconia* understory in the *Hyeronima* polycultures, after initially being suppressed beneath the overstory canopy, then began substantial height and biomass growth, leading to complementarity in this mixture; this may cause the *Hyeronima* polycultures to exceed the production of all other plantations in the study at a later stage in development (J. J. Ewel, pers. comm.).

On moist, fertile sites (such as the Costa Rican sites discussed above), light is generally considered to be the principal factor in overstory-understory competitive interactions. Competition for light in this situation is one-sided in that the overstory intercepts a disproportionately large amount of solar radiation relative to its leaf area, with the understory using transmitted and diffuse radiation without interfering with overstory light capture. In contrast, below-ground competition between overstory and understory plants is two-sided, and appears to be more proportional to the root area of individual species (Casper and Jackson 1997). Thus, understory plants may decrease growth of overstory trees on droughty sites, but the removal of understory vegetation generally does not affect overstory growth rates on sites with plentiful water. However, if productivity of the entire system is great enough, water or nutrients may become limiting even on sites with abundant supplies of both. Evidence for such a situation was found in the model ecosystem study (Haggar and Ewel 1997) where the palm/herb understory was found to markedly decrease the production of overstory *Cordia* trees on the fertile lowland rain forest site.

In arid or semiarid climates and on excessively drained sites in moister climates, soil moisture is often insufficient to support the development of dense tree canopies. This promotes the natural development of the shade tree-crop structure, with drought-tolerant understory species having access to adequate light beneath an open canopy. This stand structure may lead to facilitative interactions in several ways. Trees may reduce evapotranspiration and soil evaporation through shading. In addition, there is evidence that deep-rooted trees and shrubs may take up water from deep soils during the day, and then lose a portion of it through shallow roots to upper soil horizons at night; this water is available to all plants the next day (Horton and Hart 1998). This phenomenon of "hydraulic lift" is analogous to the input of nutrients through litter fall of nitrogen-fixing plant species—a resource accessible to only a single species is made available to the entire plant community; the importance of this interaction in affecting productivity has not been studied, however.

These effects would improve the water balance of crop species only if they more than compensate for the added competition for soil moisture from the trees. If a net increase in soil moisture does occur, there may be related benefits in improved nutrient status. Soil microorganisms may stay active for longer periods, increasing nutrient mineralization and availability. Low net primary production in arid climates results in low rates of soil organic matter and nutrient accumulation, so trees may improve site conditions by gradually accumulating nutrients in their biomass and in soil through litter fall. The cultivation of annual crops in the vicinity of *Faidherbia (Acacia) albida* trees in West Africa illustrates the balance between competition and facilitation for a tree species with phenology that makes it complementary with herbaceous species (Sanchez 1995). *F. albida* is a nitrogen-fixing species with leaves that decompose rapidly, and has deep tap roots that access ground water. Leaves are maintained through the nine-month-long dry season and are then dropped at the onset of the rainy season when crops are grown. Yields of millet (*Pennisetum americanum*) or sorghum grown within a 5-m radius around a *F. albida* tree have two to three times the yield in adjacent open areas. The effects are hypothesized to result from the improved nutrient conditions from tree litter production, plus the moderate shading effect of the bare *F. albida* branches during the rainy season, when there is no competition from the trees for water. However, the interactions have not been quantified, so it would not be possible to design an optimum arrangment of trees and crops to improve upon traditional practices (Sanchez 1995).

The balance between competition and facilitation will vary with climatic conditions for shade-tree-crop plantations, as for alley cropping. A physiological modeling approach has been used to define the range of climates where complementary production is likely to occur (Mobbs et al. 1998). In its first use (Cannell et al. 1998), the model was applied to a generic broadleaved tropical tree species (at varying density) combined with a crop of sorghum. The application of this model suggests that the overstory would not be beneficial in many situations because of competitive effects, similar to the conclusions of Sanchez (1995) for alley cropping. The simulation predicted that in climates with less than 800 mm annual precipitation, a monoculture sorghum crop would have the highest grain yield; any competition from trees would reduce the production of sorghum, and could not compensate with equal production because of moisture limitations to the trees. As precipitation was increased in the simulation, the polyculture equalled and then (above 1,000 mm precipitation) exceeded the production of the monoculture sorghum crop, but sorghum was a very small part of total production, even with a sparse overstory tree density. Cannell and coworkers concluded that unless tree products have higher value than the food crop, the complementary interaction in the shade tree-crop polyculture which gives the highest total production in humid climates will not be helpful in practice because the trees dominate resources and production. They further concluded that the polyculture would be beneficial in these climates only if the species access water supplies at different rooting depths or if soil nutrient status is improved by the presence of the overstory. This conclusion is limited in its generality because the simulation was conducted only with a shade-intolerant crop species. Complementary interactions may have greater potential for producing high yields in polyculture where the understory crop species is more tolerant of competition, as shown in the study of Haggar and Ewel (1997), described above. However, the problem with the value of the combined yield will often remain—the understory species is reduced in productivity compared to monocultures, and it generally has the highest value of yield per unit of total biomass.

## 9.8 OLD-GROWTH STAGE—HOMEGARDENS

Certain aspects of the structure of the old-growth stage of forest development can be seen in the traditional form of agroforestry known as the homegarden. These gardens consist of small plots of land surrounding farmhouses, on which a large variety of trees, shrubs, vines, and herbaceous crops are grown. Diversity of commodity production is the most common objective, with fruits, nuts, spices, medicines, vegetables, fuelwood, and timber being grown on plots usually of one hectare or less. The best-known examples of homegardens are those of Indonesian farms (Wiersum 1982), but they have been developed in many parts of the world. Their structure varies along a gradient from nearly closed tree and shrub canopies with little herbaceous vegetation to open canopies with well-developed vegetable crops.

The wide range of homegarden structures means that the relationship to a single forest structural stage is not as strong as for alley cropping, tree gardens, or shade tree-crop combinations. However, descriptions of homegardens from various parts of the world (Wiersum 1982; Raynor and Fownes 1991; Gillespie et al. 1993; Jensen 1993a; Jose and Shanmugaratnam 1993; Wojtowski 1993; Kumar et al. 1994) allow some general principles to be identified: 1) homegardens are characterized by multiple canopy layers and a patchy horizontal structure, which includes gaps in the overstory canopy where shrubs, tree seedlings, and herbaceous plants can be grown; 2) vegetation is continuously dominated by perennial species with large woody crown and root structures, providing good soil protection and nutrient retention; 3) a range of tree diameters and ages are maintained, ensuring continuous regeneration and recruitment of trees into the upper canopy.

These are characteristics associated with old-growth forest structure, but clearly the processes by which this structure is created differ between the two situations.

Old-growth stands develop through the death of large, old canopy trees, whereas planting, thinning, and harvesting are used to maintain homegardens in a similar structure but with relatively young plants. In this regard, homegardens bear a closer resemblance to silviculturally managed uneven-aged forest stands maintained by the selection method. The structural characteristics listed above apply to selection forests, and the similarity is not coincidental—the selection method is modeled partly on old-growth structure and dynamics to maintain good site protection and produce a continuous supply of wood products from a single stand (Oliver and Larson 1996, Smith et al. 1997). In selection stands, small patches are periodically harvested and regenerated; each patch grows to financial maturity (in the late stem exclusion or understory reinitiation stage), and then is harvested. Thus, a stand is comprised at any time of patches of different developmental stages.

A study of indigenous agroforestry on the South Pacific island of Pohnpei (Raynor and Fownes 1991) identified a set of successional stages in the development of homegardens. Early successional patches were first established near the farmhouse by girdling overstory trees and planting crops; these were later gradually extended toward the boundaries of the landholding. The farm entered the "early agroforest" stage as planted tree crops became dominant; production then declined with increasing tree age, a stage referred to as "late agroforest." It was hypothesized that farms could remain in this stage indefinitely if trees are harvested and crops replanted in the oldest parts of the farm. However, the pattern of tree maturation followed by reestablishment of seedlings in patches was not neatly cyclic (as envisioned in selection-forest silviculture). In some cases, entire farms would enter a fallow stage with reduced management, the intensity of management being controlled by social and economic factors.

The nature of species interactions in homegardens has not been studied in detail. Stability of production and sustainability of soil resources have been linked to high species diversity and the small amount of clearing and tilling done is relative to the biomass remaining on the site. However, there is nothing inherent in this structure that assures sustainability. Homegardens have a high fraction of stored nutrients cycled annually, compared to natural forests, as weeding and pruning supplement natural litter fall (Jensen 1993b), which may make them vulnerable to nutrient losses if disturbed. The diversity and structural complexity may guard against disturbance, but even so, the loss of nutrients to harvest and leaching may require balancing by the addition of compost or inorganic fertilizers to be sustainable. Support for the sustainability of traditional homegarden practices comes from the knowledge (often from oral history sources) that some areas have been managed and harvested under this system for hundreds of years (Wiersum 1982, Raynor and Fownes 1991), although they may have gone through periods of fallow at various times, as was found in the Pohnpei study.

## 9.9 CONCLUSIONS

A number of species interactions have been identified in the papers reviewed here as being potentially important mechanisms of facilitation or complementarity in agroforestry systems. In some cases, these mechanisms have only been hypothesized to affect production, based on observations of changes in light, temperature, nutrients, soil structure, or water availability resulting from the introduction of trees to an agricultural system. In other cases, agroforestry systems have been judged successful based on yield results from practice or experimental trials (e.g., *Faidherbia albida*/sorghum in semiarid West Africa, and in coconut plantations with understory crops in many regions of the world, but interactions have not been analyzed in detail in these systems to allow application to other species and sites. Few studies have examined the net result of the combined species interactions on productivity and sustainability in a system, particularly the balance between facilitation and competition effects. Nor have many studies made a rigorous examination of the developmental changes in agroforests as they age, and how these changes affect productivity. This review has focused on those studies that have either analyzed species interactions or studied developmental patterns in agroforestry, in order to identify some principles concerning how these factors influence the probability of success in achieving productivity and sustainability goals.

### 9.9.1 Importance of Complementary Yield as a Goal

Two systems—alley cropping and shade tree-crop plantations—have received much of the attention in recent agroforestry research. In a number of studies (using both empirical and modeling approaches), the performance of these systems has been found to be disappointing because the competitive reduction of crop yield is greater than the facilitative benefit obtained from the presence of the trees. This negative assessment is clearest where trees have been incorporated mainly or entirely for their facilitative effect on nontree food crop production. This is particularly the situation for alley cropping, where in many cases, no harvest is obtained from the hedgerow trees. Some shade tree-crop systems are similar, as in the case of *Erythrina poeppigiana* used as an overstory above coffee or cacao, where trees are pollarded to supply mulch, with timber or other products being harvested from them. However, many shade tree species produce high-value timber, fruit, or other yields.

The determination of success in these applications depends greatly upon the goal. These systems may have greater total productivity relative to monocultures as a result of the balance of interactions in mixtures (i.e., $LER > 1$), measured in total plant biomass or in biomass of harvestable fraction of all species. However, if the agroforestry system is being adopted solely to improve the current yield in nontree crops over monoculture crop systems, and analyses are made using a form of the tree-crop interaction equation, the positive result depends on $RY$ of the crop $> 1$, which is a difficult goal to attain. The facilitation benefit in many (perhaps most) cases will not compensate for the competition from trees and the land area occupied by trees.

These systems are more likely to be successful if a moderately reduced crop yield (compared to monoculture) is acceptable when accompanied by one or both of the following: 1) yield of a high value product from the tree component; 2) sustainability of the moderately reduced crop yield over long periods, with reduced costs of fertilizer, mulch, pesticides, or erosion control efforts compared to those necessary to maintain production of monoculture fields. An additional compensation for the reduced yield may be found in an environmental service provided by the tree component, such as provision of bird habitat in shade trees of coffee plantations (Perfecto et al. 1996), but this would lead to increased adoption of the system only if the landowner received some financial return for this service. It appears that if short term increase of nontree crop yields is the only goal, these practices will be deemed unsuccessful in many situations.

### 9.9.2 Importance of a Developmental Approach to System Design

Many agroforestry studies have followed the progress of a system only through the establishment and early high-production phases. This is understandable in a relatively new field of research, but it provides only a partial assessment of the performance of a system. Agroforestry stands develop further into a late stage in which trees dominate and become too large and competitive to continue the cultivation of the same set of nontree species in the stand. A decision must then be made about whether to: 1) harvest the trees and reestablish the entire system, 2) attempt to maintain the stand structure by limiting the tree competition through pruning or thinning, or 3) shift the system to a new structure and composition that is compatible with the development pattern of the overstory trees.

The forest structural model described in this paper adds some insights into the possibilities of maintaining various agroforestry plantation structures. Systems that mimic the stand initiation stage will be most difficult to hold as a stable system, because it is naturally an ephemeral stage, which ends when trees dominate the stand. Alley cropping encounters such difficulties when the competitive effects of hedgerow trees increase, and considerable labor may be involved in limiting tree size to prevent them from beginning to develop into an overstory; root competition is particularly difficult to control. Problems are less severe with shade tree-crop combinations, because the understory reinitiation stage is less dynamic in structural changes. Trees already form the overstory and grow more slowly than in the seedling/sapling stage, and overstory density can be controlled through timber and fuelwood harvests. The old-growth structure is the most permanent in that it provides for regeneration and continuity of structure within the system, as gaps are created and filled with new vegetation. No such provision exists as part of the alley cropping or shade tree-crop systems. At some point, perennial species in these systems will become too large or too poor in vigor, and the plantation must be cleared and reestablished. In contrast, the structure of homegardens allows for the continual rejuvenation of small patches of the stand. Each patch goes through developmental stages from establishment to high production in a stem exclusion or understory reinitiation structure, and is then harvested to allow for regeneration.

The incorporation of developmental or successional changes in agroforestry systems need not be confined to patchy uneven-aged stands; it could be applied to an entire stand that is kept synchronized in developmental stage. One example is the sequence of different crops grown with coconut on a single site over several decades, as the coconut palms develop from seedlings into a tall overstory (Nair 1993). Another consists of the modifications of taungya practiced in Indonesia

(Kartasubrata and Wiersum 1995), in which taungya stands develop into tree gardens with a mixed tree canopy of timber and fruit trees, rather than a monoculture timber plantation. Using these examples as a basis, a hypothetical system could be defined which uses a succession of agroforestry structures. Taungya could be applied intensively during an extended cropping period, with some trees or shrubs being pruned for mulch application to crops—thus incorporating aspects of alley cropping design. After progressing through a mixed-canopy tree garden stage, shade-tolerant understory crops could be added beneath a partially open canopy of timber and fruit trees. Eventually the stand could shift to a patchy structure as groups of overstory trees were harvested, essentially taking on the structure of a large homegarden, but management efficiency requirements may be better met by ending the timber rotation to begin again with a modified taungya planting. A succession of agroforestry structures of this kind would mimic the changes in structure and competitive interactions among life forms that occurs naturally in forests. Ideas about successional approaches
to agroforestry system design are presented in more detail by Ashton and Ducey (this volume).

## 9.10 ACKNOWLEDGMENTS

My thanks to James Fownes for a helpful review of an earlier version of this manuscript, and to Martha Roldán for preparing the figures.

## 9.11 REFERENCES

Anderson, L. S. and F. L. Sinclair. Ecological interactions in agroforestry systems, *For. Abstr.* 54(1993): 489–523.

Beer, J. Advantages, disadvantages, and desirable characteristics of shade trees for coffee, cacao, and tea, *Agroforestry Syst.* 5(1987): 3–13.

Beer, J., A. Bonneman., W. Chavez, H. W. Fassbender, A. C. Imbach and I. Martel. Modelling agroforestry systems of cacao (*Theobroma cacao*) with laurel (*Cordia alliodora*) or poro (*Erythrina poeppigiana*) in Costa Rica, *Agroforestry Syst.* 12(1990): 229–249.

Blanford, H. R. Highlights of 100 years of forestry in Burma, *Empire For. Rev.* 37, 1(1958): 33–42.

Berish, C. W. and J. J. Ewel. Root development in simple and complex tropical successional ecosystems, *Plant and Soil* 106(1988): 73–84.

Cannell, M. G. R., D. C. Mobbs and G. J. Lawson. Complementarity of light and water use in tropical agroforests, II., Modelled theoretical tree production and potential crop yield in arid to humid climates, *For. Ecol. and Manage.* 102(1998): 275–282.

Casper, B. B. and R. B. Jackson. Plant competition underground, *Annu. Rev. of Ecol. and Syst.* 28(1997): 545–570.

Dix, M. E., R. J. Johnson, M. O. Harrell, R. M. Case, R. J. Wright, L. Hodges, J. R. Brandle, M. M. Schoeneberger, N. J. Sunderman, R. L. Fitzmaurice, L. J. Young and K. G. Hubbard. Influences of trees on abundance of natural enemies of insect pests: a review, *Agroforestry Syst.* 29(1995): 303–311.

Ewel, J. J., M. J. Mazzarino and, C. W. Berish. Tropical soil fertility changes under monocultures and successional communities of different structure, *Ecol. Appl.* 1(1991): 289–302.

Ewel, J. J. and S. W. Bigelow. Plant life-forms and tropical ecosystem functioning, in: Biodiversity and Ecosystem Processes in Tropical Forests. Eds. G. Orians, D. Dirzo and C. Cushman. (*Ecol. Stud.* vol. 122. Verlag, Berlin: Springer-Verlag 1996), 101–106.

Fassbender, H. W., J. Beer, J. Heuveldop, A. Imbach, G. Enriquez, and A. Bonneman. Ten-year balances of organic matter and nutrients in agroforestry systems at CATIE, Costa Rica, *For. Ecol. and Manage.* 45(1991): 173–183.

Gillespie, A. R., D. M. Knudson, and F. Geilfus. The structure of four homegardens in the Peten, Guatemala, *Agroforestry Syst.* 24(1993): 156–170.

Haggar, J. P. and J. J. Ewel. Primary productivity and resource partitioning in model tropical ecosystems, *Ecology* 78(1997): 1211–1221.

Harper, J. L. *Population Biology of Plants.* New York: Academic Press, 1997.

Herzog, F. Multipurpose shade trees in coffee and cocoa plantations in Côte d'Ivoire, *Agroforestry Syst.* 27(1994): 259–267.

Hooper, D. U. The role of complementarity and competition in ecosystem responses to variation in plant diversity, *Ecology* 79(1998): 704–719.

Horton, J. L. and S. C. Hart. Hydraulic lift: A potentially important ecosystem process, *Trends in Ecol. and Evol.* 13(1998): 232–235.

Jama, B. A., P. K. R. Nair and M. R. Rao. Productivity of hedgerow shrubs and maize under alley cropping and block planting systems in semiarid Kenya, *Agroforestry Syst.* 31(1995): 257–274.

Jensen, M. Soil conditions, vegetation structure, and biomass of a Javanese homegarden, *Agroforestry Syst.* 24(1993a): 171–186.

Jensen, M. Productivity and nutrient cycling of a Javanese homegarden, *Agroforestry Syst.* 24(1993b): 187–201.

Jose, D. and N. Shanmugaratnam. Traditional homegardens of Kerala: A sustainable human ecosystem, *Agroforestry Syst.* 24(1993): 203–213.

Kartasubrata, J. and K. F. Wiersum. Traditions and recent advances in tropical silvicultural research in Indonesia, *Unasylva* 46, 2(1995): 30–35.

Kelty, M. J. Comparative productivity of monocultures and mixed-species stands, in: *The Ecology and Silviculture of Mixed-Species Forests,* Eds. Kelty, M. J., B. C. Larson, and C. D. Oliver (Dordrecht, The Netherlands: Kluwer Academic Publishers, 1992), 125–141.

Kumar, B. M., S. J. George, and S. Chinnamani. Diversity, structure, and standing stock of wood in the homegardens of Kerala in peninsular India, *Agroforestry Syst.* 25(1994): 243–262.

Mayhew, J. E. and A. C. Newton. *The Silviculture of Mahogany.* Oxford, U.K.: CABI Publishing, 1998.

Mobbs, D. C., M. G. R. Cannell, N. M. J. Crout, G. J. Lawson, A. D. Friend, and J. Arah. Complementarity of light and water use in tropical agroforests, I, Theoretical model outline, performance, and sensitivity, *Forest Ecol. and Manage.* 102(1998): 259–274.

Naeem, S., L. J. Thompson, S. P. Lawler, J. H. Lawton, and R. M. Woodfin. Declining biodiversity can alter the performance of ecosystems, *Nature* 368(1994): 734–737

Nair, P. K. R. An introduction to agroforestry. Dordrecht, The Netherlands: Kluwer Academic Publishers, 1993.

Oliver, C. D. and B. C. Larson. *Forest Stand Dynamics.* New York: John Wiley and Sons, 1996.

Ong, C. K. A framework for quantifying the various effects of tree-crop interactions, in: *Tree-Crop Interactions: a Physiological Approach,* Eds. C. K. Ong and P. Huxley (Wallingford, U.K.: CAB International, 1996), 1–23.

Ong, C. K., C. R. Black, F. M. Marshall, and J. E. Corlett. Principles of resource capture and utilization of light and water, in: *Tree-Crop Interactions: A Physiological Approach,* Eds. C. K. Ong and P. Huxley (Wallingford, U.K.: CAB International, 1996), 73–158.

Palm, C. A. Contribution of agroforestry trees to nutrient requirements of intercropped trees, *Agroforestry Syst.* 30(1995): 105–124.

Perfecto, I., R. A. Rice, R. Greenberg and M. E. Van der Voort. Shade coffee: A disappearing refuge for biodiversity, *Bioscience* 46(1996): 598–608.

Raynor, W. C. and J. H. Fownes. Indigenous agroforestry of Pohnpei, 2. Spatial and successional vegetation patterns, *Agroforestry Syst.* 16(1991): 159–165.

Sanchez, P. A. Science in agroforestry, *Agroforestry Syst.* 30(1995): 5–55.

Schroth, G. Tree root characteristics as criteria for species selection and systems design in agroforestry, *Agroforestry Syst.* 30(1995): 125–143.

Smith, D. M., B. C. Larson, M. J. Kelty, and P. M. S. Ashton. *The Practice of Silviculture, Ninth Ed.* New York: John Wiley and Sons, 1997.

Tilman, D. and J. A. Downing. Biodiversity and stability in grasslands, *Nature* 367(1994): 363–365.

Trenbath, B. R. Biomass productivity of mixtures, *Adv. in Agro.* 26(1974): 177–210.

Wiersum, K. F. Tree gardening and taungya on Java: Examples of agroforestry techniques in the humid tropics, *Agroforestry Syst.* 1(1982): 53–70.

Willey, R. W. The use of shade in coffee, cocoa, and tea, Horti. Abstr. 45(1975): 791–798.

Wojtkowski, P. A. Toward an understanding of tropical homegardens, *Agroforestry Syst.* 24(1993): 215–222.

Vandermeer, J. *The Ecology of Intercropping.* Cambridge, U.K.: Cambridge University Press, 1989.

# 10 Agroforestry Systems as Successional Analogs to Native Forests

*Mark S. Ashton*[1] *and Mark J. Ducey*[2]

## TABLE OF CONTENTS



## 10.1 INTRODUCTION

In this paper we first describe the patterns and processes that facilitate the growth of species as compatible mixtures in moist and dry temperate and tropical forests. We then provide a rationale for their simplistic reconstruction as agroforestry analogs on abandoned agricultural lands. Lastly this paper describes models for potential testing and development of mixed plantations and provides preliminary information on experiments that have established them based on our knowledge of forest dynamics.

Many studies have documented tree species mixtures to be stratified both over time and in vertical space for natural forests of moist temperate and tropical climates. In these regions, soils are moist enough during the growing season to reduce the importance of water acquisition by the plant roots as a limiting growth process. This has promoted the expansion of below-canopy environments within which shade-tolerant plants can grow and survive. The complexity of vertical stratification is

[1] School of Forestry and Environmental Studies, Yale University, New Haven, CT 06511; Tel.: (203) 432-9835; Fax: (203) 432-8903; e-mail: mark.ashton@yale.edu.

[2] Department of Natural Resources, University of New Hampshire, 215 James Hall, Durham, NH 03824; Tel.: (603) 862-4429; Fax: (603) 862-4976.

0-8493-2206-5/00/$0.00+$.50

largely limited by the degree to which the availability of water and other edaphic resources to the roots limits the ability of the plant to allocate resources for light capture. The degree of vertical complexity in stratification varies at resource scales that are often based on light-water interactions. Such changes can be seen primarily between different topographic sites within a watershed such as between ridge and valley, or between different soil or geological types such as between a shallow till and deep outwash. Other examples of change can be observed across more regional, physiographic landscapes. For example floristic changes between the leeward and windward sides of a mountain range, or between northern and southern aspects of mountain slopes. Watershed influences on vertical canopy stratification can largely be attributed to below-ground differences in the nature of hydrological flow pathways and storage, and entrained flows and deposition of nutrients and soil structural components. Regional physiographic influences on vertical canopy stratification are often associated with climatic differences in precipitation or incident radiation, but may also be associated with changing water or nutrient availability due to differing parent material of the soil or dominant surficial geology.

Water is more limiting in regions with seasonal dry forest and woodland. In these regions root competition among trees restricts canopies to single homogeneous strata with understories of woody shrubs and grasses (Figure 10.1). In more severely limiting circumstances root competition among trees for water is enough to promote open canopied savanna woodland. Though tree canopies are open above-ground, below-ground growing space is fully occupied by roots. In these situations vertical stratification of roots is prevalent with canopy tree species having deep root systems that seek groundwater aquifers, while understory grasses and shrubs more effectively acquire surface water from seasonal rains and often avoid droughts by dormancy and dieback.

Conceptually, canopy stratification in moist forest regions can be characterized in two modes: dynamic and static. Both exist and occur together but vary in their importance depending on the nature and disturbance history of the site. Dynamic stratification can be described as that part of vertical complexity of forest stand structure that is most closely associated with succession. Moisture-loving tree species considered fast-growing pioneers are overtopped by slower-growing and longer-lived tree species. This process has been described as having several phases of stand development which closely parallel changes in the resource use efficiency and therefore the competitive ability of trees to grow and develop (Figure 10.2). Based primarily on North American research four phases of development have been proposed by Oliver and Larson (1990). The initiation phase can be considered the first developmental stage of stand reorganization, regeneration site colonization and/or release after disturbance. The stem-exclusion phase follows this period and can be regarded as the most active postestablishment period of sorting and self-thinning with growing space totally occupied by the stand. A stand enters into the understory initiation phase of stand development when the maintenance of larger canopy tree sizes promotes less efficient capture and use of resources, and therefore makes growing space available for the reinitiation of groundstory advance regeneration, sprout growth, and other usually clonal herbs. This phase also occurs when an increase in the spatial scale of canopy gaps, caused either by increasing tree size or by partial disturbance,

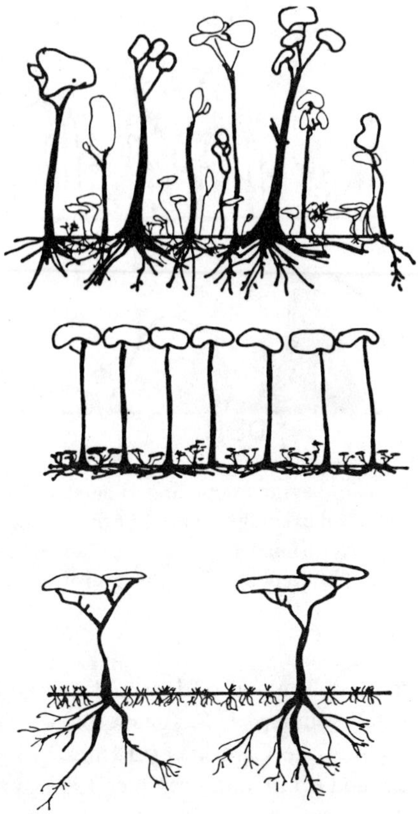

**FIGURE 10.1** A diagram depicting canopy structure (as illustrated by stratification and closure) in relation to availability of soil water. (1) Evermoist soils promote stratification and a heterogenous canopy surface. The canopy is water-wasteful because of large surface areas with a canopy roughness that promotes high evapotranspiration from air turbulance. (2) Seasonally moist or hydric (physiologically droughty) soils restrict stratification to distinct layers with an even canopy with lower surface area that reduces evapotranspiration from air turbulence. (3) Seasonally dry soils with poor stratification and open canopy conditions that are spatially limited by below-ground root competition.

exceeds the declining ability of the large canopy trees to reoccupy vacant growing space. The old growth phase is the last period of stand development. Here, the process of understory reinitiation has progressed enough to promote irregular canopy tree death and subsequent patchy release of the groundstory that eventually develops an all-aged tree canopy (See Kelty, Chapter 9, for detailed description).

A similar stand development paradigm has been described in the European temperate and old world tropical rain forest literature (Watt 1947, Whitmore 1984, 1990). Gap phase can be equated to stand initiation; building phase to stem exclusion; and mature phase to the combination of understory reinitiation and old growth. Examples of dynamic stratification of tree mixtures have been reported for the birch, oak, and

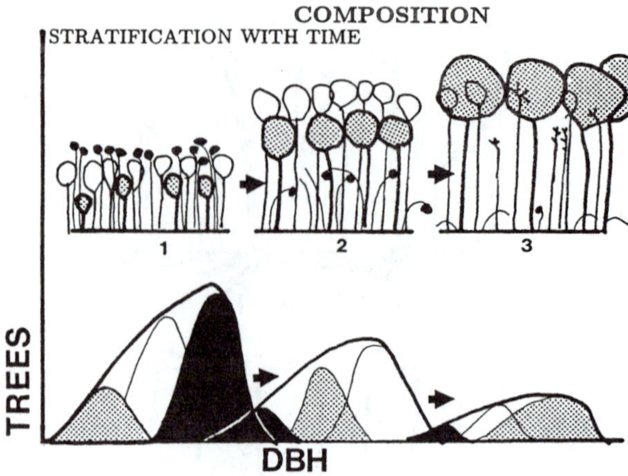


**FIGURE 10.2** A hypothetical diagram depicting diameter distribution and associated dynamic canopy stratification of an even-aged cohort of mixed-species over different phases of stand development [early successional (1); midsuccessional (2); late-successional (3)]. The different shading represents different species crown positions and diameter distributions over time.

maple forests of southern New England (Oliver and Stephens 1977, Oliver 1978); bottomland hardwood of the southern U.S. (Clatterbuck et al. 1987, Oliver et al. 1989); piedmont hardwoods of the Carolinas (O'Hara 1986); spruce-aspen of the north central states and Canada (Palik and Pregitzer 1993); and coniferous mixtures of the coastal northwest (Stubblefield and Oliver 1978, Wierman and Oliver 1977). The moist tropics have been similarly described particularly with mixtures in the neotropics of central and south America (Budowski 1961, Ewel 1980, Uhl et al. 1981, Brokaw 1985, Uhl et al. 1988), and old world tropics of southeast Asia and west Africa (Whitmore 1984, Swaine and Hall 1988, Swaine and Whitmore 1988).

Static stratification can be described as that part of vertical complexity of the forest stand that promotes the permanent existence of a subcanopy and groundstory comprised of plants that never succeed to the canopy. This pattern is most characteristic of moist forest regions with long-term disturbance intervals that allow for progressive accentuation of vertical habitat strata. Examples of this kind of stratification have been well-described in the classical tropical literature (Davis and Richards 1933, 1934, Beard 1944, Black et al. 1950, Ashton 1964, Whitmore 1974). A simple type of static stratification has been described for the southern boreal forest (Cooper 1913, 1928). Many of the more complex temperate forests also have well-defined strata particularly in the wet coastal Pacific Northwest (Wierman and Oliver 1977) and the cove forests of the southern Appalachians (Braun 1942, Lorimer 1980) (Figure 10.3). The description of this type of stratification as static is largely one of temporal perspective; some static stratification patterns, such as those of the boreal forest, may be dependent on severe disturbance for long-term maintenance at the landscape scale.

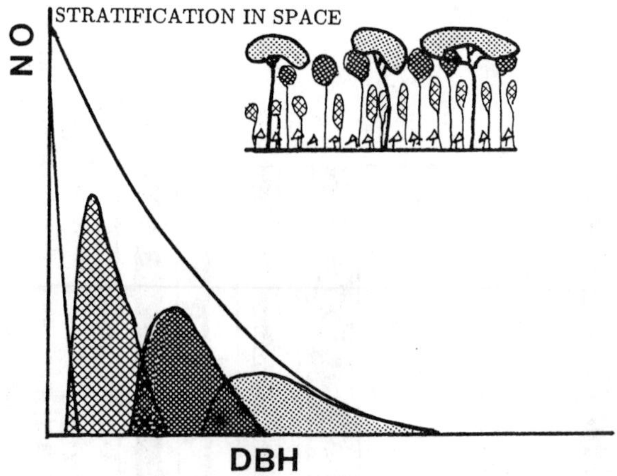


**FIGURE 10.3** A hypothetical diagram depicting diameter distribution and associated canopy stratification of an even-aged cohort of mixed-species that permanently occupy different canopy strata over time.

Intermediate between these two types of stratification are examples of long-lived canopy trees that eventually relinquish their canopy space to subcanopy trees that are still longer-lived through gradual canopy disturbances such as ice storms and branch breakage from winds. Depending on the time scale of stand development this can almost be interpreted as part of static forest stratification rather than the last part of dynamic stratification. These intermediate descriptions of forest stratification have been documented for southern New England for hemlock beneath oak (Kelty 1986), and for white oak beneath tulip poplar (O'Hara 1986).

The diversity and complexity of both the static and dynamic processes of forest stratification for these moist forest regions can largely be attributed to the wide variations in ways that these forests rely upon release of advance regeneration (Figure 10.4). This indicates the importance of those kinds of disturbance that serve to release regeneration, such as the various kinds and sizes of windthrows, as compared to disturbances that are lethal to the groundstory such as catastrophic wildfires, landslides, or volcanic eruptions. Reliance on release-type disturbances in moist forest regions means all growing space is actually occupied before, or shortly after, a disturbance occurs. The term for this type of regeneration initiation has been called allogenic succession, and is often associated with an initial floristics pathway of development (Egler 1954, Drury and Nisbet 1973, Henry and Swan 1974). Disturbance regimes lethal enough to destroy the groundstory can be considered relatively unusual in these forests, but do occur, in many instances associated with human clearance for agriculture, mining, and development. Such disturbance patterns often lead to successional processes that promote changes in regeneration dominance characterized by sequential recruitment of different species over time, often in pulses as one species modifies the groundstory environment and facilitates the regeneration of another species. Many examples of this kind of successional process existed during the period of agricultural abandonment in eastern North America in the form of old-field pine and cedar (Lutz 1928, Billings 1938, Oosting 1942, Bormann 1953, Raup 1966). This

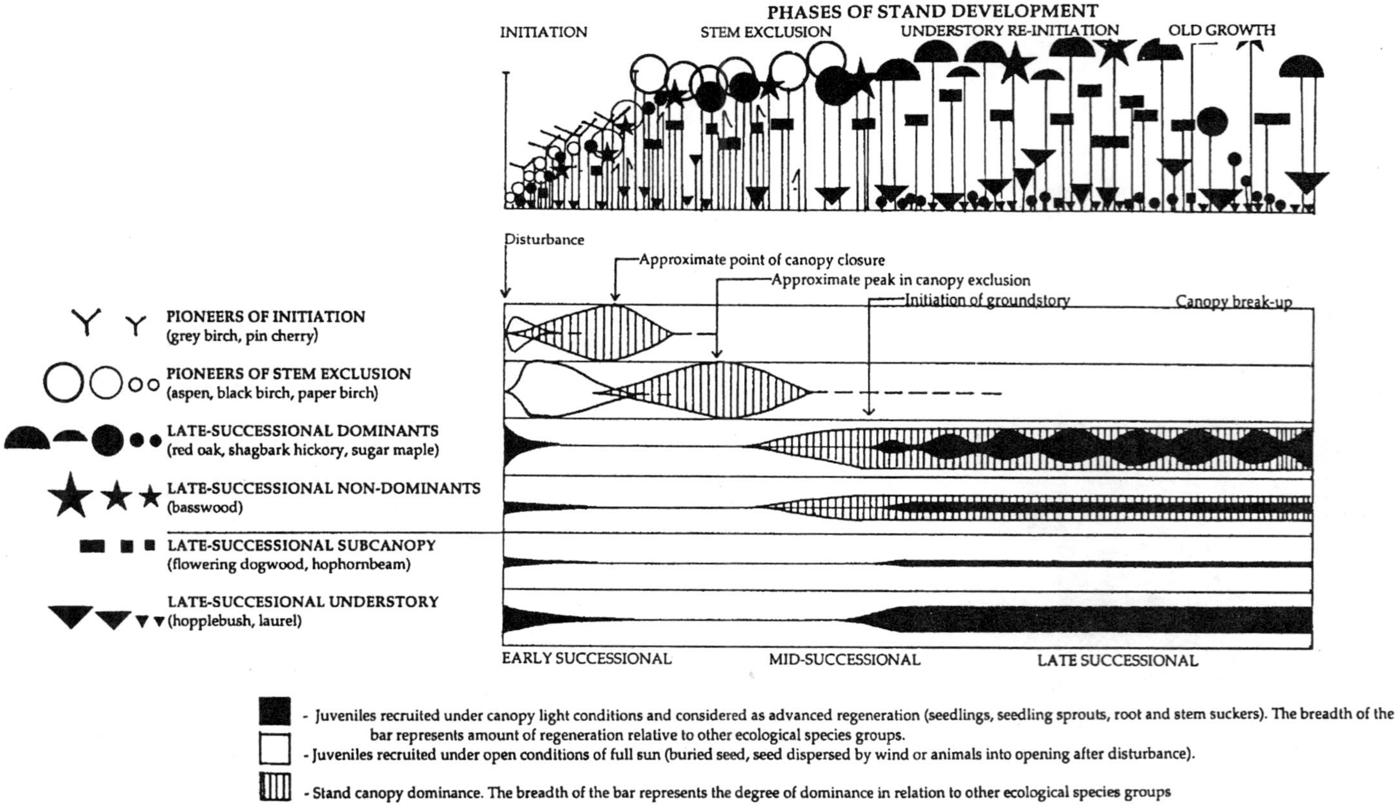


**FIGURE 10.4** Regeneration recruitment frequency and stand canopy dominance of ecological species groups over different successional stages of stand development for a mixed-hardwood forest of eastern North America. Examples of species are given for each ecological group along with codes denoting their structural position within the stand over time. Note the periodic recruitment of seedlings for tree species belonging to the late-succes-sional canopy dominants (modified after Ashton 1992).

kind of sequential regeneration initiation has been termed autogenic succession or relay floristics (Clements 1916, Daubenmire 1952, MacArthur and Connell 1966).

Dry forests and boreal woodlands have successional pathways that are simpler with fewer species combinations than moist forests (see review by Connell and Slatyer 1977). In many regions, late seral stands can become monospecific for example *Shorea robusta* (sal) in south Asia; *Pinus contorta* (lodgepole pine) and *P. ponderosa* (ponderosa pine) in the interior U.S. west; and *P. banksiana* (jack pine) in northern Canada. In dry regions, forest regeneration has adapted to establishing after more lethal disturbances that kill both the existing canopy trees as well as the groundstory vegetation. Fire and its variation in severity and frequency is a characteristic disturbance associated with such climates. Fire adapted species regenerate from buried seed banks, serotinous cones, capsules or pods, or from sprout growth of surviving roots and stumps. Seedling regeneration of some species (e.g., *P. palustris,* longleaf pine) can withstand subsequent fires through insulation of their stems from thick bark and dense foliage. Regeneration initiation in these forests can also be termed allogenic but unlike moist forest regions the majority of the vegetation physically occupies new growing space after a disturbance creates it. In the absence of catastrophic disturbance over long periods of time autogenic patterns of succession may occur in these regions. Mladenoff and Pastor (1993) describe an example of the diversity of successional pathways with and without catastrophic lethal disturbances for forests of north central U.S. (Figure 10.5).

Lastly, site specialization of tree species across below-ground resource gradients that are usually related to variations in topography of the landscape can play an important determining role in the spatial heterogeneity of tree mixtures. In general, site specialization is more characteristic of late-successional species that are dependent upon advance regeneration as compared to pioneers. Evidence suggests differences in species composition can exist due to edaphic factors: soil moisture (Bourdeau 1954, Ashton et al. 1995, Ashton and Larson 1996) and nutrient status (Denslow et al. 1987, Latham 1992, Burslem et al. 1995, Gunatilleke et al. 1996, Gunatilleke et al. 1997a); and biotic factors including density dependence of host-specific seed predation (Janzen, 1971, Condit et al. 1992, 1994), pathogen mortality (Gilbert et al. 1994), seedling herbivory (Becker et al. 1985), and microfaunal and floral symbioses (Janos 1988, Newberry et al. 1988).

## 10.2 RATIONALE FOR DEVELOPMENT OF AGROFORESTRY TREE-MIXTURES

Currently evermoist and seasonally dry mixed-species forests supply much of the world demand for quality timber (furniture, interior panelling, flooring, turnery, veneer, and other speciality woods). In most instances the mode of exploitation has led to a high-graded forest that has no quality timber production capability or value. The majority of species that are desired for quality timber in these forests are canopy tree species that are late-successional site-specialists relying mostly upon advance regeneration for their establishment and growth. Studies have shown that their real price values have dramatically increased compared with other timber values

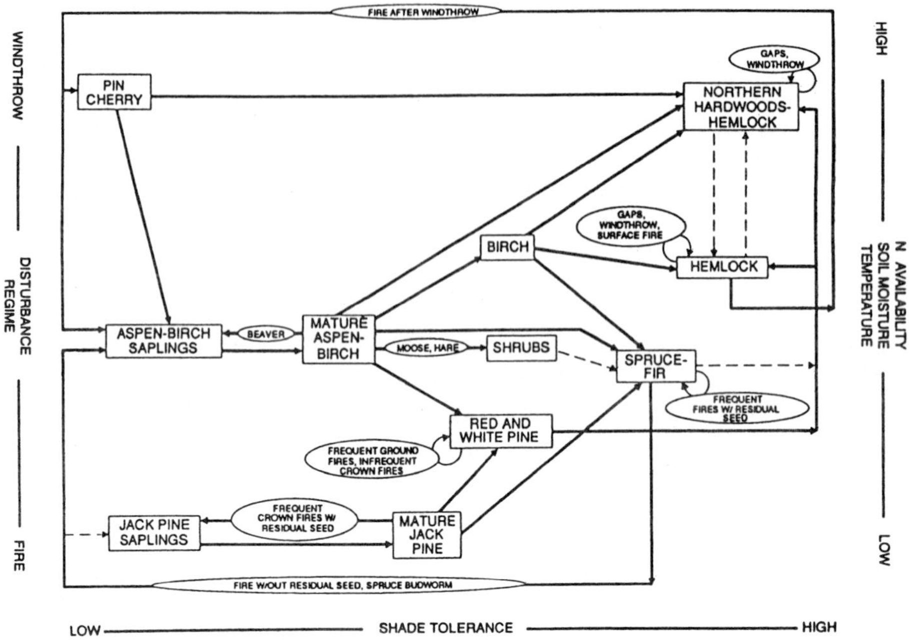


**FIGURE 10.5** Succesional pathways. Forest landscape succession in the western Great Lakes portion of the northern hardwood and conifer forest region (from Mladenoff and Pastor 1993).

such as fiber, pulp, and production sawlogs (Burgess 1993, Howard 1995, Verissimo et al. 1995). Use of tree species that produce quality timber in conventional plantation or agroforestry systems has thus far been generally poor (Wormald 1992, Ashton et al. 1993). This can be mainly attributed to: A) their inability to compete with weedy competition under full sun conditions; B) their tendency to stagnate or produce poor bole form when self-thinning amongst themselves; and C) their poor survival and establishment on soils for which such specialists are not suited.

Once established, mixtures of compatible tree species have been demonstrated to have higher yields than single-species plantation systems. Studies in North America have shown greater yields for mixed plantations of *Alnus rubra* (red alder) and *Pseudotsuga menziesii* (Douglas fir) (Binkley and Greene 1983, Binkley 1984); *Robinia pseudoacacia* (black locust), *Elaeagnus umbellata* (autumn olive), and *Juglans nigra* (black walnut) (Paschke et al. 1989, Schlesinger and Williams 1984); *Quercus rubra* (red oak) and *Tsuga canadensis* (eastern hemlock) (Kelty 1986). European literature has reported similar findings for mixtures of *Betula pendula* (silver birch) and *Pinus sylvestris* (Scots pine) or *Picea abies* (Norway spruce) (Mielikainen 1985, Tham 1988). Only a few well-documented studies have been done in the tropics; notable are those by DeBell et al. (1985, 1989) and Binkley et al. (1992) using different mixtures of *Eucalyptus saligna, Albizia falcataria,* and *Acacia melanoxylon,* and experiments comprising monospecific and mixed plantations using indigenous species in Costa Rica (Montagnini and Sancho 1990, Montagnini et al. 1995).

Nontimber species are often associated with the same problems in establishment that characterize late-successional canopy trees. They are also solely exploited from these same forests and are quickly depleted. Many of these species are subcanopy trees, lianas, and groundstory shade-demanding herbs that yield a great variety of specialized but highly desirable products (sugars, latex, spices, medicines, and fruits). However, these products are usually ecologically costly for the plant to manufacture, and are often associated with chemical compounds that the plant synthesizes for protection. In certain circumstances, the management for these products alone can generate more income and produce a higher net present value for a natural forest than any other use for the land (Peters et al. 1989, Balick and Mendelsohn 1992). Examples of single products that continue to rise dramatically in value are *Calamus* spp. (rattan) and *Taxus* spp. (yew), both of which are now multibillion-dollar industries (Manokaran and Wong 1985).

The more reliable and more frequent the yield of a crop plant, the less financial risk, and hence the more desirable for cultivation by landholders. Important examples of such crops are obviously those that are widely grown as intensive single-species plantation systems such as *Camelia sinensis* (tea) or *Hevea brasiliensis* (rubber). Products that produce at infrequent intervals or that take several years to maturity for a one-time harvest are often those plants that are still exploited from natural forests and are most prone to scarcity (Fortmann 1985, DeBeer and McDermott 1989).

We propose that the cultivation of nontimber and quality timber species in mixture, provided the complexities of their autecology are known, can make good economic sense. Mixtures can reduce the downside financial risks of crop failure, and potentially provide a crop yield at least once a year for some species (sugars, latex, fruits). At the same time, mixtures can provide for high value through the sequential yield of several one-time crops that mature over the long-term stand development process. The aggregation of multiple products over time is frequently considerably higher in net present value than if each species could be grown alone (Peters et al. 1989).

Lastly, in many instances commercial species that are currently grown in single-species plantation systems can also be incorporated into mixed plantation or agroforestry systems. They can be grown, particularly for their own reduction in risk of crop failure from certain pathogens and insects, and for reducing income risk of the whole system by providing a crop with greater market security. Under such circumstances these crops can provide a reliable, early and directly marketable product for income during the early stages of plantation establishment. These commercial species are usually food crops that are early seral annuals such as cereals (i.e. *Oryza sativa,* rice), corn (*Zea mays*), vegetables (i. e., *Brasica* spp., broccoli, cabbage), or fodder (grasses); or they could be short-lived perennial shrubs that can be repeatedly harvested for a commercial product such as *Musa* spp. (banana), *Carica papaya* (pawpaw), *Coffea arbica* (coffee), *Piper nigrum* (pepper), *Camelia sinensis* (tea) and *Hevea brasiliensis* (rubber). Certain other species have often been planted in mixture with a more commercial plant, to provide shade and to avoid sun scorch (*Milicia excelsa*—Gibson and Jones 1977); to reduce leader weevil damage on *Pinus strobus* (eastern white pine) (Boyce 1954) and *Hypsipyla* spp. stem borer damage to host genera of the Meliaceae (Africa—*Khaya, Entandrophragma;* Asia—*Toona, Chukrasia;* S. America—*Swietenia, Cedrella*).

Obviously, in drier and colder climates where resources are more limiting, mixed-plantings for agroforestry will usually have less vertical canopy complexity, with strata that are more static in time than moister tropical and temperate climates. However, because of more severe below-ground competition among plants for water, arrangements of species plantings will be more distinct and less intimate. Examples of the most successful agroforestry systems in these drier climates reflect these spatial patterns (i. e., alley cropping; contour-field cultivation with line planting; open grown orchards/pollards with grazing).

## 10.3 SUCCESSIONAL MODELS FOR THE DEVELOPMENT OF AGROFORESTRY TREE-MIXTURES

Many studies have recorded the establishment and survival of mixed-species plantations (Wormald 1992). Mixtures have been organized in different ways. Mixtures vary in degrees of composition dominance, their spatial arrangements and their age structures. Mixtures of plantings often imply intimate, tree by tree, or line by line establishment, but arrangement can be in blocks such as some of the first mixed planting trials in North America (Hawley and Lutz 1943) and conifer-hardwood mixtures in the U.K. (Evans 1984). Mixtures need not always be the result of actual planting. Many studies have now measured the recruitment of natural regeneration beneath the canopy of single-species plantations (Guariguata et al. 1995, Parrotta 1995).

The models that we propose are based largely on knowledge gained of mixed-species forest dynamics described in the introduction to this paper. It can be condensed into a list of key principles derived from this understanding:

1. Where soil resources have not degraded or diminished beyond a threshold that affects the ability of more site-specific tree species to establish, or where the radiation environment does not exceed the light tolerance of certain tree species, then plantation establishment of species can be completed at approximately the same time (initial floristics). In situations where site-specific species cannot establish on a plantation site because conditions for their establishment are too severe, then plantation establishment should be sequential over time (relay floristics).
2. To obtain satisfactory growth without continuous thinning it is important to select species that are successionally compatible with each other (Table 10.1).
3. The spatial arrangement of mixtures should be consistent with the differential degree of self-thinning that occurs between tree species that are of different successional status (Table 10.2). There should be more

**TABLE 10.1**
**Examples of successionally compatible mixtures based on some stand development studies in moist forest regions of North America.**

| Region | Nurse Phase | Training Phase | Tree Crop Phase |
|---|---|---|---|
| | Successional stages of plantation development | | |
| Southern New England[1] | | *Acer rubrum*<br>*Betula lenta* | *Quercus rubra* |
| Piedmont[2] | | *Liriodendron tulipifera* | *Quercus alba* |
| New Hampshire[3] | *Prunus pensylvanica* | *Betula alleghaniensis* | *Acer saccharum*<br>*Fagus grandifolia* |
| Northern Michigan[4] | | *Populus grandidentata* | *Quercus rubra*<br>*Acer rubrum* |
| Coastal Washington[5] | *Alnus rubra* | *Pseudotsuga menziesii* | *Tsuga heterophylla* |

1–Oliver 1978, 2–O'Hara 1986, 3–Bormann and Likens 1979, 4–Palik and Pregitzer 1993, 5–Stubblefield and Oliver 1978.

early-successional species per unit area than late-successional at time of planting. For example this means that in a temperate mixed-species plantation that includes mid-successional *Q. rubra* (red oak) and and early-successional *B. papyrifera* (paper birch) there should be a higher number of *B. papyrifera* planted than *Q. rubra*.

**TABLE 10.2**
**Differences in self-thinning over a 10-year-period (1986–1996) for northern hardwood stands on thin till (xeric) and swale till (mesic) sites at the Great Mountain Forest in northeastern Connecticut. Both stands are now 27 years old and would be considered within the stem exclusion stage of stand development (Liptzin and Ashton 1998).**

| Species | Tolerance Rank | % Stem Mortality | | Mortality of Stems/ha | |
|---|---|---|---|---|---|
| | | Till | Swale | Till | Swale |
| *Betula populifolia* | very intolerant | −65.4 | −100.0 | −345 | −198 |
| *Prunus pensylvanica* | very intolerant | −50.0 | −46.2 | −223 | −170 |
| *Betula papyrifera* | intolerant | −49.1 | −47.6 | −1,704 | −694 |
| *Betula lenta* | intermediate | −40.6 | −10.1 | −872 | −127 |
| *Quercus rubra* | intermediate | −25.9 | −57.9 | −609 | −992 |
| *Fagus grandifolia* | tolerant | −13.0 | +42.4 | −183 | +198 |
| *Acer rubra* | tolerant | −5.9 | −6.7 | −20 | −14 |
| *Tsuga canadensis* | very tolerant | 0.0 | 0.0 | 0.0 | 0.0 |

4. The spatial arrangement of trees should be compatible with their crown spatial requirements as they grow (Figure 10.6).
5. Careful selection of the late-successional more shade tolerant tree species needs to be made to insure their site compatibility.

There are many more considerations that need to be taken into account in selecting species mixtures that can concern such factors as susceptability to pathogens, herbivory, nutrients, and water use. These factors have been discussed elsewhere in previous chapters. The principles that we propose above are based primarily on our understanding of successional patterns of native forests.

Our proposed models consist of three phases (Ashton et al. 1993). These phases of plantation growth represent the same phases described for the natural forest dynamic but can be more complex for wetter more moderate climates (Figure 10.7) than for ones that are drier and more extreme. The nurse phase can be equated to stand regeneration and initiation; the training phase is the period

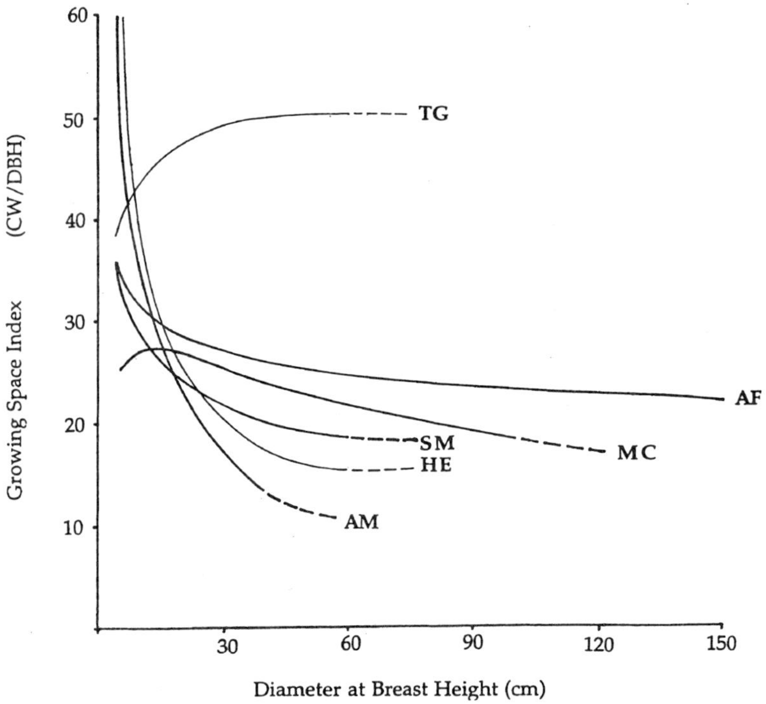


**FIGURE 10.6** Relationships depicting changes in use efficiency of growing space as measured by the ratio of crown width/diameter at breast height with increase in diameter at breast height for some tropical tree species of different successional status. [TG—*Tectona grandis* (mid-successional); AF—*Albizia falcataria* (mid-successional); MC—Michelia champaca (late-successional); SM—*Sweitenia macrophylla* (mid-successional); HE—*Hibiscus elatus* (early-successional); AM—*Alstonia macrophylla* (early-successional). (Samarasingha et al. 1995).

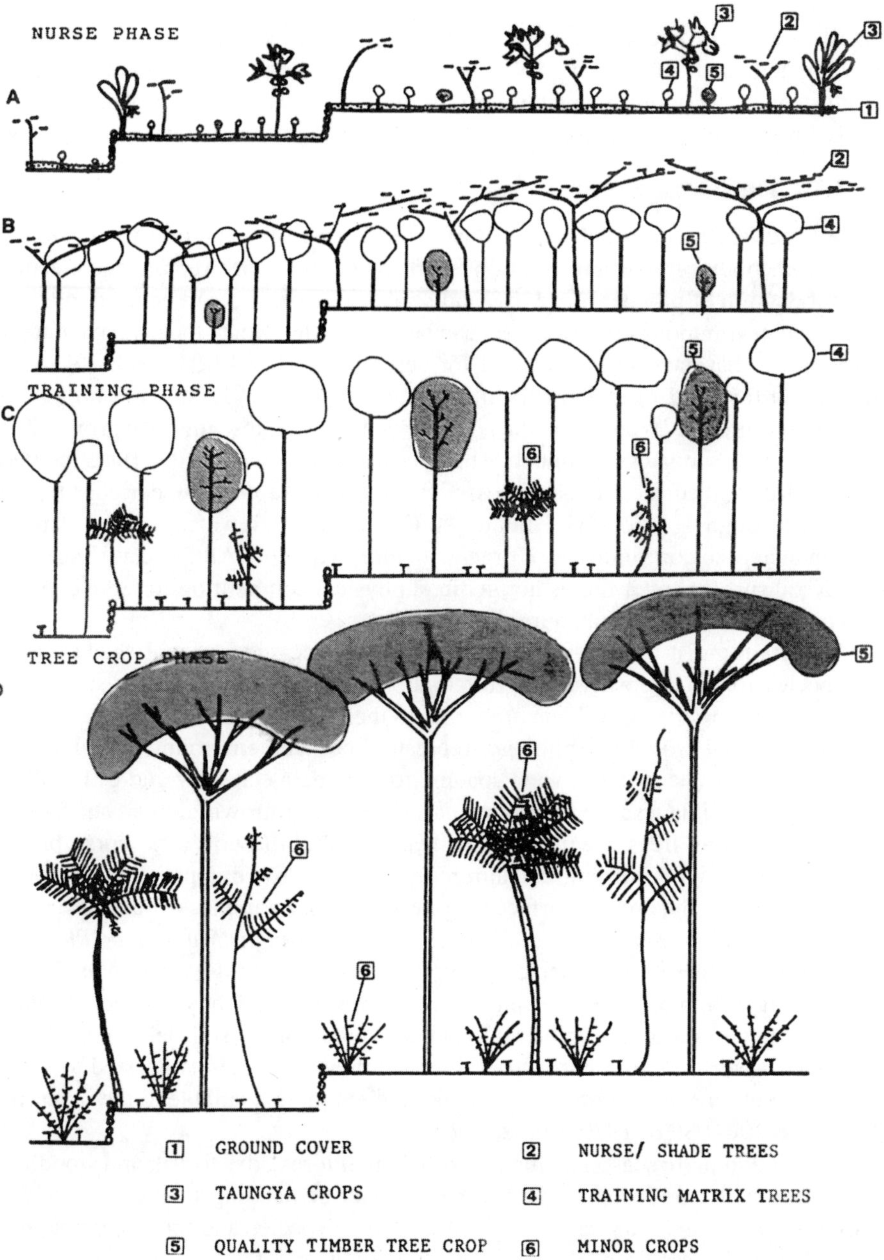


**FIGURE 10.7** A representation of an agroforestry system that illustrates the progressive development of a mixed-planting showing the different phases of development for a moist climate (Ashton et al. 1993).

described as stem exclusion; and the tree crop phase is analogous to the understory initiation and old-growth phase of stand development. All species can be planted simultaneously at establishment, but each species that is selected for the plantation mixture is representative of a different part of the plantation's successional development.

### 10.3.1 The Nurse Phase

This period can be considered the most critical time of plantation establishment. Where soil conditions merit restoration and then protection of structure and nutrition, the introduction of pioneer nurse tree and herbaceous species can be used to advantage. Many leguminous ground covers can be selected to imitate the invasive but protective roles that have been recorded for certain early seral herbaceous forbs and grasses (Safford and Filip 1974, Bormann and Likens 1979, Ashton et al. 1997b). Leguminous ground covers have the added advantage of a low-uniform growth habit that can form a compact live mulch. This cover greatly lowers soil surface temperatures, retards soil surface moisture losses from direct evaporation, enriches top soil nitrogen through fixation, and builds soil structure and water-holding capacity through increased contributions of organic matter to the soil. All these attributes promote a soil surface cover that is not in direct physical competition for above-ground growing space with the planted tree seedlings.

In certain moist tropical circumstances where late-successional shade-tolerant tree species are very sensitive to high radiation, early seral pioneer tree species can be established to serve as a shade umbrella. In these regions many pioneers have the rapid growth and crown morphology to create this shade environment. Nurse trees can be seeded or planted at a wide spacing to provide as cheaply and as rapidly as biologically possible a shade environment for the slower growing, more shade-tolerant or -demanding trees species planted beneath. Within two years, nurse pioneer trees can develop a spreading but thin mono-layered crown approximately four to five meters above the ground surface (Figure 10.7a, b). Examples of such species are in the genera *Cecropia, Gliricidia, Leucaena, Macaranga, Mussanga, Rhus,* and *Trema.* Most of these nurse species are very short-lived and die as soon as they are over-topped (Whitmore 1984). Some studies have suggested they may act as temporary nutrient sinks, trapping mineral ions that would otherwise leach out of the soil after a disturbance. As the nurse trees die back they have been suggested to slowly relinquish nutrients back into the soil, making them more available to the other trees (Budowski 1961, Stark 1970, Marks 1974).

Colder and more seasonal climates with boreal forest, dry forest, and woodlands do not have such nurse pioneers because their water-demanding attributes make them susceptible to drought and cold temperatures. Such species and the ecological roles that they can play are probably irrelevant in these circumstances.

Leaves of nurse tree species often have fewer toxic compounds (exceptions exist for such genera as *Gliricidia*—rotinone) or protective characteristics such as surface wax, hairs, or coriaceousness than slower-growing tree species (Ewel 1980). Many of these species also have the capability to coppice or pollard and can have high nitrogen contents because they are able to fix atmospheric nitrogen. All these attributes make them suited to producing arboreal fodder for livestock.

During this period of plantation establishment, other very short-lived species can be inter-planted to bring direct tangible benefits to the local community, as long as this does not interfere with future stand dynamics (Figure 10.7a). A successful example of this type of planting for a food crop is the taungya system used to establish teak timber plantations in south Asia (Brandis 1897, Champion and Seth 1968). Use of this system encourages protection and care of the tree species that otherwise could be neglected or under-appreciated by the local community. Examples of species that are often socially desirable for local communities include light-demanding high carbohydrate food/fruit crops (*Musa* spp., *Manihot esculenta, Carica papaya*), and even some annual crops that require tillage (e.g., *Zea mays,* corn; Cucurbitaceae, squashes, melons; Papilionoideae, beans, peas), especially in the first two years. If grown as a commercial crop, the income generated can offset some or all of the costs of plantation establishment.

Because water is more limiting, canopy closure of trees never really occurs in drier climates. Trees that are planted are usually widely spaced to insure adequate availability of water and nutrients and to avoid root competition particularly during droughts. This can promote more intensive cultivation for arable crops and pasture during the wetter periods of the year that would ordinarily not be possible because of understory shade in moist climates. This phenomenon of cultivation at the groundstory in seasonally dry climates can be a permanent part of agroforestry systems, because temporal successional processes of tree canopy stratification are much less dynamic and because spatial windows of growing space can exist during the wet periods of the year that can only be used by seasonally active plants.

### 10.3.2 The Training Phase

When other tree species overtop the shade trees of the nurse phase, fully occupy the plantation canopy, and begin shading out the groundstory cover and crop plants, then the mixed planting can be considered to have entered the training phase of its development. Tree species that dominate the canopy at this stage typically have autecological characteristics that make them upwardly fast-growing with strong epinastic control. Their crowns are small but compact making them efficient users of growing space. In their native forests they often grow in dense stands and in most circumstances readily self-thin amongst each other (Figure 10.7a, b). Examples of species that fit this description include many from the genera *Acacia, Albizia, Alstonia, Betula* (birches), *Eucalyptus* (eucalypts), *Pinus* (pines), and *Populus* (poplars). Many species in these genera are planted as single-species plantations because they can produce some of the world's highest yields of sawtimber and fiber (Shepherd 1986, Evans 1990).

Species characteristic of this group are usually considered pioneers, and produce abundant seed regularly almost every year. Their regeneration is dependent upon seed that germinates on new growing space that has been created by a forest canopy disturbance. Their purpose in the kind of mixed-species plantation or agroforestry system that we propose is to provide the same kind of support, stem training and rapid ability to self-thin that improves yield and quality of the slower growing, longer-lived tree species in moist mixed-species forests. If the trees are planted at a dense spacing, thinning regimes can be adjusted to capture their timber and fiber values, or if no markets are available, reliance can be made on their own self-thinning. During the training period of stand development these species create the matrix within which other more shade-tolerant and late-successional species grow that eventually create the canopy and subcanopy. At this time the mixed planting, like the stem exclusion stage of a moist mixed-species forest, is undergoing considerable self-thinning in the canopy with little to no growing space available to other plants at the groundstory.

On sites with very favorable environmental conditions, or with planted species mixtures that do not require immediate partial shade for survival, the nurse phase of establishment can be very brief. In some moist temperate climates and in most dry tropical climates there are no good examples of large-leaved pioneers that germinate and grow. Instead the ground cover is quickly dominated by forbs such as *Rubus* or species in the Compositae, which do not interfere substantially with seedling growth. However, on other sites species such as ferns or members of the Ericaceae may proliferate rapidly, and can interfere with stand development over long time periods. This is particularly evident in Mediterranean climates where tree competition with grasses and woody shrubs can be severe. In these circumstances, scarifying the topsoil and then direct seeding pioneer species typically dominant during the training phase along with a temporary ground cover might be a more satisfactory protocol. If a buried seed bank exists within the soil, then scarification and exposure of mineral soil might be the only necessity for release of succession (Cohen et al. 1995).

Studies have documented the invasive role of many pioneer tree species on abandoned agricultural lands. These are also the species that usually dominate the stand canopy of the training phase of our model. In moist temperate circumstances of eastern North America many species of pines have facilitated the understory initiation of late-successional angiosperms (Lutz 1928, Billings 1938, Oosting 1942, Bormann 1953). Plantations of species with similar growth rate have been documented to facilitate secondary rain forest vegetation in the moist tropics (Guariguata et al. 1995, Parrotta 1995). On sites that have soils and above-ground radiation environments that are too extreme for the immediate establishment of site-sensitive, late-successional species, then plantings should be established by sequential introduction over time. Using species like *Pinus* as an establishment matrix, the site can quickly be occupied, shading out the groundstory. Afterwards more site-sensitive late-successional species can be underplanted or line-planted beneath the thinned canopy (Ashton et al. 1997a).

In seasonally dry climates pioneer species considered to belong to the training phase should not be planted at a close spacing because under these circumstances self-thinning can stagnate. In these climates, fires serve as an important thinning mechanism for forest stands. Examples of overly dense and stagnant stands that have been protected from fire for many years exist in the interior western U.S. (*P. contorta,* lodgepole pine; *P. ponderosa,* ponderosa pine) and Australia (*Eucalyptus* spp.).

### 10.3.3 The Tree Crop Phase

The initiation of the tree crop phase of the mixed planting begins when the late-successional tree species begin to overtop the fast-growing pioneers that dominate the canopy of the training phase (Figure 10.7d). The late-successional species, because they are more shade-tolerant, are able to grow steadily through the stratum of training phase pioneers. During this process, the late-successional species often change canopy morphology dramatically from a crown structure that is monopodial and columnar in shape, to one that upon receiving full sun becomes broadly branched and expansive. These species often do not perform well in competition for growing space with each other, but a more shade-intolerant matrix of pioneers allows for their crown expansion. Although diameter increments might prove low during the training phase, increments should increase substantially as these species attain canopy status during the tree crop phase. These are the species that would be harvested at the end of the rotation for high quality timber products (veneer, furniture, interior panelling, flooring, turnery). Late-successional species of the tree crop phase in dry climates are slower growing trees than their counterparts in the moist climates, but they can have very dense and durable woods used for specialty purposes such as ornate furniture, carving, and turnery (e.g., *Diospyros,* ebony; *Guaiacum* lignum-vitae).

Other species that represent true below-canopy strata of mature evermoist forests can be grown for the production of various non-timber crops. Because they are adapted to relatively poor light environments, many of these species have morphological and physiological adapations that make them efficient at "harvesting" light. Their leaves are usually broad and often variegated and arranged in single-layers that are either A) in planar whorls that make crowns deep and monopodial; or B) shallow crowns that are flat and spreading. Their conservative use of resources promotes greater allocation to production of secondary compounds for their protection from pathogens and herbivores. These attributes make them desirable for use as flavorful beverages (obvious commercial examples are the original tea and coffee plants), and medicines. In dry climates, understory conditions for shade tolerant plants never occur because water limitations at certain periods always promote canopy openness. These situations can therefore allow the continuous cultivation of seasonal light-demanding crops that can only be done in moister climates during the first two years before canopy closure occurs.

The rotation lengths of these agroforestry systems are dependent upon the size and maturity of the canopy late-successional timber trees. The progress of sequentially moving through these phases of agroforestry or mixed plantation development from start to finish could be anywhere between 40 to 100 years depending on the successional dynamic of the mixture chosen for planting, the integrated economic value of the products obtained, and the nature of the soil and climate. Because the subcanopy species, like the late-successional timber species, are slow-growing and site-specific, no satisfactory plantation has been developed to date. However, this aspect of the model deserves testing.

## 10.4 CONCLUSIONS

The growth and development of mixed plantings in agroforestry systems can be understood and managed using an analog to the development of natural mixed-species stands. Each of the three phases of development corresponds to one of the stand development stages of Oliver and Larson (1990): The nurse phase to the stand initiation stage, the training phase to the stem exclusion stage, and the tree crop phase to the understory reinitiation and old-growth stage. Ecologically, each stage is dominated by a different mix of life forms and successional species, while facilitating the regeneration, growth, and development of late-successional or site-sensitive species. Economically, each stage is dominated by a different mix of product yields, offering the possibility of frequent and dependable income compatible with the long-term production of high-quality timber. These features suggest that agroforestry systems that reflect these mixed-species plantings can offer a variety of social benefits and considerable silvicultural flexibility, while reducing elements of risk often associated with single-species systems.

## 10.5 REFERENCES

Ashton, P. M. S. Establishment and early growth of advance regeneration of canopy trees in moist mixed-species broadleaf forest, In: eds. M. D. Kelty, B. C. Larson, and C. D. Oliver. *The Ecology and Silviculture of Mixed-Species Forests*. (Dordrecht, The Netherlands: Kluwer Academic Publishers, 1992), 101–125.

Ashton, P. M. S. and B. C. Larson. Germination and seedling growth of *Quercus* (Section *Erythrobalanus*) across openings in a mixed-deciduous forest of southern New England, USA. *For. Ecol. and Manage.* 80(1996): 81–94.

Ashton, P. M. S., C. V. S. Gunatilleke, and I. A. U. N. Gunatilleke. A case for the evaluation and development of mixed-species even-aged plantations in Sri Lanka's lowland wet zone, in: *Ecology and Landscape Management in Sri Lanka,* eds. W. Erdelen, C. Preu, N. Ishwaran, and C. M. Madduma Bandara. Margraf Scientific, D-97985 Weiersheim, 1993), 275–288.

Ashton, P. M. S., C. V. S. Gunatilleke, and I. A. U. N. Gunatilleke. Seedling survival and growth of four *Shorea* species in a Sri Lankan rain forest. *J. of Trop. Ecol.* 11(1995): 263–279.

Ashton, P. M. S., S. Gamage, C. V. S. Gunatilleke, and I. A. U. N. Gunatilleke. Restoration of a Sri Lankan rain forest: using caribbean pine *Pinus caribaea* as a nurse for establishing late successional tree. *J. of Appl. Ecol.* 34(1997a): 915–925.

Ashton, P. M. S., S. J. Samarsingha, I. A. U. N. Gunatilleke, and C. V. S. Gunatilleke. Role of legumes in release of successionally arrested grasslands in the central hills of Sri Lanka. *Restoration Ecol.* 5(1997b): 36–43.

Ashton, P. S. Ecological studies in the mixed dipterocarp forests of Brunei State. *Oxford For. Memoirs* 25, 1964.

Balick, M. J. and R. Mendelsohn. Assessing the economic value of traditional medicines from tropical rain forests. *Conserv. Biol.* 6(1992): 128–130.

Beard, J. S. Climax vegetation in tropical America. *Ecol.* 25(1944): 127–158.

Becker, P., L. W. Lee, E. D. Rothman, and W. D. Hamilton. Seed predation and coexistence of tree species: Hubbell's models revistited. *Oikos* 44(1985): 382–390.

Billings, W. D. The structure and development of old-field pine stands and certain associated physical properties of the soil. *Ecol. Monogr.* 8(1938): 437–499.

Binkley, D. Ecosystem production in Douglas fir plantations: interaction of red alder and site fertility. *For. Ecol. and Manage.* 5(1983): 215–227.

Binkley, D. Importance of size-density relationship in mixed stands of Douglas fir and red alder. *For. Ecol. and Manage.* 9(1984): 80–85.

Binkley, D., K. A. Dunkin, D. DeBell, and M. G. Ryan. Production and nutrient cycling in mixed plantations of *Eucalyptus* and *Albizia* in Hawaii. For. Sci. 38(1992): 393–408.

Black, G. A., A. Dobzhansky, and C. Pavan. Some attempts to estimate species diversity and population density of trees in Amazonian forests. *Bot. Gaz.* 111(1950): 413–425.

Bormann, F. H. Factors determining the role of loblolly pine and sweetgum in early old-field succession in the Piedmont of North Carolina. *Ecol. Monogr.* 23(1953): 339–358.

Bormann, F. H. and G. E. Likens. *Pattern and Process of a Forested Ecosystem.* New York: Springer-Verlag, 1979.

Bourdeau, P. F. Oak seedling ecology determining segregation of species in Piedmont Oak Hickory Forests. *Ecol. Monogr.* 24(1954): 297–320.

Boyce, J. S. Forest plantation protection against disease and insect pests. FAO Forestry Development Paper 3, 1954.

Brandis, D. *Forestry in India.* London: Empire Press, 1897.

Braun, E. L. Forests of the Cumberland Mountains. *Ecol. Monogr.* 12(1942): 413–447.

Brokaw, N. V. L. Gap-phase regeneration in a tropical forest. *Ecology* 66(1985): 682–687.

Budowski, G. Studies on forest succession in Costa Rica and Panama. Ph.D. thesis, Yale University, New Haven, CT, 1961.

Burgess, J. C. Timber production, timber trade, and tropical deforestation. *Ambio* 22(1993): 136–143.

Buslem, D. F. R. P., P. J. Grubb, and I. M. Turner. Responses of nutrient addition among shade-tolerant tree seedlings of lowland tropical rain forest in Singapore. *J. of Ecol.* 83(1995): 113–122.

Champion, H. G. and S. K. Seth. *General Silviculture of India.* Delhi: Government of India Press, 1968.

Clatterbuck, W. K., C. D. Oliver, and E. C. Burkhardt. The silvicultural potential of mixed stands of cherrybark oak and American sycamore: spacing is the key. *South. J. of Appl. For.* 11(1987): 158–161.

Clements, F. E. Plant succession: an analysis of development of vegetation. Carnegie Washington DC: Carnegie Institute, 1916.

Cohen, A., B. M. P. Singhakumara, and P. M. S. Ashton. Releasing rain forest succession: a case study in the *Dicranopteris linearis* fernlands of Sri Lanka. *Restoration Ecol.* 3(1996): 261–270.

Condit, R., S. P. Hubbell, and R. B. Foster. Recruitment near conspecific adults and the maintenance of tree and shrub diversity in a neotropical forest. *Am. Naturalist* 149(1992): 261–286.

Condit, R., S. P. Hubbell, and R. B. Foster. Density dependence in two understory tree species in a neotropical forest. *Ecology* 75(1994): 671–680.

Connell, J. H. and R. D. Slatyer. Mechanisms of succession in natural communities and their role in community stability and organization. *Am. Naturalist* 111(1977): 1119–1144.

Cooper, W. S. The Climax Forest of Isle Royale, Lake Superior, and its development, I. *Bot. Gaz.* 55(1913): 1–44.

Cooper, W. S. Seventeen years of successional change upon Isle Royale, Lake Superior. *Ecology* 9(1928): 1–5.

Davis, T. A. W. and P. W. Richards. The vegetation of Moraballi Creek, British Guyana: an ecological study of limited area of tropical rain forest, Part I. *J. of Ecol.* 21(1933): 350–384.

Davis, T. A. W. and P. W. Richards. The vegetation of Moraballi Creek, British Guyana: an ecological study of limited area of tropical rain forest, Part II. *J. of Ecol.* 22(1934): 106–155.

DeBeer, J. H. and M. J. McDermott. *The Economic Value of Nontimber Forest Products in Southeast Asia.* Amsterdam: Netherlands Committee for I. U. C. N., 1989.

DeBell, S. D., C. D. Whitesell, and T. H. Schubert. Mixed plantations of eucalyptus and leguminous trees enhance biomass production. Pacific southwest forest & range experiment station, USDA Research Paper PSW-175, 1985.

DeBell, D. S., C. D. Whitesell, and T. H. Schubert. Using $N_2$-fixing *Albizia* to increase growth of *Eucalyptus* plantations in Hawaii. *For. Sci.* 35(1989): 64–75.

Denslow, J. C., P. M. Vitousek, and J. C. Schultz. Bioassays of nutrient limitation in a tropical rain forest soil. *Oecologia* 74(1987): 370–376.

Drury, W. H. and I. C. T. Nisbet. Succession. *J. of the Arnold Arboretum.* 54(1973): 331–368.

Egler, F. E. Vegetation science concepts: initial floristic composition a factor in old field vegetation development. *Vegetatio* 4(1954): 412–417.

Evans, J. Plantation forestry in the tropics. Oxford: *Oxford Sci.* 1990.

Evans, J. *Silviculture of Broadleaved Woodland, Bulletin No. 62* London: Forestry Commission, 1984.

Ewel, J. J. Tropical succession: manifold routes to maturity. *Biotropica* 12(1980): 2–7.

Fortmann, L. The tree tenure factor in agroforestry with particular reference to Africa. *Agroforestry Syst.* 2(1985): 229–251.

Gibson, I. A. S. and T. Jones. Monoculture as the origin of major forest pests and diseases, especially in the tropics and southern hemisphere, *Origins of Pest, Parasite, Disease, and Weed Problems,* eds. J. M. Cherrett and G. R. Sagar. (Oxford: Blackwell Scientific, 1977), 139–161.

Gilbert, G. S., S. P. Hubbell, and R. B. Foster. Density and distance-to-adult effects of a canker disease of trees of a moist tropical forest. *Oecologia* 98(1994): 100–108.

Guariguata, M. R., R. Rheingans, and F. Montagnini. Early woody invasion under plantations in Costa Rica: implications for forest restoration. *Restoration Ecol.* 3(1995): 252–260.

Gunatilleke, C. V. S., G. A. D. Perera, P. M. S. Ashton, P. S. Ashton, and I. A. U. N. Gunatilleke. Seedling growth of *Shorea* section *Doona,* (Dipterocarpaceae) in soils from topographically different sites of Sinharaja Rain Forest in Sri Lanka, in: M. D. *Tropical Tree Seedling Ecology,* ed. M. D. Swaine. (Paris: Parthenon Press, UNESCO, 1996), 124–141.

Gunatilleke, C. V. S., I. A. U. N. Gunatilleke, G. A. D. Perera, D. F. R. P. Burslem, P. M. S. Ashton, and P. S. Ashton. Responses to nutrient addition among seedlings of eight closely related species of *Shorea* on Sri Lanka. *J. of Ecol.* 85(1997): 301–311.

Hawley, R. C. and H. J. Lutz. Establishment, development, and management of conifer plantations in the Eli Whitney Forest. New Haven, CT: *Yale Sch. of For. Bull.* 53, 1943.

Henry, J. D. and J. M. A. Swann. Reconstructing forest history from live and dead plant material—an approach to the study of forest succession in southwest New Hampshire. *Ecology* 55(1974): 772–783.

Howard, A. F. Price trends for stumpage and selected agricultural products in Costa Rica. *For. Ecol. and Manage.* 75(1995): 101–110.

Janos, D. P. Mycorrhiza applications in tropical forestry: are temperate zone approaches appropriate? in: *Trees and Mycorrhiza,* ed. F. S. P. Ng. (Kuala Lumpur: Forest Research Institute Malaysia, 1988), 133–188.

Janzen, D. H. Seed predation by animals. *Annu. Rev. of Ecol. and Syst.* 2(1971): 465–492.

Kelty, M. J. Productivity of New England hemlock hardwood stands as affected by species composition and canopy structure. *For. Ecol. and Manage.* 28(1986): 237–257.

Latham, R. E. Co-occurring tree species change rank in seedling performance with resources varied experimentally. *Ecology* 73(1992): 2129–2144.

Lipzin, D. and P. M. S. Ashton. Early successional dynamics of single-aged mixed hardwood stands in a southern New England forest, USA. *For. Ecol. and Manage.* 116(1998): 141–150.

Lorimer, C. G. Age structure and disturbance history of a southern Appalachian virgin forest. *Ecology* 61(1980): 1169–1184.

Lutz, H. J. Trends and silvicultural significance of upland forest successions in southern New England. *Yale Sch. of For. Bull. No. 22,* 1928.

MacArthur, R. H. and J. H. Connell. *The biology of Populations.* New York: John Wiley, and Sons 1966.

Manokaran, N. and K. M. Wong. Proceedings of the rattan seminar 2–4 October, Forest Research Institute, Kuala Lumpur, Malaysia, 1985.

Marks, P. L. The role of pin cherry (*Prunus pensylvanica* L.) in the maintenance of stability in northern hardwood ecosystems. *Ecol. Monogr.* 44(1974): 73–88.

Mielikainen, K. The structure and development of pine and spruce stands in birch mixture, in: *Broadleaves in Boreal Silviculture, Swedish University of Agricultural Sciences, Report 14,* eds. B. Hagglund and G. Petterson. n.p., 1985, 12–27.

Mladenoff, D. J. and J. Pastor. Sustainable forest ecosystems in the northern hardwood and conifer forest region: concepts and management, in: *Defining Sustainable Forestry.* eds. G. H. Aplet, N. Johnson, J. T. Olson, and V. A. Sample. (Covelo, CA: Island Press, 1993), 145–180.

Montagnini, F. and F. Sancho. Impacts of native trees on tropical soils: a study in the Atlantic lowlands of Costa Rica. *Ambio* 19(1990): 386–390.

Montagnini, F., E. Gonzalez, C. Porras, and R. Rheingans. Mixed and pure forest plantations in the humid neotropics: a comparison of early growth, pest damage, and establishment costs. *Common. For. Rev.* 74(1995): 306–314.

Newberry, D. M., I. J. Alexander, D. W. Thomas, and J. S. Gartlan. Ectomycorrhizal rain forest legumes and soil phosporus in Korup National Park, Cameroon. *New Phytologist* 109(1988): 433–450.

O'Hara, K. L. Development patterns of residual oak and oak and yellow poplar regenerations after release in upland hardwood stands. *South. J. of Appl. For.* 10(1986): 244–248.

Oliver, C. D. The development of northern red oak in mixed species stands in central New England. *Yale Sch. of For. and Environ. Stu. Bull.* No. 91, 1978.

Oliver, C. D. and E. P. Stephens. Reconstruction of a mixed-species forest in central New England. *Ecology* 58(1977): 562–572.

Oliver, C. D., E. C. Burkhardt, and W. K. Clatterbuck. Spacing and stratification patterns of cherrybark oak and American sycamore in mixed, even-aged stands in the southeastern U.S. *For. Ecol. and Manage.* 29(1989): 214–222.

Oliver, C. D. and B. C. Larson. *Forest Stand Dynamics.* New York: John Wiley and Sons, 1990.

Oosting, H. J. An ecological analysis of the plant communities of Piedmont, North Carolina. *Am. Midland Naturalist* 28(1942): 1–26.

Palik, B. J. and K. S. Pregitzer. The vertical development of early successional forests in northern Michigan, U.S. *J. of Ecol.* 81(1993): 271–285.

Parrotta, J. A. Influence of overstory composition on understory colonization by native species in plantations on a degraded tropical site. *J. of Vegetation Sci.* 6(1995): 627–636.

Paschke, M. W., O. D. Jeffrey, and M. B. David. Soil nitrogen mineralization in plantations of juglans regia interplanted with actinorhizal *Elaeagnus umbellata* or *Alnus glutinosa. Plant and Soil* 118(1989): 33–42.

Peters, C. M., A. H. Gentry, and R. O. Mendelsohn. Valuation of an Amazonian rain forest. *Nature* 339(1989): 655–656.

Raup. H. M. The view from John Sanderson's farm. *For. Hist.* 10(1966): 2–11.

Samarasinghe, S. J., P. M. S. Ashton, I. A. U. N. Gunatilleke, and C. V. S. Gunatilleke. Thinning guidelines for tree species of different successional status. *J. of Trop. For. Sci.* 8(1995): 44–52.

Safford, L. O. and S. M. Filip. Biomass and nutrient content of a four-year-old fertilized and unfertilized northern hardwood stand. *Can. J. of For. Res.* 4(1974): 549–554.

Schlesinger, R. C. and R. D. Williams. Growth responses of black walnut to interplanted trees. *For. Ecol. and Manage.* 9(1984): 235–243.

Shepherd, K. R. *Plantation Silviculture.* Dordrecht, The Netherlands: Kluwer Academic Publishers, 1986.

Stark, N. The nutrient content of plants and soils from Brazil and Surinam. *Biotropica* 2(1970): 51–60.

Stubblefield, G. W. and C. D. Oliver. Silvicultural implications of the reconstruction of mixed alder-conifer stands, in: *Utilization and Management of Red Alder, USDA Forest Service General Technical Report PNW-70,* eds. W. A. Atkinson, D. Briggs, and D. S. De Bell. p., 1978, 307–320.

Swaine, M. D. and J. B. Hall. The mosaic theory of forest regeneration and the determination of forest composition in Ghana. *J. of Trop. Ecol.* 4(1988): 253–269.

Swaine, M. D. and T. C. Whitmore. On the definition of ecological species groups in tropical rain forests. *Vegetatio* 75(1988): 81–86.

Tham, A. Yield prediction after heavy thinning of birch in mixed stands of Norway spruce (*Picea abies* (L) Karst.) and birch (*Betula pendula* roth and *Betula pubescens* ehrh.). *Swed. Univ. of Agric. Sci. Rep. No. 23,* 1988.

Uhl, C., K. Clark, H. Clark, and P. Murphy. Early plant succession after cutting and burning in the upper Rio Negro region of the Amazon basin. *J. of Ecol.* 69(1981): 631–649.

Uhl, C., K. Clark, N. Dezzeo, and P. Maquirino. Vegetation dynamics in Amazonian treefall gaps. *Ecology* 69(1988): 751–763.

Verrissimo, A., P. Barreto, R. Tarifa, and C. Uhl. Extraction of a high-value natural resource in Amazonia: the case of mahogany. *For. Ecol. and Manage.* 72(1995): 39–60.

Watt, A. S. Pattern and process in the plant community. *J. of Ecol.* 35(1947): 1–22.

Whitmore, T. C. Change with time and the role of cyclones in tropical rain forest of Kolombangara, Solomon Islands. *Oxford For. Memoirs No. 46,* 1974.

Whitmore, T. C. *Tropical Rain Forests of the Far East.* Oxford: Clarendon Press, 1984.

Whitmore, T. C. *An Introduction to Tropical Rain Forests.* Oxford: Oxford University Press, 1990.

Wierman, C. A. and C. D. Oliver. Crown stratification by species in even-aged mixed stands of Douglas fir/western hemlock. *Can. J. of For. Res.* 9(1979): 1–9.

Wormald, T. J. Mixed and pure forest plantations in the tropics and subtropics, FAO Forestry Paper No. 103, 1992.

# Part III

## The Silvicultural Framework for Agroforestry Systems

# Part III Introduction

This part of the book can be considered the conclusion. Chapter 11 concerns site classification and the appropriate scale of silvicultural treatment for agroforestry systems. How are resource gradients accounted for? Reinterpretations are made of traditional forestry applications to account for the biological and social complexity of agroforestry systems. The last chapter, Chapter 12, summarizes the main points made in previous chapters. The chapter is organized within a framework that enables the reader to better think about silvicultural application in an agroforestry setting.

# 11 Site Classification and the Management of Landscape Pattern in Agroforestry Systems

*Mark J. Ducey[1], Mark S. Ashton[2], and Bruce C. Larson[2]*

## TABLE OF CONTENTS



## 11.1 INTRODUCTION

In this chapter, we delve into the spatial aspects of agroforestry systems. At the simplest level, we can ask what species and what types of systems are appropriate in different portions of the landscape, both at regional and local scales. There are several ways of addressing this seemingly elementary question, and, as we will see below, none is fully satisfactory in and of itself. Indeed, many of the site assessment and classification techniques employed in traditional forestry applications fare poorly when confronted with the biophysical, ecological, and social complexity of the agroforestry setting. We propose here some reinterpretations of familiar concepts such as the phytometer, and discuss how they can be used to guide thinking about appropriate agroforestry systems.

[1] Department of Natural Resources, University of New Hampshire, 215 James Hall, Durham, NH 03824; Tel.: (603) 862-4424; Fax: (603) 862-4976.

[2] School of Forestry and Environmental Studies, Yale University, New Haven, CT 06511; Tel.: (203) 432-9835; Fax: (203)-432-8903.

0-8493-2206-5/00/$0.00+$.50

A more complex set of questions arises when we consider that the bits of land on which we practice different agroforestry systems do not exist in isolation. Rather, they participate in a broader system with physical, biological, and social components. By considering this broader system, we can gain insight into what types of systems might coexist together, and how they should be arrayed across the landscape. Here, we will use the silvicultural concept of a stand to define our atomic units of analysis. How can stands of different types be combined to achieve goals at larger scales? When is their spatial arrangement important?

Finally, we will take up the question of planning over long time frames. Here, the dynamic nature of agroforestry systems will be emphasized. We will build upon the successional framework outlined in Chapter 9 to discuss how a group of stands will change through time. We will take a look at traditional forestry approaches to sustained yield in this context, and explore possible techniques for planning under multiple objectives and risk.

## 11.2 A SYSTEMS APPROACH TO THE LANDSCAPE

When we confront agroforestry problems at large scales, we immediately encounter complexity. The dominant question at the scale of the individual field or stand, that of whether or not a specific approach is more or less productive than some other alternative (Cannell et al. 1996), gives way to more subtle questions about how a particular approach meshes with broader physical, biological, and social systems. For example, Gupta (1994) emphasizes that ecological geography in isolation may not be the best determinant of sustainable farming strategies. We are forced to contend with the spatial and temporal aspects of agroforestry systems, and the ways in which multiple values can be balanced against each other. Decisions about selecting and planning appropriate agroforestry systems become much more relevant and difficult than simply optimizing some measure of yield over time. A systems approach allows us to grapple with this complexity and develop insight about the kinds of solutions which may be effective.

As an example of the need for a systems approach, we can take the central question of whether or not agroforestry should be considered as part of an overall farm strategy. Some authors (e.g., Cannell et al. 1996) have questioned whether agroforestry systems can provide greater social or economic benefits than an admixture of pure agricultural and forestry systems, in the absence of any biophysical yield increase. This seems particularly probable if the primary benefits under consideration are the direct biological products of plant growth, and if those benefits are to be exchanged in a market which is reasonably efficient. However, the same conclusion seems less obvious if we consider true economic externalities, or broader social objectives such as sustainability (Price 1995). A physical admixture of perennial and annual crops may lead to changes in the timing or quality of products, leading to increases in per-unit price (Hoekstra 1990), increases in efficiency of on-farm use, or reductions in labor or other binding resources associated with harvest or conversion into market or nonmarket goods and benefits. Furthermore, in many farm or village situations access to other goods and services, through either a barter or cash economy, is restricted. Land availability itself may impose operational constraints which restrict or influence farmer choices. This does not obviate the need to consider the biophysical aspects of yield; it simply forces us to consider agroforestry solutions in a broader systemic context.

A *sine qua non* of the systems approach is an emphasis on hierarchical structures in natural, social, and ecological phenomena, and we will employ an hierarchical strategy in this chapter (Figure 11.1). However, we should also note that hierarchies are often more characteristic of conceptual models than of actual processes. Indeed, hierarchy theory has been criticized for its continuing failure to lead to testable ecological hypotheses, in part because natural systems rarely reflect the strict structure of information and feedback found in designed social and institutional systems, such as corporations or military forces (Steele 1989, Schneider 1994). The decision to impose a hierarchical analysis on social structures themselves may obscure subtle connections and patterns which are loosely organized, but nonetheless of critical importance. Here, we will use an hierarchical approach in the spirit of the comment, often attributed to statistician George Box, that all models are wrong, but some models are useful. The use of a specific model in describing a general situation does not free the professional from the responsibility for identifying discrepancies between the model and reality in a specific application. Furthermore, we believe it is absolutely essential to consider multiple scales and hierarchical levels simultaneously in any successful landscape management problem.

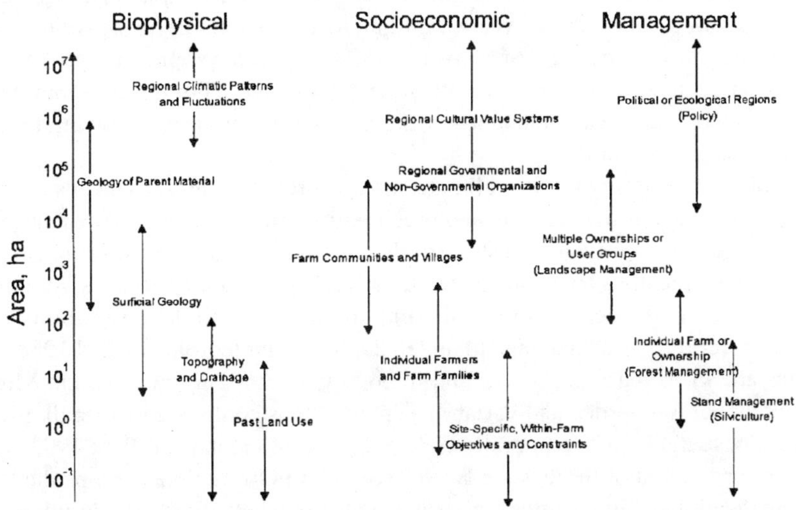


**FIGURE 11.1** Parallel hierarchies of biophysical characteristics, socioeconomic characteristics, and management structures. Biophysical characteristics are important for possibilities and constraints, socioeconomic characteristics determine values, and management structures help translate values into actions. Within the management hierarchy, names for the forestry disciplines conventionally operating at each scale are shown. Note, however, that in all cases the scales of different characteristics and structures are imprecise and usually overlap.

## 11.3 REGIONAL CLASSIFICATION

At a broad spatial scale, we are confronted with the problem of regional classification. Classification at regional scales is often related to the objectives of research planning, as exemplified by the meso- and macro-scale methods of the diagnosis and design, or D&D, approach (Raintree 1983, 1987, 1990, Rocheleau et al. 1989). An awareness of regional context is also important for decisions at a local scale, particularly when novel systems are to be designed or tested. Regional-scale information can offer important insights into opportunities currently being missed, or challenges that must be confronted, by local-scale solutions. While many factors are potentially important at the regional scale, we will focus our attention in three areas: climate, soils, and broad social and economic issues.

The climate of a region is defined as its characteristic patterns of temperature, precipitation, and air speed and humidity. It is important to distinguish between climate, which is a predictable feature of broad regions and long time scales, from individual weather events, which may be characterized by short spatial and temporal scales, and are often relatively unpredictable. The fact that an individual weather event, such as a thunderstorm, may be only a few kilometers across, last only a few minutes, and may or may not pass over a particular valley, does not obviate the fact that the valley lies in a region characterized by such storms, or that the expectation of such storms should be considered in an agroforestry planning process. The climate can be viewed as the statistical ensemble of its weather events.

A wide variety of climatic data sources are available for regional assessment of agroforestry options, and many are familiar to the practicing professional. Maps based on the life-zone concept of Merriam (1898), or simple approaches to the annual water balance such as that of Thornthwaite (1948) and Thornthwaite and Mather (1955), have proven invaluable in assessing regional differences in potential productivity and species suitability.

While it is tempting to view climate strictly through long-term means of important variables such as temperature and precipitation, such a view obscures variability which bears directly on agroforestry problems. Because the trees in an agroforestry system must typically survive and grow for several years, if not decades, interannual variability in climate becomes critically important. While climatic variability at just such scales represents a challenge to understanding in meteorology (Karl 1988), such changes are an increasingly active and important area of current research. Much of the interannual to interdecadal variability in climate is contained in a small group of large-scale, spatially and temporally coherent patterns (Mann and Park 1993, 1994). The most prominent of these is the El Niño Southern Oscillation, perhaps the largest signal in the global climate after the annual cycle (Mitchell 1976). The North Atlantic Oscillation affects climatic patterns from the Arctic to Saharan Africa (Lamb and Peppler 1987, Hurrell 1995), and is associated with patterns of drought and tree growth in northeastern North America (Ducey 1996). An important implication of these patterns is that climatic variability, like mean climatic expectations, can be regionalized.

Qualitatively, the relatively large contribution of interannual and interdecadal variability to climatic variability overall has important implications for the design and implementation of agroforestry systems. Because of such variability, the apparent means and variances of climatic variables calculated from short-term (i.e., years to decades) meteorological records may fail to capture important climatic behavior, especially near the extremes. Likewise, even relatively long-term experiments in which an agroforestry system is monitored through a rotation may fail to include important climatic events which may be probable in succeeding rotations. As Lamb (1995) points out, the history of human civilization is rife with tales of institutions and systems that failed to adapt to climatic variability and change. To be both sustainable and productive, agroforestry systems must be adaptable to changes in the physical climate of a region.

To be complete, regional classification must include a consideration of soils. It should not be surprising that regional soils classifications often partly mirror regional climatic classifications, as climate is an important factor in determining soil development over time. However, as Nair (1993) points out, regional soil classifications based on the five soil-forming agents of climate, vegetation, topography, parent material, and age often describe soils as they might be rather than as they actually are. Thus, while regional differences do exist among soils, and these do provide some guidance about what types of agroforestry systems might be appropriate, we should expect that variability within a region will be great. Nonetheless, soils maps such as the UNESCO *Soil Map of the World* (FAO/UNESCO 1974) often prove useful in understanding the common limitations, opportunities, and objectives in maintaining and enhancing soil productivity for agroforestry systems.

A critical point of interaction between soils and climate, the hydrological cycle, remains problematic. Solomon and West (1985) list the possible effects of hydrological variation on tree and forest processes, including changes in growth, foliage turnover, disease, mortality, and nutrient availability, frequently tied to excesses or deficiencies of soil moisture. However, despite its importance, soil moisture remains difficult to quantify, and few long-term observational records exist. As a result, variation in soil moisture is frequently inferred from models or indices, which often present difficulties in interpretation (Alley 1984a, b). Although techniques are emerging to make such indices regionally and temporally comparable, no clear consensus on methodologies has emerged (Guttman 1998). Thus, we should expect continuing advances in the hydrological sciences to change how we view drought and moisture surplus at regional scales.

Finally, we must consider the regional, social, and economic context for agroforestry systems. Although social and economic factors have often been considered outside the domain of silviculture, they are critically important to successful implementation. While physiographic and climatic factors are causally related to potential biological productivity, social and economic factors are intimately linked to realized productivity and sustainability. In an agroforestry context, local inhabitants must be considered not merely as beneficiaries of some outside design process, but as participants. Thus, identifying who the participants are is a critical first step. When considering regional-scale goals, equity of costs and benefits among those participants is an important consideration (Ascher and Healy 1990, Gowen et al. 1994, Miller et al. 1994). Patterns of property rights, land tenure, and social regulation vary widely, may dictate roles for different participants, and often lead to controversy over resource use and allocation (Fortman and Bruce 1989, Peluso 1992, Padoch 1994). Placing present social, economic, and biophysical patterns in historical context should be a prerequisite, and not an afterthought (Mendelbaun 1991). Finally, patterns of resource production, consumption, and transportation may suggest opportunities or limitations to agroforestry systems. As an extreme example, regional rates of deforestation vary dramatically, with differing consequences for future price and supply of many potential agroforestry products. Thus, while the regional, social, and economic context may provide considerable impetus and insight into possible agroforestry solutions, solutions designed in ignorance of that context are doomed to failure.

## 11.4 SITE CLASSIFICATION

Site classification in traditional silviculture plays a role similar to that of land evaluation in agriculture and agroforestry. The primary goal of site classification, like land evaluation, is to assess the suitability and relative desirability of a parcel of land for a set of potential management or utilization strategies (Duchhart et al. 1990). This assessment can be considered in two parts. The suitability of a parcel of land has much to do with the constraints imposed by the site on what vegetation structures and management styles *might* be imposed. Desirability has to do with what values are at stake, and the sensitivity of those values to intervention on that site. Desirability involves both social and biophysical factors. The interaction of the suitability and desirability should determine what agroforestry strategy, if any, should be employed. Site classification plays an important secondary role in simplifying the complexity of an available land base, to facilitate subsequent systems analysis at scales involving multiple parcels (Jansens 1990). Site classification in forestry, as in agriculture, has often focused primarily on the biophysical with little attention to social and economic aspects (Duchhart et al. 1990). Here, we will begin by examining some of the traditional approaches to site classification in forestry, to see what they can offer in an agroforestry context. As we will see, many of the traditional approaches in forestry are not well-suited to the combination of complexity and intensive management often encountered in agroforestry systems. However, a reexamination and expansion of the concept of the phytometer suggests promising directions.

It is tempting to believe that site classification could be accomplished by the direct measurement of those factors influencing plant growth, such as light, water, and nutrients. While the mechanistic aspects of resource capture by crops have been studied intensively (Monteith et al. 1994), direct measurement remains difficult on an operational basis, especially over the long time scales of variability which trees experience in a single rotation. This is particularly true in a limited-resource environment; many techniques for the direct measurement of water or nutrient availability which are quite practicable on demonstration or research plots are quite out of reach for a farm family attempting to improve their management strategy. The complex spatial pattern imposed by many agroforestry systems creates sampling difficulties, even when direct measurements are available and appropriate. We can view the site as a statistical ensemble of its microsites, much as the climate is the statistical ensemble of its constituent weather events. Precise description of that statistical ensemble requires detailed measurement of its components, as well as the summary of those measurements in such statistical terms as their means, variances, and measures of pattern such as spatial and temporal correlation. In the case of climate, detailed records are often unavailable or insufficient for precise characterization of the statistical ensemble, and yet we are able to make good decisions with coarse, qualitative descriptions, supplemented with such quantitative data as can be brought to bear. Likewise, in the case of site classification and description, we can substitute indirect measurements and qualitative evaluations of resource characteristics, while still maintaining high confidence in our decisions.

A good starting point in site classification is the physiographic characteristics of the site. These characteristics include soil texture, drainage class, and potential rooting depth for crops and trees; slope, aspect, and elevation; and, where significant impacts on soil chemistry or physical properties may have occurred, past human land use. Useful distinctions can be drawn between those aspects of site which are, for all practical purposes, fixed, such as texture and composition of the mineral soil; and those which can be modified by agroforestry practices or other interventions, such as surface organic matter content, structure, and nutrient availability. Depending on the context, water availability may also be modifiable through irrigation or drainage; topographic factors may be modifiable through bedding or terracing.

Physiographic site characteristics bear directly on two of the goals of site classification. Physiographic factors often impose constraints on vegetative structure which cannot be alleviated by any economically feasible management action. As an extreme example, it is usually not possible to grow a complex, multistoried home garden with high levels of leaf area index on a bottomless, droughty, and nutrient-deficient sand. At the opposite extreme, structural simplicity and low species diversity may be difficult to maintain, and may even be wasteful of both effort and resources, on moist but well-oxygenated, nutrient rich soils with moderate texture. Physiographic factors also have much to say about values that may be at stake in the planning process. For example, steeply sloping areas with fine-textured soils present an almost certain erosion risk. Because of their direct importance, physiographic factors should play a leading role in any site classification scheme for agroforestry. However, one must be careful not to design schemes which presume too much detail or measurement expense for the problem at hand, or which are intolerant of missing information.

As an alternative to the direct measurement of site resources, or as a complement to physiographic resources in site assessment, one might consider using a measure of resource capture by the plants themselves to assess site quality. One of the most common approaches to site classification in forestry involves the use of *site index:* the height obtained at a reference age by a crop of trees which have never experienced suppression. In practice, the heights of dominant canopy trees are measured, their ages are obtained, and this height-age pair is compared to a set of species-specific curves. A higher site index value is taken to indicate higher productive capacity on the site. Methods for constructing site index curves are detailed in Clutter et al. (1983). A closely related set of techniques, collectively known as *growth-intercept methods,* use measurements of the annual or periodic growth rate of trees at some convenient reference height. As appealing as the site index concept is, it suffers from some serious drawbacks. The most serious of these is the assumption that trees are already on the site, and that their ages can be determined. For much of the potential agroforestry land base, trees are simply absent. In many tropical areas, accurate assessment of tree age is difficult to impossible. Finally, much of the site index concept's original appeal lies in the insensitivity of height growth in many tree species to the density of lateral competitors, at least within the range of density often encountered in forest stands (Oliver and Larson 1996). Unfortunately, even in an established agroforestry setting, wide spacings, the direct manipulation of belowground resources through tillage or fertilization, and indirect manipulation through management of other crop species, can all impact tree height growth, compromising either site index or growth intercept methods as measures of inherent site quality. Finally, if the "central agroforestry hypothesis" that trees are acquiring resources the surrounding crops cannot acquire (Cannell et al. 1996), proves true for a site, then measurement of tree growth may provide a poor indicator of the potential performance of agroforestry stands at the system level. Certainly, if multiple resources are involved, reduction of variability in the landscape to a single site quality axis is unlikely to prove satisfying.

A more complex approach to site classification involves the use of the plant community as an indicator; the plant community represents a phytometer (Major 1951, Daubenmire 1976, Spurr and Barnes 1980). The tradition of using forest understory species as indicators is quite old in forestry; the best known example is Cajander's work in Finland (1926), although the approach has been widely adapted and modified in other parts of the world. Daubenmire's (1952, 1966) approach, also widely adopted in forestry, uses understory and overstory composition to infer the potential climax community on a site. An ecologically oriented review of many of the statistical techniques currently used is found in Jongman et al. (1995). Although it is possible for rational and well-informed professionals to amuse themselves with bitter disputes on the subject, the decision to develop discrete species groups, or to consider the species as occupying points along one or more axes, ought to depend as much in practice on the data available, the goals of the project, and the tradeoff between conceptual sophistication and ease of use in the project at hand, as it does on one's ecological perspective. Clearly, these factors differ between a purely scientific phytosociological investigation, and a prescriptive, management-oriented situation.

The success of plant community-oriented classification and ordination schemes in traditional silviculture is precluded in the agroforestry setting by the relative absence of forests on the sites in question. Indeed, the investigator will be fortunate to have examples of reasonably intact forest communities on similar sites, much less the elusive phantasm of the climax forest. We may certainly question the relevance of the Clementsian climax concept in a cultured agroforestry context. Where a well-developed indigenous agricultural tradition is present, the species planted on a site in that tradition may well be taken as a sort of evolved community, and the use of this knowledge in assessing site capabilities should be encouraged. However, use of the forest community as a measure of site capability in agroforestry will require a conceptual shift. This shift involves changing our focus from the specific constituents of the plant community and their growth rates, to the dynamic structures and patterns found in forested and semiforested sites.

Given a regional context, and a physiographic characterization of a site (however imperfect), we are in a position to ask what sorts of communities arise on similar sites, and how those communities change through time. Our purpose is to develop a framework for identifying possible agroforestry communities, with particular reference to their structural characteristics, and the ways in which those characteristics change through time. Our goal is not simply to describe what sort of community might arise in the absence of disturbance over long periods. We must pay particular attention to those disturbances which initiate community development, and those which modify developmental pathways. It is these disturbances which will serve as a model for silvicultural interventions in agroforestry systems. The community structures and developmental pathways which arise across a gradient of disturbance frequencies can provide models for alternative agroforestry solutions. To a large degree, our site-level solutions should be designed to capture the dynamics associated with disturbance intensities and frequencies consonant with the values at stake on the site. Examples of these solutions are given in the previous chapter; the process of site classification is a means to make such possibilities operational. In this sense, we are using the forest ecosystem itself as a metaphytometer, not so much to meter the productive capacity of the site, but the structures and dynamics we may choose to bring forth upon it.

## 11.5 ARRANGEMENT OF ELEMENTS WITHIN AN AREA

The allocation of areas to different strategies, and the spatial arrangement of that allocation, is critical to solving the three central challenges posed to agroforestry systems. Can we increase total production? Can we help farmers through more diversified production? Can we develop more sustainable productivity (Lundgren 1982)? At one extreme, we might elect to allocate some areas to pure agriculture, and others to pure forestry. At the other extreme, we might opt everywhere for an intimate mixture of trees with herbaceous crops. A more realistic, yet complex, situation arises when some areas are allocated to pure agriculture, some to pure forestry, and some to uses which span the difference between the two. Regardless of our eventual choice, those final allocation decisions cannot be made solely on the basis of our classification. Classification indicates what a site might support biologically; it does not indicate what it ought to support in the context of the farm or the farm community as a human activity system.

At this point, we require more specific definitions of the terms site, stand, and agroforestry system which we have heretofore used somewhat loosely. Smith (1986) describes a stand as a contiguous group of trees of similar species composition, age class distribution, and condition which can be recognized as a manageable unit; Oliver and Larson (1996) include the requisite of similar climatic and soil conditions in their definition. Here, we extend the definition of a stand to the agroforestry setting. A stand is a spatially contiguous vegetative community, with a consistent but not necessarily uniform admixture of species in vertical and horizontal structure, bound together by a shared developmental pathway. Movement along that pathway is driven by competition for shared resources, and by response to a shared history of disturbance and management. The physical locus of a stand is its site. The intersection of a stand on a site with social activity through cultivation and management over time is precisely what is meant by an agroforestry system. As a stand is a dynamic entity, an agroforestry system must also be dynamic. A static structure describes an agroforestry system only in a brief moment in time. Our management decision is how to allocate dynamic, not static, structures in space.

The simplest analytical situation in which multiple stands occur is that of the individual farm. The situation is simple because there decisions can be made in a somewhat centralized fashion. This does not imply that the objectives for those decisions are simple, or that conflict does not exist. Indeed, we should expect some competition among objectives. We will assume for the time being that those objectives are not in such extreme conflict as to preclude any satisfactory solution.

Decisions about allocation of agroforestry systems to sites are essentially decisions about choosing alternative strategies. An appropriate problem-solving approach involves six steps (Miller et al. 1994):

1. A clear statement of the decision to be made. It may not be possible or desirable to allocate or reallocate systems to all sites in a farm simultaneously. Alternately, it may be advantageous to consider revamping the entire farm as a system, rather than considering solutions piecemeal. The exact nature of the decision depends on the needs, goals, and social context of the farm and the farm family.
2. Enumeration of criteria and constraints. Criteria may be farmwide or site specific. For example, improving cash income is likely to be a farmwide criterion, while protecting the banks of an intermittent stream is a site-specific criterion. Explicit assignment of site-specific criteria to locations will facilitate the planning process. Likewise, constraints may be farmwide or site-specific. Though a farm family may have certain needs for fuelwood, those needs are at least potentially satisfiable from any site or sites on the farm. However, a specific erodible bank may require constant vegetative cover for protection, a need which can only be met *in situ*. Miller et al. (1994), emphasize the subjective, creative nature of this step, and suggest that criteria not be eliminated from consideration during the initial stages. Care should be taken to avoid specifying solutions as criteria. For example, the assertion "it would be good to have a living fence on this site" is premature at this stage of the process. It will be far more revealing to ask why a living fence might be a good solution. Are there forage or pole needs at the farmwide level? Is there a need to keep livestock away from a specific site? Enumeration of criteria should involve all interested parties to capture gender and role differences in values. Once criteria and constraints have been generated, they should be refined and clarified.

3. Identification of alternatives. Again, this is a subjective and creative process (Miller et al. 1994). In the farm planning context, this can be conceived as a three-stage process. The first stage involves identifying agricultural, agroforestry, and pure forestry alternatives for each site. This identification will draw on information from the site classification process, from indigenous farming knowledge, and from any available studies. By exploring the range of anthropogenic disturbance regimes which might be employed on each site, we can envision dynamic successional structures which might be employed to satisfy the constraints and criteria previously outlined. We may also be able to identify probable species compositions for those structures. At this first stage, we should be extremely careful not to eliminate alternatives prematurely. An agroforestry system need not satisfy all criteria simultaneously; it should remain in contention so long as it contributes to at least one criterion, and does not violate any site-specific constraints. The result is a menu of agrosilvicultural options for each site.

   The second stage involves integration of the site-specific options into suites of strategies at the farm level. The goal of this second stage is to creatively and heuristically explore combinations, to create strategies which satisfy farmwide constraints, while advancing site-specific and farmwide criteria. If additional criteria emerge at this stage, they should be refined and incorporated into the planning process. As the goal of this stage is to generate several feasible alternatives, it is important not to rule out or discard strategies prematurely. Prejudging alternatives during this exploratory process can obscure potentially successful or illuminating alternatives (Miller et al. 1994).

   The third stage, refinement of alternatives, may proceed once the several alternatives are available. Here, the goal is to eliminate or modify those alternatives which fail to meet important constraints, or which clearly perform worse than other alternatives on one or more criteria, without compensating gains in other criteria (Figure 11.2). Here, a considerable amount of give-and-take should be expected between parties to the decision. Where unavoidable tradeoffs are found, the participants may wish to return to the creative process of the second stage to identify further, better alternatives, or defer judgement until Step 5.
4. Evaluate the impact of each farm-level strategy on farmwide and site-specific criteria. Here, each of the farm-level strategic alternatives identified in Step 3 is evaluated for its impact on the criteria enumerated in Step 2. This evaluation may be quantitative, or it may reflect qualitative preference. It is at this stage that much of our objective scientific information becomes most relevant. Evaluation of alternative strategies for meeting goals at the ownership level must consider the dynamic nature of agroforestry systems, including the timing and variability in the yield of products, services, and amenities as stand structure changes through time. Evaluation of strategies should also include fundamental operational considerations. For example,

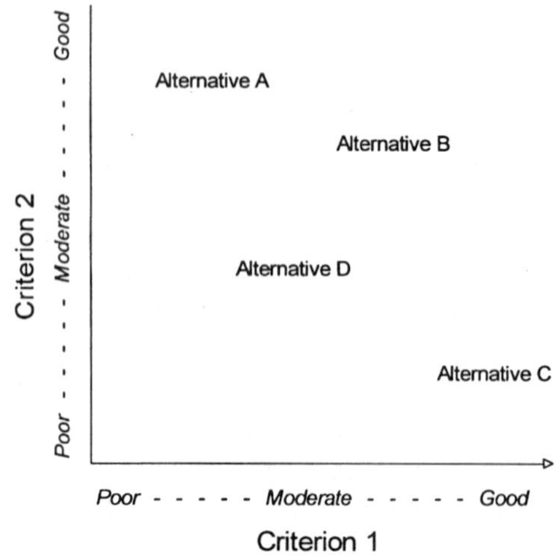


**FIGURE 11.2** Identifying inferior alternatives during the refinement process, for a problem with two criteria. Because switching from Alternative D to Alternative B would improve performance on both criteria, without resulting in decreased performance on any criteria in the problem, we can safely eliminate Alternative D from further consideration. In choosing between Alternatives A, B, and C, increases in performance on one criterion result in decreases in performance on another. This family of alternatives, in the language of decision science, is the "nondominated set." In real problems, there may be many criteria, and a very large number of possible alternatives. Simple matrix methods or more sophisticated mathematical programming techniques can help reduce the number of alternatives which need to be considered.

goods which are difficult to transport, stand structures which require intensive maintenance, or particularly valuable crops requiring close surveillance, should be placed close to dwellings or other areas of intensive use. By contrast, goods that are easy to transport in bulk, which grow in stand structures which require infrequent attention and that do not require defense against animals or people may be more conveniently placed on more distant sites. Spatial arrangement may also contribute to ecological objectives, particularly where maintaining (or avoiding) connectivity of forest or semiforest cover is important. While tempting, we should avoid automatic acceptance of the idea that "nature knows best" where spatial pattern is concerned. Analysis of pattern and scale in natural communities may indeed provide insight into arrangements which might sustain ecological processes at the landscape scale (Forman and Godron 1986). However, not all such processes are desirable. For example, certain types of contiguous habitat can contribute to the spread of mosquito-borne diseases, or may tend to place dangerous organisms in unnecessarily close contact with human habitation. Such considerations depend strongly on the local and regional ecological setting, and are likely to be reflected in indigenous knowledge. In any case, we must remember that an agroforestry system is a cultured ecosystem, and that sustaining the human inhabitants is an important component of sustainability.

5. Select the best alternative. Again, this step involves a subjective expression of values, albeit clarified by the potentially more objective stages in Step 4. Where a clear ranking of criteria is present, this decision may be relatively straightforward. However, where no clear ranking exists, or where participants share different ranking, the choice will undoubtedly require shared discussion and political compromise.
6. Identify steps that must be taken to translate the strategy chosen in Step 5 into action. The purpose of such a process, deliberate and cumbersome as it may seem, is to encourage full use of creative as well as analytical capabilities in the problem solution. As simple as decisions about a single farm may seem, we must remember that most farms involve a plurality of decision-makers, with different social roles, objectives, and perceptions. It is not necessarily the case that these roles, objectives, and perceptions can be collapsed into a single performance scale that is both achievable and meaningful. The complexity of the decision, in terms of the number of possible alternatives and site-specific constraints and objectives, grows dramatically as the spatial domain of the problem increases. In such a case, we may be forced to take a hierarchical approach to the problem itself. Analyses at the multihousehold level, as typefied by the meso-scale application of the D&D procedure, are often expensive both in terms of information and personnel (Rocheleau and van den Hoek 1984). The design of methods of analysis and decision support appropriate to these scales, and the development of facilitating institutions and social structures, present ongoing challenges in the development of agroforestry systems.

## 11.6 DYNAMIC PATTERN AND THE TIMING OF HARVESTS

Decisions about the allocation of agroforestry systems to different areas conclude with the problem of developing an action plan. In situations where all available land is already allocated to one system or another, there remains the need to translate intention into action. Of particular concern are harvests of woody crops which change the yield of other components, change the structure of the system, or change its successional dynamics. Perhaps the most extreme example of such a harvest is the final removal of trees at the end of a rotation. However, intermediate operations, such as thinning, pruning, or lopping may have important effects on system behavior. In silviculture, such decisions are traditionally made, or at least first considered, at the stand level. However, stand-level decisions must interact to produce desirable outcomes at the farm level. The analysis of multiple stand systems in conventional forestry falls generally into the subject of forest management. Here, however, is a case where too rigid an adherence to hierarchical separation of problems may lead to poor decisions. We will take up the problem of harvest scheduling in agroforestry for the single-stand case first, but we must bear in mind that successful solutions need to satisfy both stand-level and farm-level goals and constraints.

The conventional tool for setting harvest schedules in single-stand problems is net present value analysis. In this method, all goods and services produced by the stand are first converted to currency values. These currency values are discounted back to the present by removing the effects of an appropriately chosen compound interest rate. Hoekstra (1990) details issues of methodology for evaluation and net present value analysis in an agroforestry context. Such an analysis is, in principle, straightforward to perform provided high-quality information exists about the potential yields of the system over time. Where such information is lacking, net present value analysis becomes problematic. Its use grows increasingly difficult as novel systems or manipulations are contemplated.

Despite these difficulties, net present value analysis remains important, as it provides an important basis for comparing the economic effects of different strategies. Net present value analysis and optimization is a classic example of a hard systems approach to the harvest scheduling problem. A hard systems, optimization approach may be quite appropriate in an *ex ante* economic analysis, which aims to assess or demonstrate the technical and economic feasibility of a proposed system (Hoekstra 1990). Such an analysis would most likely be undertaken on a pilot project level, as part of an integrated research, demonstration, and extension effort.

Price (1995) points out that relatively few studies have considered economic aspects of agroforestry systems beyond an elementary discounted cash flow analysis, and suggests theoretical difficulties with using net present value as the sole criterion for evaluating strategies. First, the exponential discounting of future yields tends to downplay the impacts of harvesting behavior on sustainability. Since sustainability is often a major objective for agroforestry systems, net present value analysis may require modification to meet sustainability goals or constraints.

As an alternative to maximization of net present value, we might consider maximization of forest rent. Here, the mean yields over time are maximized without discounting. As with net present value analysis, it is customary to convert yields to a single currency or utility function, and both the yields and the conversion are presumed known through the length of the rotation. Rotations under this system are universally longer than under maximization of net present value, and there is a tendency to retain more growing stock on the site. We should not make the facile leap to believe that longer rotations increase sustainability, however. Whether or not longer rotations with increased growing stock contribute to or detract from sustainability depends on the biophysical and social situation. While longer rotations usually result in fewer nutrient removals and reductions in the duration and frequency of soil erosion and compaction, they may also lead to difficult cash flow constraints, exposure to variability in the biophysical or economic environment, or reduced adoptability if income is deferred beyond an acceptable planning horizon for the farmer. The relative importance of these factors must be examined on a case-by-case basis.

A second difficulty with net present value analysis, and one which is shared by maximization of forest rent, is that the reduction of all objectives to a single currency is unlikely to capture the complexity of harvesting decisions. This difficulty is not so much theoretical as it is practical. Even within a household there are often strong differences in economic role, with consequences for individual values and conflicts. For example, responsibility for the intramural subsistence economics of a farm, such as fuelwood production, and the extramural, cash, or barter economics are often split along gender lines, leading to differences in goals, constraints, and perceptions among household members. Some values, such as species diversity or multiplicity of uses for their own sake, are often difficult to quantify and not fully captured in simple measures of either production or conservation values (Padoch 1994). Multicriteria methods may be used to elicit from the decisionmakers the relative importance of such values, to assess consistency of value judgements, and to discriminate between alternatives (Saaty 1980). However, such methods are computationally complex, and elicitation can prove difficult where responsibility is not clear or where conflict is severe. Analyses which are multiobjective (Cohon 1978) or pluralistic (Ducey and Larson in press) may lead both to greater insight and improved adoptability.

In pure forestry, the counterpart to stand analysis, as we move from the scale of the individual stand to a group of stands under unified control and ownership, is forest planning or forest analysis (Baskerville 1992). In the case of agroforestry, we might call this problem farm analysis, and use the term landscape analysis to denote analysis and planning at the scale of multiple farms. The farm analysis problem is analogous to an investment analysis problem, in which a portfolio of assets (stands) must be managed together through time to achieve overall goals and satisfy overall constraints. In many cases, this portfolio can best be conceived as organized at the household level, while recognizing that important assets may be drawn from other households, or from open-access or common property resources (Gupta 1994). Such planning must be complex enough to be realistic. Failure to recognize the ecological, economic, and informational constraints characteristic of farm families or groups of farmers has often led to fall-down when translating experimental results into on-farm practices (Duchhart et al. 1990). At the same time, to be adoptible, the planning process itself must be computationally simple, and must respect those same economic and informational constraints. Many techniques which are quite manageable for a large industrial concern, or even for a graduate student in a university setting, are wholly inappropriate for a limited-resource farmer because of their computational and informational demands.

Forest analysis and planning has traditionally been approached from a hard systems, optimization standpoint in forestry, and some techniques translate at least conceptually to the agroforestry domain. Perhaps the most common mathematical approach to managing multiple stands has been linear programming. In essence, the linear programming approach seeks to maximize the additive contributions of the individual stands to some objective, such as net present value. The optimization is constrained by the area available, as well as labor, capital, and other constraints. Sustainability is most often represented by constraints on the decline of yield over time, or by constraints on levels of growing stock at the end of the planning period. Linear programming problems are solved by an elegant solution scheme known as the simplex method, which allows modeling extremely complex decision problems with many stands and constraints on an ordinary personal computer. Johnson and Scheurman (1977) provide a detailed treatment of linear program specification; Clutter et al. (1983), Dykstra (1984), and Davis and Johnson (1986) provide more general discussion and examples. Mendoza and Sprouse (1989) discuss alternative objective functions motivated by fuzzy set theory.

Despite its mathematical elegance, linear programming has its difficulties. The first of these, in an agroforestry context, is the sheer expense in terms of time and expertise in formulating and solving the linear programming problem. The second is that accurate yield functions must be specified for every management alternative for every stand over the course of the planning horizon. Due to the biophysical complexity of agroforestry systems, this seems problematic, especially at a site-specific level. Some authors have severely criticized the linear programming approach for reducing all goals to a single performance measure (Iverson and Alston 1993). To some degree, this shortcoming can be addressed by adopting a true multiobjective linear programming model, in which decisions about the relative importance of different objectives are deferred until tradeoffs in performance of the alternatives are established (Cohon 1977). Finally, because of its insistence on accurate yield functions and deterministic outcomes, the linear programming problem can be extremely difficult to implement where either novel objectives or novel stand structures are contemplated.

A mathematical alternative to linear programming is dynamic programming. Here, the assumption of deterministic outcomes is relaxed; outcomes of management activities are viewed as probabilistic and conditional. This is extremely advantageous in dealing with the perennial component of agroforestry systems, as their long time frame creates inherent uncertainties about future resource conditions, markets, and nonmarket desires. Dykstra (1984) treats dynamic programming in a forestry context, and Gong (1992) presents an extension to multiobjective problems. While dynamic programming resolves some of the difficulties with linear programming, and perhaps presents a more faithful representation of the decision problem of a farmer at a single instant in time, it does so at some cost. Dynamic programming is not a single technique like linear programming, and as such, it can require considerable expertise in developing problem-specific mathematical formulation and solution schemes. Also, while dynamic programming allows uncertainty in the sense of having probabilistic outcomes, those probabilities are presumed precise and specifiable, which may present difficulties in practice.

A difficulty with both linear programming and dynamic programming methods is the requirement that performance be measurable on a precise, continuous scale (Ducey and Larson in press). However, sustainability in agroforestry systems requires consideration of aspects which can be difficult to quantify, and which change over time. Stability of production depends on the biophysical stability of an agroforestry system, the constancy and predictability of harvests, and the stability and adaptability of management (Harwood 1979, Wiersum 1990). Sustainability is best conceived in a dynamic, rather than a static sense, complicating its assessment at a single decision point. While levels of production and extraction must not degrade site resources or jeopardize the ability to achieve similar levels of production in the future (York 1988), methods of production must allow the capacity to respond to changes in the social and ecological environment (Wiersum 1990). No particular agroforestry system, or arrangement of systems on a farm, will remain fixed both in spatial pattern and management style over multiple rotations. For example, agroforestry using *Acacia mearnsii* in Java has decreased in importance following a shift toward a commercial farm economy based on cash crops (Berenschot et al. 1988), while population pressures and development of markets have led to profound changes in traditional multi-storied village gardens in Bogor (Michon and Mary 1994). Although these difficulties do not render a hard systems, optimization approach to agroforestry problems valueless, they do suggest that they must be complemented by a qualitative, soft systems approach.

## 11.7 CONCLUSIONS

As we have seen, the process of characterizing an area, and translating that characterization into practical, operational prescriptions can be quite complex. In a pure forestry context, Baskerville (1992) points out that while fixed, "cookbook" methods of making stand prescriptions are often attractive because of their simplicity, they often have long histories of failure because the biophysical complexities and unique attributes of a stand are not taken into account. We hypothesize that these difficulties can only be exacerbated in an agroforestry context, where stand development and product yields are intimately and inextricably intertwined with the multiple objectives of the members of a household or other social unit. At the same time, a more thoughtful, analytical approach to stand and landscape management in an agroforestry context is unlikely to lead to the sorts of determinate dynamics, hard yield figures, and crisp outcomes upon which a mathematical programming approach to agrosilviculture would depend. This does not devalue the results of careful studies of yield relationships in different agroforestry systems; indeed, such studies remain valuable as ever. However, to the degree that these circumstances require us to change our analytical perspective, we should not expect those changes to come smoothly. Changing approaches has also proven difficult in traditional timber-oriented silviculture (O'Hara et al. 1994). Addressing multiple objectives and sustainability in a complex biophysical and socioeconomic system ought to be, and is, a challenging and exciting enterprise.

## 11.8 REFERENCES

Alley, W. M. On the treatment of evapotranspiration, soil moisture accounting, and aquifer recharge in monthly water balance models. *Water Resour. Res.* 20(1984a): 1137–1149.

Alley, W. M. The Palmer Drought Severity Index: limitations and assumptions. *J. of Clim. and Appl. Meteorol.* 23(1984b): 1000–1009.

Ascher, W. and R. Healy. *Natural Resource Policy Making in Developing Countries.* Durham, NC: Duke University Press, 1990.

Baskerville, G. L. Forest analysis: linking the stand and forest levels, in: *The Ecology and Silviculture of Mixed-Species Forests,* M. J. Kelty, B. C. Larson, and C. D. Oliver, eds. (Dordrecht, The Netherlands: Kluwer Academic Publishers, 1992), 257–277.

Berenschot, L. M., B. M. Filius, and S. Hardjosoediro. Factors determining the occurrence of the agroforestry system with *Acacia mearnsii* in Central Java. *Agroforestry Syst.* 6(1988): 119–135.

Cajander, A. K. The theory of forest types. *Acta For. Fenn.* 29(1926): 108

Cannell, M. G. R., M. van Noordwijk, and C. K. Ong. The central agroforestry hypothesis: the trees must acquire resources that the crop would not otherwise acquire. *Agroforestry Syst.* 34(1996): 27–31.

Cohon, J. L. *Multiobjective Programming and Planning.* (New York: Academic Press, 1978), 333.

Clutter, J. L., J. C. Fortson, L. V. Pienaar, G. H. Brister, and R. L. Bailey. *Timber Management: A Quantitative Approach.* (New York: John Wiley & Sons, 1983.)

Davis, L. S. and K. N. Johnson. *Forest Management.* (New York: McGraw-Hill, 1986), 790.

Daubenmire, R. F. Forest vegetation of northern Idaho and adjacent Washington, and its bearing on concepts of vegetation classification. *Ecol. Monogr.* 22(1952): 301–330.

Daubenmire, R. F. Vegetation: identification of typal communities. *Science* 151(1966): 291–298.

Daubenmire, R. F. The use of vegetation in assessing the productivity of forest lands. *Bot. Rev.* 42(1976): 115–143.

Ducey, M. J. Scale and structure in forest response to climatic change. Ph.D. dissertation, Yale University, New Haven, CT, 1996.

Ducey, M. J. and B. C. Larson. A fuzzy set approach to the problem of sustainability. *For. Ecol. and Manage.,* in press.

Duchhart, I., R. Van Haeringen, and F. Steiner. Introduction: integrated planning for agroforestry, in: *Planning for Agroforestry,* W. W. Budd, I. Duchhart, L. H. Hardesty, and F. Steiner, eds. (Amsterdam: Elsevier, 1990), 1–17.

Dykstra, D. P. *Mathematical Programming for Natural Resources Management.* New York: McGraw-Hill, 1984.

Food and Agriculture Organization/United Nations Educational, Scientific, and Cultural Organization. *Soil Map of the World.* 9 vols. Paris: UNESCO, 1974.

Forman, R. T. T. and M. Godron. *Landscape Ecol.* New York: John Wiley & Sons, 1986.

Fortman, L., and J. Bruce. *Whose Trees? Proprietary Dimensions of Agroforestry.* Boulder, CO: Westview Press, 1989.

Gong, P. Multiobjective dynamic programming for forest resource management. *For. Ecol. and Manage.* 48(1992): 43–54.

Gowen, M. M., W. R. Bentley, and E. Stijfhoorn. Tropical forest management and wood-based biomass energy as development assets, in: *Forest Resources and Wood-based Biomass Energy as Rural Development Assets,* W. R. Bentley, and M. M. Gowen, eds. (New Delhi: Oxford & IBH Publishing, 1994), 28–63.

Gupta, A. K. Generating sustainable rural development options around wood-based biomass energies: The Case of Local Knowledge Systems, in: *Forest Resources and Wood-based Biomass Energy as Rural Development Assets,* W. R. Bentley and M. M. Gowen, eds. (New Delhi: Oxford & IBH Publishing, 1994), 109–126.

Guttman, N. B. Comparing the Palmer Drought Index and the Standardized Precipitation Index. *J. of Am. Water Res. Assoc.* 34(1998): 113–121.

Harwood, R. R. *Small Farm Development: Understanding and Improving Farming Systems in the Humid Tropics.* Boulder, CO: Westview Press, 1979.

Hoekstra, D. A. Economics of agroforestry, in: *Agroforestry: Classification and Management,* K. G. MacDicken and N. T. Vergara, eds. (New York: John Wiley & Sons, 1990), 310–331.

Hurrell, J. W. Decadal trends in the North Atlantic oscillation: regional temperatures and precipitation. *Science* 269(1995): 676–679.

Iverson, D. C. and R. M. Alston. Ecosystem-based forestry requires a broader economic focus. *J. of Sustainable For.* 1(1993): 97–106.

Jansens, J. W. Landscape development scenarios for planning and implementing agroforestry: a case study in the semiarid lands of Eastern Kenya, in: *Planning for Agroforestry,* W. W. Budd, I. Duchhart, L. H. Hardesty, and F. Steiner, eds. (Amsterdam: Elsevier, 1990), 267–292.

Johnson, K. N. and H. L. Scheurman. Techniques for prescribing optimal timber harvest and investment under different objectives: discussion and synthesis. *For. Sci. Monogr.* 18(1977): 31.

Jongman, R. H. G., C. J. F. Ter Braak, and O. F. R. Van Tongeren. *Data Analysis in Community and Landscape Ecology.* Cambridge: Cambridge University Press, 1995.

Karl, T. R. Multivear fluctuations of temperature and precipitation: the gray area of climate change. *Climate Change* 12(1988): 179–197.

Lamb, H. H. *Climate, History, and the Modern World.* London: Routledge, 1995.

Lamb, P. J. and R. A. Peppler. North Atlantic oscillation: concept and an application. *Bull. Am. Meteorol. Soc.* 68(1987): 1218–1225.

Lundgren, B. Introduction. *Agroforestry Syst.* 1(1992): 3–6.

Major, J. A functional factorial approach to plant ecology. *Ecology* 32(1951): 392–412.

Mann, M. E. and J. Park. Spatial correlations of interdecadal variation in global surface temperatures. *Geophys. Res. Lett.* 20(1993): 1055–1058.

Mann, M. E. and J. Park. Global-scale modes of surface temperature variability on interannual to century timescales. *J. of Geophys. Res.* 99(1994): 25819–25833.

Mendelbaun, S. The past in service to the future, in: *Hist. and Public Policy,* D. B. Mock, ed. (Malabar: Krieger, 1991), 39–53.

Mendoza, G. A. and W. Sprouse. Forest planning and decision making under fuzzy environments: an overview and illustration. *For. Sci.* 35(1989): 481–502.

Merriam, C. H. Life Zones and Crop Zones. *USDA Div. Biol., Survey Bull. No. 10,* 1898.

Michon, G. and F. Mary. Conversion of traditional village gardens and new economic strategies of rural households in the area of Bogor, Indonesia. *Agroforestry Syst.* 25(1994): 31–58.

Miller, S. E., C. W. Shinn, and W. R. Bentley. *Rural Resource Management: Problem Solving for the Long Term.* Ames, IA: Iowa State University Press, 1994.

Mitchell, J. M. Jr. An overview of climatic variability and its casual mechanisms. *Quaternary Res.* 6(1976): 481–493.

Monteith, J. L., R. K. Scott, and M. H. Unsworth. *Resource Capture by Crops.* Bonnington, U.K.: Sutton Nottingham University Press, 1994.

Nair, P. K. R. *An Introduction to Agroforestry.* Dordrecht, The Netherlands: Kluwer, Academic Press, 1993.

O'Hara, K. L., R. S. Seymour, S. D. Tesch, and J. M. Guldin. Silviculture and our changing profession: leadership for shifting paradigms. *J. of For.* 92(1994): 8–13.

Oliver, C. D. and B. C. Larson. *Forest Stand Dynamics.* New York: John Wiley and Sons, 1996.

Padoch, C. The woodlands of Tae: traditional forest management in Kalimantan, in: *Forest Resources and Wood-Based Biomass Energy as Rural Development Assets,* W. R. Bentley and M. M. Gowen, eds. (New Delhi: Oxford & IBH Publishing, 1994), 307–314.

Peluso, N. *Rich Forests, Poor People: Resource Control and Resistance in Java.* Berkeley: University of California Press, 1992.

Price, C. Economic evaluation of financial and nonfinancial costs and benefits in agroforestry development and the value of sustainability. *Agroforestry Syst.* 30(1995): 75–86.

Raintree, J. B., ed. Resources for agroforestry diagnosis and design, ICRAF Working Paper No. 7. Nairobi, Kenya: ICRAF, 1983.

Raintree, J. B. *D&D Users Manual.* Nairobi, Kenya: ICRAF, 1987.

Raintree, J. B. Agroforestry diagnosis and design: overview and update, in: *Planning for Agroforestry,* W. W. Budd, I. Duchhart, L. H. Hardesty, and F. Steiner, eds. (Amsterdam: Elsevier, 1990), 33–57.

Rocheleau, D. E. and A. van den Hoek. The application of ecosystems and landscape analysis in agroforestry diagnosis and design. ICRAF Working Paper No. 11. Nairobi, Kenya: ICRAF, 1984.

Rocheleau, D., F. Weber, and A. Field-Juma. *Agroforestry in Dryland Africa.* Nairobi, Kenya: ICRAF, 1989.

Saaty, T. L. *The Analytical Hierarchy Process: Planning, Priority Setting, Resource Allocation.* New York: McGraw-Hill, 1980.

Schneider, D. C. *Quantitative Ecology: Spatial and Temporal Scaling.* New York: Academic Press, 1994.

Solomon, A. M. and D. C. West. Potential responses of forests to $CO_2$-induced climate change, in: *Characterization of Information Requirements for Studies of $CO_2$ Effects: Water Resources, Agriculture, Fisheries, Forests, and Human Health,* M. R. White, ed. (Washington, D. C.: U.S. Dept. of Energy, DOE/ER-0236, 1985), 145–170.

Spurr, S. H. and B. V. Barnes. *Forest Ecology,* 3rd ed., New York: John Wiley & Sons, 1980.

Steele, J. H. Discussion: scale and coupling in ecological systems, in: *Perspectives in Ecological Theory,* J. M. Roughgarden, R. M. May and S. A. Levin, eds. (Princeton, NJ: Princeton University Press, 1989), 177–180.

Thornthwaite, C. W. An approach toward a rational classification of climate. *Geogr. Rev.* 38(1948): 55–94.

Thornthwaite, C. W. and J. R. Mather. The water balance. *Publ. Climatol., Lab. Climatol., Drexler Institute Technology* 8(1955): 1–104.

Wiersum, K. F. Planning agroforestry for sustainable land use, in: *Planning for Agroforestry,* W. W. Budd, I. Duchhart, L. H. Hardesty, and F. Steiner, eds. (Amsterdam: Elsevier, 1990), 18–32.

York, E. T. Improving sustainability with agricultural research. *Environment* 30(1988): 37–40.

# 12 Defining Silvicultural Systems within Agroforestry

*Mark S. Ashton,[1] Florencia Montagnini,[2] and Matthew J. Kelty[3]*

## TABLE OF CONTENTS



[1] Mark S. Ashton, 360 Prospect St, Yale University, New Haven, CT 06511. Tel.: (203) 432-9835; Fax: (203) 432-3809; email mark.ashton@yale.edu.

[2] Florencia Montagnini, Centro Agonómico Tropical de Investigación y Enseñanza (CATIE), Turrialba, Costa Rica.

[3] Matthew J. Kelty, Department of Forestry and Wildlife Management, University of Massachusetts, Amherst, MA 01003. Tel.: (413) 545-1799; Fax: (413) 545-4358; e-mail: kelty@forwild.umass.edu.

0-8493-2206-5/00/$0.00+$.50

## 12.1 INTRODUCTION

### 12.1.1 Concepts Behind System Definition

Silviculture systems have been designed to create and maintain desired combinations of plants and animals in space and over time. The purpose of classifying these combinations is chiefly to insure communication to other people of the silvicultural activities undertaken on a particular tract of land. The simplest classification of a system can be defined upon the regeneration method and spatial arrangement of age classes and species composition. What is important to understand, however, is that there is a multitude of ways of understanding different combinations of plants and animals in time and space. All too often managers revert to systems that are "cookbook" prescriptions selected from a set of ready-made recipes obtained from texts on agroforestry classification. Following such recipes blindly can lead to disastrous consequences of ignoring the specific social conditions and biological constraints of the site under scrutiny. Other reasons that should caution reliance on cook book prescriptions are: (i) that it makes it easier for the manager (a person who should be the most knowledgeable about the biological and social circumstances of the site) to relinquish responsibility in system design; (ii) that it does not cater to continued evolution that must occur if systems are to successfully accommodate new inputs of knowledge and changes in human values; and (iii) that it often evolves into legislation which if applied across a wide variety of situations, forces particular set combinations of plants and animals that are imperfect, and in many instances harmful to the eventual integrity of the ecosystem process.

Given the qualifications we have described above various considerations make up the construction and evolution of silvicultural systems. First, a system for a particular tract of land should be designed to fit well within the overall goals and objectives of the surrounding lands. Second, it should represent the best combination of silvicultural activities that will satisfy the following major objectives that have been listed as follows (based on Smith et al. 1997):

1. Harmony with goals and characteristics of ownership
2. Provision for regeneration
3. Efficient use of growing space and site productivity
4. Control of damaging agencies
5. Protection of soil and water resources
6. Provision for sustained yield
7. Optimum use of capital and growing stock
8. Concentration and efficient arrangement of operations
9. Maintenance of desired plant and animal populations
10. Execution of policies about landscapes, scenery, and aesthetic considerations.

In this chapter we do not present a rigid design or blueprint for planning silvicultural systems. Our intentions are to provide a loose set of descriptors for various biological factors that could be considered when resource managers intend to describe the system of silvicutural activities that are carried out on their site. The factors that we intend to outline for classification have been considered in previous chapters of this book. This includes consideration of plant uses of nutrients (see Berlyn and Cho, Chapter 2; Montagnini et al. Chapter 7), water (see Berlyn and Cho, Chapter 2; Smith, Chapter 6) and light (see Berlyn and Cho, Chapter 2; Ashton and Ducey, Chapter 10) across sites of differing spatial (physiographic, regional) and temporal (succession) scale. These factors have been considered together as a combination of effects that can promote facilitative (see Bryan, Chapter 3, for a review on nitrogen-fixation) or competitive interactions among plants (see Ashton, Chapter 4, for review on plant diversity theory) and productivity (see Kelty, Chapter 9), and trophic interactions among producers, consumers and predators (see Beckerman for review on predation and grazing, Chapter 5). We use stand dynamics theory (Oliver and Larson 1996; outlined by Kelty in Chapter 9; and used by Ashton and Ducey in Chapter 10) as a basis for arranging these factors within a temporal and spatial framework that can make sense from an agroforesty perspective.

### 12.1.2 The Primary Factor Defining Scale of Management: Site Productivity

Identifying the nature of the physiography within the landscape is the best starting point for defining appropriate management scales for agroforestry (see Ducey et al. this volume, Chapter 11). We start with physiography because of the close association between physiographical attributes of soil texture, soil nutrition, drainage, aspect, and slope that bear directly on site productivity. The ability to determine how site productivity changes across different spatial and temporal scales provides the first set of guidelines for defining the thresholds of management practice from one site to another (Berlyn and Ashton 1996, Toman and Ashton 1996). In reality threshold limits in site productivity, and their associated physiographic attributes, change across landscapes as continua rather than as abrupt transitions (Toman and Ashton 1996). For example a droughty ridge, with a bedrock soil sensitive to erosion, has obviously a lower site productivity with a narrower set of value options and a more sensitive set of ecological constraints, as compared to a fertile and deep soil of a valley — but the transition zone on the midslope between the two sites can be diffuse. Management demands that the two sites at the ends of the gradient be managed differently and that they be clearly identified for the purpose of different treatments. In such circumstance, management promotes abrupt edges and transition zones that cannot be avoided.

Physiographic measures of site productivity should be considered carefully. Some are very static measures of site productivity such as aspect, topographic position, and slope. Other measures are more sensitive to change because of landuse impacts or long-term changes in climate (e.g., surface soil organic matter, nutrient availability in the soil surface). Both kinds of measures need to be taken account of, particularly in soils that have been severely impacted by past land use.

Silvicultural treatments are frequently used to change the allocation of resources from one plant to another, by cutting or killing some plants. It is also possible through human ingenuity to manipulate the overall level of resources at a site. From a management perspective the three fundamental components of all attribute measures of site productivity, namely light, soil water, and soil nutrition, can all be manipulated to some degree by human ingenuity. The limit to human efforts of manipulation is economic — for instance no farmer would irrigate the ridge site mentioned previously, as compared to a mesic area that has deeper less erodeable soils. Upon examination of those three variables, however, there are differing abilities to change each. Changing a resource such as solar energy (light), which, unlike nutrients, and to a lesser extent soil water, cannot be stored easily for later use by plants, makes this component of site productivity the least attractive choice for changing site productivity by humans. Irrigation and drainage are manipulations by humans that can have dramatic impacts on site productivity. These manipulations can be expensive, however, and often require society and governments in general to support construction of large water storage and transport schemes. Fertilization is the most available, and often the cheapest option for changing site productivity just by the application of manure, forest litter, or inorganic fertilizer. The order of ability to change site productivity from least to most is therefore light, soil water, and then soil nutrition. This trend is tied to the nature of how these factors are available to plants. Light is generally available with amounts that can be extremely variable across the day and season, but it has no ability to be stored so plants have to be both opportunistic and responsive. Soil water varies over the course of seasons, a time scale whereby water can be more reasonably manipulated both by people and nature. Soil fertility and weathering processes occur over years making this the least variable but most manipulative factor.

### 12.1.3 A Framework for Identifying Scale of Stand Management

Three baseline considerations can help define the appropriate scale of management that an agroforestry stand should be delineated by. The first is a clear understanding of how site productivity changes across the landscape (Figure 12.1a; Figure 12.2a). The ability to construct a site productivity map provides the baseline information for determining the biophysical constraints to options of what can be done and where.

The second baseline consideration is defined by property rights and human access to the various products and services garnered from the land. Management is surely going to be different even if the sites are similar in productivity if management objectives are different. This usually comes down to who owns, or has rights to, the particular site in question. In certain cases sites can be managed for multiple values that are compatible with each other, allowing for coexistence of values within the same site. More often than not values are incompatible. Under such circumstances the only alternative, if both values are to coexist, is to divide the site in such a way that meets agreement with both users (see Figure 12.1b; Figure 12.2b).

The third and last consideration for identifying a management unit is the nature of past disturbance and the spatial composition of species and distribution of age-classes across the site. Predetermined mixes of species and age-classes force man-

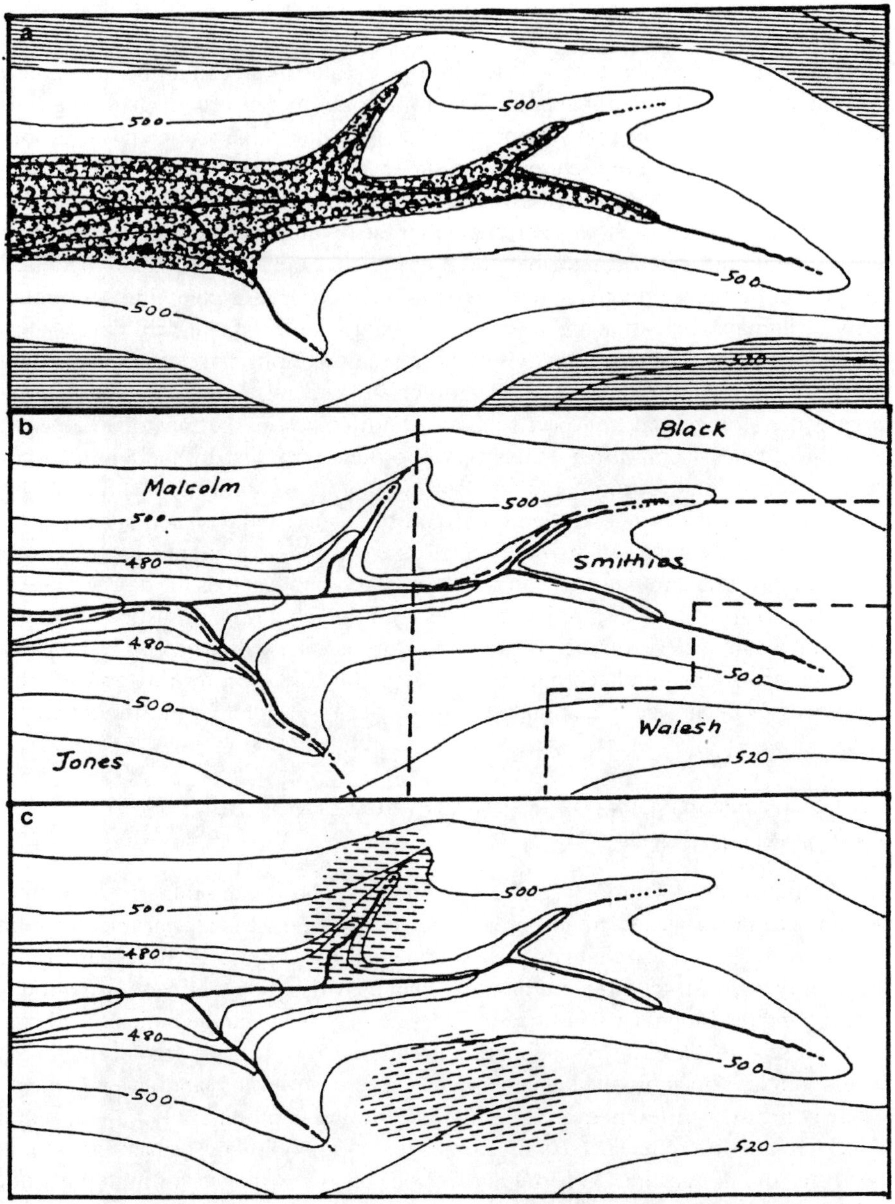


**FIGURE 12.1** The three baseline considerations that help define an appropriate scale of stand management: A) Changes in site productivity as depicted by differences in shading of an imaginary landscape. Changes in site productivity can be related to in soil type and underlying geology with reflective differences in floristics; B) Changes in resource ownership as depicted by property boundaries; C) Changes in age class distribution, species composition, and stocking of a forested landscape depicted by shading that denotes young regeneration from recent windthrow disturbances.

**FIGURE 12.2** A) Changes in site productivity across the valley-ridge landscape of a Sri Lankan rain forest. B) Changes in property ownership across a Bavarian forest hillside identified by plantation boundaries. C) Changes in age class distribution of spruce-aspen forest from disturbances caused by clear cutting.

agers to accommodate and fit existing vegetation composition and structure into current planning for agroforestry management. A spatial area that defines vegetation with similar composition, stocking, structure (as defined by such measures as stem density, basal area, vertical distribution of foliage), and age distribution would be classified as a stand from purely a biological perspective (Figure 12.1c; Figure 12.2c). The biological definition of a stand is therefore usually the finest scale for defining a management unit.

Products such as wood, fruit crops, fodder, and fuel wood (direct values) are often most appropriately managed as defined by the biological stand definition because of their direct dependence on the nature of composition, structure, and age-class distribution of the existing vegetation. However, though the biological definition of a stand may be the most refined scale for delineation of a spatially explicit management unit, if the driving value is not relevant to this scale, such as service values like water quality and yield, then stands need to be aggregated up to larger spatial scales that can more realistically account for costs of management.

## 12.2 PRINCIPAL DESCRIPTORS FOR SILVICULTURAL SYSTEMS

### 12.2.1 Succession and Age-Class Distribution

Probably the most important descriptors for characterizing the nature of silviculture with agroforestry relates to how the vegetation originates and changes in both composition and growth habit over time. For this section we will start by categorizing the main types of **regeneration origin,** then we describe the nature of **age-class** distributions that can occur for **single species** stands, we elaborate these themes with **species mixtures** belonging to the same **growth habit,** and lastly we consider different growth habits.

#### 12.2.1.1 Regeneration Origin

Natural vegetation arises after the occurrence of one, or a combination of many kinds of **disturbances.** It is necessary to know which appears after what event as no species is adapted to all kinds of disturbance.

Geological events that promote erosion constitute the most severe kind of natural disturbance. In ecological terms, the only true primary succession starts with catastropic disturbances such as landslides or the formation of new land by water or volcanic eruption. Erosion or earth-moving caused by people can also expose mineral soil that are free of organic matter and deficient in nutrients. These disturbances would be considered **lethal,** killing all vegetation previously found on-site. It is vacancies such as these that are colonized by true **pioneer** vegetation. These plants are more likely to be herbaceous than woody, but there are some tree species, such as the true poplars and alders, that can be carried by the wind or water and can colonize without some initial herbaceous stage. Ordinarily, in most forest regions, trees do not start on such exposed surfaces until some other vegetation has begun to build up the organic matter or ameliorate the microlimate enough for establishment of shade for tree seedlings.

The next most severe natural disturbance is the very hot fire. Most forest fires, even crown fires in living stands, are fueled by the litter of the forest floor. In such circumstances often not all this litter is consumed and rootstocks of many living plants may survive. However, if there is a large quantity of dead, organic debris on the ground as a result of blowdowns or mortality from pathogen or insect outbreaks, fires can burn so hot that they can be as lethal as a landslide. The main difference is that much of the organic matter incorporated in the mineral top soil is actually untouched by even a very hot fire, whereas the entire upper soil horizon is lost in landslides.

Most sprouting species are killed from fires of this kind, providing open space for establishment of completely new vegetation from seed. Subsequent succession would be considered as secondary because the remaining organic matter enables some species to start earlier than they would otherwise.

Artificial disturbances often can have significant similarities to severe fire. Bare soil with incorporated organic matter but without woody perennials capable of sprouting can be created by agricultural cultivation, and by the most intensive kinds of silvicultural site preparation. The old-field pine stands that colonized abandoned agricultural lands represent a stage in this kind of succession. This legacy was so productive that these old field stands have been perpetuated particularly on the coastal plains of the southeastern United States where intensive site preparation can be achieved. The silviculture that simulates these cases therefore depends on either natural seeding or planting of species adapted to colonize vacant areas. The tree species suited to such environments grow rapidly in diameter and height, regardless of how productive they are at the stand level in terms of stemwood volume or biomass. Consequently these species are very important in silviculture because of the greater economic advantages of reaching merchantable size quickly.

The other general category of regenerative disturbance is where forests, and the trees in them, die "from the top downward"; that is, the larger trees are killed but most of the plants of the lower strata are spared. The most common example of this category is damage from hurricanes and other windstorms; some insects and pathogens also kill overstory trees without affecting lower strata. The species that are adapted to regenerate in such circumstances are those that are tolerant of shade. Their seedlings are usually not adapted to exposed microclimates and their juvenile growth is slow.

These species can survive as understory plants for many years beneath old stands. After years of suppression they can retain the capacity to initiate rapid height growth whenever some disturbance event releases them (Ashton 1992). Most of these species can maintain height growth for long periods once released. This characteristic, along with an ability to hold deep canopies of foliage, leads to highly productive stands. However, such species tend to be limited to sites and regions that are continuously moist enough to reduce exposure to fire or drought stress. These species have been termed "advance-growth-dependent" because they are best regenerated naturally under some sort of protective cover before they are released by the creation of a canopy opening. The most complex forests of wet temperate and tropical climates are usually dependent upon this form of regeneration (see Ashton and Ducey, Chapter 10, Figure 10.4).

We have assembled species into autecological groups based primarily on the origin of their natural regeneration. In reality there are numerous ways in which plants can be dispersed and established. Attempting to set up detailed classifications that differentiate species ecological adaptations are fruitless. In the end each species of a given locality seems to have its own separate regeneration niche. Similar sets of adaptations for ecologically similar groups of species, which can be called **guilds,** can usually be found among representatives from geographically separated regional flora. However, one species seldom exhibits exactly the same adaptations throughout its natural range. The only real solution to the problem is to find out as much as possible about the regeneration requirements of each species in each locality. For purposes of simplification we have used the Smith et al. (1997) classification that divides propagule regeneration into six broad guilds (Table 12.1).

#### 12.2.1.2 Age-Class Distribution

There are three general types of age-class structure within stands: single-cohort, double-cohort, and multiple-cohort. We will describe the simplest age structures first and then progress to the more complex. The simplest stands are those that are pure single-species stands. All trees are the same age or have been released after a disturbance at the same time in a **single-cohort** stand. Intimately mixed groups of trees consisting of at least three age-classes within a site are considered to be a **multiple-cohort** stand. A **double-cohort** represents an intermediate category in which two age-classes exist, and in which one of the cohorts may be considered temporary or a permanent part of the stand. All gradations of age-class can be found in nature and can be created through silvicultural treatments (Figure 12.3).

Distinctions can also be made between **balanced** and **unbalanced** or **irregular** age-classes. When all age-classes are uniformly spaced throughout the stand and occupy an equal amount of basal area the stand is considered balanced. Multi-cohort stands that do not keep to this constraint are considered to have irregular age-class distributions.

#### 12.2.1.3 Stages of Stand Development

We base our descriptors for the successional development of agroforestry mixtures on Oliver and Larson's (1996) paradigm for stand development. This has served as a basic guideline for understanding silvicultural manipulations for forests in North America and we believe this can be appropriately applied to silvicultural aspects of agroforestry systems (see Kelty this volume, Chapter 9, for review of stand dynamics model relevant to agroforestry). In their model, regeneration cohorts that successfully establish, and that are subsequently released, represent the first stage of stand development. This stage is called **stand initiation.** This period can be an exceedingly dynamic window of time that results in plant community reorganization that is partial or complete, depending on the nature and scale of the disturbance.

Once plants fill above- and belowground space, competition between plants excludes further colonization of new individuals. This is a phase of stand development whereby the crowns of individuals vigorously compete amongst each other such that the weaker ones die or become more suppressed. The crowns of the young trees recede because of shading by the upper foliage, beneath which light, soil

**TABLE 12.1**
**Autecology of tree propagules of forests and woodlands (reproduced with permission from Smith et al. 1997).**

| Regeneration Guild | Mode of Dispersal | Life History Traits | Disturbance Type | Species Examples |
|---|---|---|---|---|
| Seed dispersed into the stand from outside. | Wind, small birds, mammals. | Seed that is abundant and small; light-demanding pioneers that can be long-lived but usually short-lived. | Severe — landslides, sediments from floods, exposed mineral soils | Birch, poplar. |
| Seed stored on the tree or on the ground surface. | Gravity, wind, and explosive dehiscence. | Borne within protective wood cones or pods; small-seeded; light-demanding pioneers that can be long-lived. | Severe — promotes release of seed — fire. | Jack pine, some eucalypts and acacias. |
| Buried seed stored in the mineral soil. | None — accumulating seed bank from small mammals and birds. | Medium-sized seed; light-demanding pioneers. | Severe — promotes germination of seed from changes in soil environment. | Pin cherry, raspberries. |
| Vegetative propagation that arises from the stems, roots and branches of residual parent plants. | None. | Fast-growing tree species that are stress adapted. | Releasing — promotes death of above-ground parts of residual trees but allows survival of roots. | Oaks, maples, aspen. |
| Advance growth of seedlings, seedling sprouts and saplings that originate as cohorts during mast years. | Hoarding mammals, hoarding birds, and gravity. | Seed successfully germinates sporadically as a distinct cohort. Seedlings accumulate beneath closed forest canopy as a seedling bank. Large-seeded. Intermediate to very shade-tolerant; clumped distribution. | Releasing — promotes release of groundstory regeneration. Windthrows of all kinds, canopy herbivory, and pathogens. | Dipterocarps, Oaks, hickories. |
| Advance growth of seedlings, seedling sprouts and saplings that originate almost continuously from year to year in small amounts. | Bats, primates, and large birds. Seeds require ingestion by animal. | Seed successfully germinates seasonally but regularly in small amounts beneath a closed forest canopy. Large-seeded, shade-tolerant, and as a seedling susceptible to herbivory. Scattered distribution. | Releasing — as above. | Mangos, durians, nutmegs. |

**FIGURE 12.3** Age-class distributions. A) A single-cohort single-species stand in the stem exclusion stage of stand development (old-field white pine in Southern New England). B) A wildfire origin interior cedar-hemlock stand in British Columbia. This stand would be considered multicohort multispecies in the stem exclusion phase of stand development. Note the large residual trees that survived the fire in the background. C) A single-cohort multi-species stand after logging in a Sri Lankan rain forest. The stand would be considered in the stem exclusion phase. D) A multi-cohort multi-species stand in a Belizean second growth rain forest. Most of the canopy originated after agricultural abandonment several hundred years ago. The stand would be considered in the understory reiniation with a dominant palm understory. Several canopy disturbances from wind have promoted realease of new cohorts including fast growing pioneers like *Cecropia*.

nutrition, and soil moisture are so limiting that the groundstory is barren. All these events signify a period of vigorous competition that has been termed **stem exclusion.**

Over the course of many years competition and the processes of self-thinning between individuals allows taller trees to be better competitors for capturing growing space. However, as trees become larger, their body size increases and the ratio between photosynthetic to nonphotosynthetic tissues declines. Tree leaves are forced to have higher compensation points, primarily because of increases in respiration maintenance, necessary for the functioning of the greater amounts of living nonphotosynthetic tissue (roots, branches, stem). This balance forces trees of large size to lose proportionately larger and larger amounts of their foliage in the lower and inner parts of their crowns, where low light environments cannot allow leaves to create a net photosynthetic gain. Excess light filters lower in the canopy enabling smaller plants with more efficient ratios of photosynthetic to nonphotosynthetic tissues to establish and persist at the groundstory. This stage of stand development has been termed **understory reinitiation.** Height growth is probably also limited by reduced photosynthesis as a result of moisture stress in foliage at greater heights above the roots. Also, just the increased height of the canopy, even without changes in canopy density, gives greater proportion of penumbral shade. Lastly, crown abrasion from wind and ice can be a factor.

Over still longer periods of time the actual canopy trees succumb to disease and that promotes the break up of the canopy. These processes facilitate the release of the groundstory. Most likely if disturbances are small but periodic, the growth of the groundstory into the canopy will probably occur through a process of periodic release and suppression events as the canopy opens and then closes in. When the trees that originated during the understory reinitiation period eventually move into the canopy, the stand can be considered in the **old-growth** stage of stand development. It should be recognized that the old-growth stage is never acheived in many forest regions with frequent catastrophic disturbances.

#### 12.2.1.4 Age-Class Distribution: Single-Cohort Single Species Stands

Single species that establish or are planted as a monocrop plantation will move through stages of stand development that have been described in the previous section (Figure 12.4a). Understanding the developmental pathway of single-cohort single-species stands is therefore relatively simple. In silviculture individuals competing amongst each other in these circumstances can easily be classified by their crown position relative to their competitors. Five crown classes are generally recognized as depicted by Figure 12.5.

**Emergent:** trees with crowns that extend completely above the general level of the canopy and that receive light from above and from all sides.

**Dominant:** trees with crowns extending well above the general level of the canopy so that crowns receive light at least partly from the sides and from above.

**Codominant:** trees with crowns that are part of the general level of the canopy and with crowns that receive light from above.

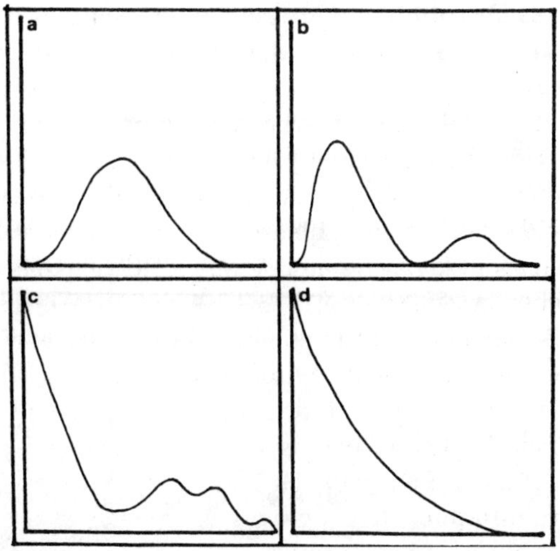


**FIGURE 12.4.** Age class distributions as denoted by differences in tree size (diameter) for different single-species stands. A) Single-cohort single-aged stand; B) double-cohort stand with two age classes; C) An unbalanced age class distribution where several cohorts that have been released from multiple disturbances; and D) A balanced all-age class distribution where progressive (almost yearly) cohorts of regeneration establish, grow and represent an equal ground area of all size classes.

**Intermediate:** trees that reach the general level of the canopy but barely so, receiving a little light from above and none from the sides.
**Overtopped:** trees with crowns that are clearly beneath the general level of the canopy and therefore receive only partial shade. For trees that tolerate shade but grow much more slowly this crown position has also been termed suppressed.

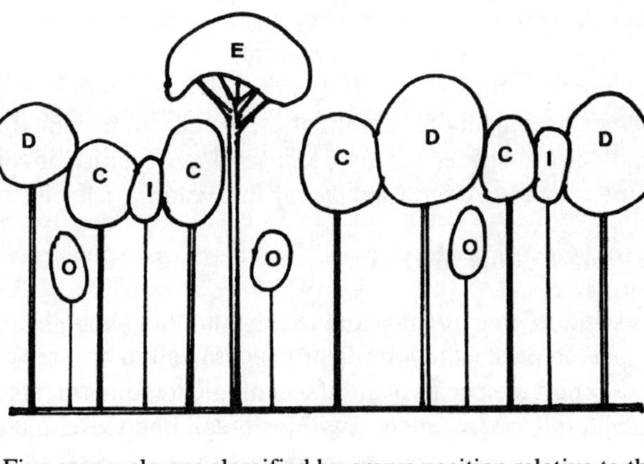


**FIGURE 12.5** Five crown classes classified by crown position relative to their competitors. E — emergent; D — dominant; C — codominant; I — intermediate; O — overtopped.

#### 12.2.1.5 Age-Class Distribution: Double-Cohort and Multiple-Cohort Single-Species Stands

Double-cohort stands can frequently occur where a new age-class of trees has started beneath an older one that has partially or nearly entirely been eliminated by cutting or natural disturbance (Figure 12.4b). Multiple-cohort single-species stands can really be regarded as an elaboration of the same process, only several to many age classes of trees are present but enough growing space is open to allow the release of another new age-class. The number of age-classes represented within the stand and the degree to which the age-classes are balanced is reflective of the periodicity and nature of the disturbance regime (Figures 12.4c, d).

#### 12.2.1.6 Single-Cohort Mixtures

Ordinarily, combinations of species, often belonging to different regeneration guilds, will vertically stratify with one species tending to suppress its associates. Small differences in height at the beginning can become accentuated and noticeable as the stand matures. Two stratification processes (i) dynamic; (ii) static) have been defined by Ashton and Ducey this volume (Chapter 10), both of which can exist in differing combinations depending on the amount of light and availability of the soil resources.

**Dynamic stratification** can be described as that part of vertical complexity of forest stand structure that is most closely associated with succession. Fast-growing pioneer species are overtopped by slower-growing but eventually taller and longer-lived tree species. This process has been described as having several phases of stand development which closely parallel changes in the resource use efficiency and therefore the competitive ability of trees to grow and develop (see Ashton and Ducey this volume, Chapter 10, Figure 10.2).

**Static stratification** can be described as that part of vertical complexity of the forest stand that promotes the permanent existence of a subcanopy and groundstory comprised of plants that never succeed to the canopy. This pattern is most characteristic of moist forest regions with long-term disturbance intervals that allow for the progressive accentuation of vertical habitat strata.

Single-aged mixtures therefore generally comprise both kinds of stratification but when soil resources become more limiting these processes become less evident both in horizontal and vertical space, usually because of availability of soil moisture (see Smith this volume, chap. 6.; Ashton and Ducey this volume, Chapter 10, Figure 10.1).

#### 12.2.1.7 Multiple-Cohort Mixtures

The intimate mixing of trees of different ages complicates both the structure and dynamics of stands. In most situations disturbances need to be large enough to establish and release a cohort of species of similar ecological requirements that, given continued thinning and release treatments, will eventually be represented by a single or couple of individuals when reaching the canopy of the stand (Figure 12.6). In these circumstances each cohort undergoes the same stand development processes as dis-

**FIGURE 12.6** A balanced age-class stand depicting a multiple-cohort mixture of Norway spruce with some European beech in Germany.

cussed previously (initiation, stem exclusion, etc.), but the scale is defined by the size of the periodic disturbance regime within the stand.

##### *12.2.1.7.1 Spacing*

Spacing between individuals and species is obviously driven by limitations in the abiotic components of the resource. Quantitative spacing and planting guidelines for light and soil moisture have been constructed for numerous species and species mixtures across different sites. Much of this has been published in the agroforestry and forestry literature (Evans 1984, Nyland 1996, Savill et al. 1997). For reviews in this volume Berlyn and Cho (Chapter 2) and Smith (Chapter 6) consider soil moisture as an important factor effecting plant spacing, Montagnini et al. (Chapter 7) consider soil nutrition, and Ashton and Ducey (Chapter 10) consider self-thinning and plant successional status.

The effects of **density dependence** of predators, pests, and pathogens as an influencing factor on spacing for agroforestry is less clear, but evidence reviewed by Ashton (Chapter 4.) strongly suggests that this may be another important contributing factor affecting the construction of intimate plant mixtures. Recent studies by

Wills et al. (1997) have shown that populations of the more common species of a Panamanian rain forest show density dependence effects of predators and pathogens. Examples in agroforestry systems might be that of mahogany and the shoot borer. In any event, the more specific the pathogen is to the host crop plant, likely the greater necessity of cultivating the host species in mixture with other less susceptible plants. Little work has examined these effects however, but Worwald (1992) in a review concluded that the level of plant mixtures that would be practical to manage would probably not be enough to achieve a reduction in host plant susceptibility to damage. Suffice to say there have been no well-designed experiments that have tested for these effects.

#### *12.2.1.7.2 Facilitation Versus Competition*

In most studies competition among plants for limiting resources is the driving mechanism for spacing, but an increasing number of studies are demonstrating the importance of facilitative interactions among plants, and between plants and animals (see review by Kelty, Chapter 9). However, studies investigating nitrogen-fixation have shown certain amounts of facilitation among plant mixtures that can alter spacing and increase overall productivity (see review by Bryan this volume, Chapter 3; and application to agroforestry by Bryan, Chapter 8). This practice can be sequential, such as the use of clover to enrich the soil before planting another crop, but it is most often practiced simultaneous to crop cultivation, such as the close integration of nitrogen-enriching trees with the crop plants. Also, though unproven, ectomycorrhizal plants that grow well on acid and low fertility soils, may indirectly benefit other plants through the release of nutrients (in particular phosphorus), brought up from deeper horizons, or made more available from the surface soil (see Ashton this volume, Chapter 4).

#### *12.2.1.7.3 Growth Habit*

Clearly, it is easy to understand the compatibility of tree mixtures particularly over successional time, as outlined in this volume by Ashton and Ducey (Chapter 10), or compatability associated with facilitative interactions within some different tree mixtures relating to nutrient use, directly or indirectly (see Montagnini et al. this volume, Chapter 7; and Bryan, this volume, Chapter 8). These authors provide many examples of compatible woody mixtures that require low levels of human input because shade-tolerant understory species can exist beneath more light-demanding ones both in temporal (successional) space, and at one time in different vertical strata.

Plants of distinctly different **growth habits** (i.e., **annual** agricultural crops and wood **perennials**) that both require high light regimes can give a completely different story. Such intimate mixtures cannot be promoted indefinitely without repeated control of in-growth by the woody species into growing space allocated for the annuals. Another such mixture is the compatibility of cultivating pasture grasses with tree crops — in dry climates such combinations need considerably less human tending to control woody in-growth as compared to wet climates. This is borne out by the success of these mixtures in more seasonally dry climates (see Beckerman, Chapter 5, for underlying theoretical basis; also see Smith, Chapter 6 for application).

In all circumstances when mixtures of plants that have different growth habits have to be maintained through continual human intervention this can only be regarded as a successionally **arrested system.** Examples of such treatments (**alley cropping, live fencing,** orchards, and **pollarding** within agricultural fields and pastures) are numerous in modern agroforestry. They have also been widely described in the literature (Nair 1989, MacDicken and Vergara 1990, Nair 1993), as well as in this volume by Bryan (Chapter 8) and Smith (Chapter 6). Such systems can afford to have high human inputs in controlling growing space allocation chiefly because in most cases the main component of the mixture (agricultural crop) benefits by the partial shade, mulch, and/or nutrition provided by the woody component.

## 12.3 CONCLUSION

There are a multitude of ways of comprehending the different combinations of plants and animals in agroforestry. In summary, we have provided a series of sequential steps upon which to first conceptualize, and then to methodically construct silvicultural treatments within agroforestry. We first evaluate scale of stand mangement appropriate for silvicultural treatment that considers site productivity, disturbance, and social boundaries. We then consider the origin of regeneration as a determinator of species suitability, composition, and dynamic of the agroforest stand. This provides the basis for consideration of age-class distribution and species composition (single, mixed) within the context of the stand development model of Oliver and Larson (1996). Using the stand development model we consider density dependence and possible effects on spacing, the processes that promote competition versus facilitation, and the differences in growth habit that can occur in mixtures. We hope that this will also prove useful to the reader when evaluating the pattern and dynamic of tree mixtures within agroforestry.

## 12.4 REFERENCES


Ashton, P. M. S. Establishment and early growth of advance regeneration of canopy trees in moist mixed-species broadleaf forest, in: *The Ecology and Silviculture of Mixed-Species Forests,* M. D. Kelty, B. C. Larson, and C. D. Oliver, eds. Kluwer (Dordrecht, The Netherlands: Kluwer Academic Publishers, 1992), 101–125.

Berlyn, G. P. and P. M. S. Ashton. Forests and the ecosystem paradigm. *Journal of Sustainable For.* 7(1998): 141–157.

Oliver, C. D. and B. C. Larson. *Forest Stand Dynamics.* New York: John Wiley and Sons, 1996.

MacDicken, K. G., and N. T. Vergara, eds. *Agroforestry: Classification and Management.* (New York: John Wiley and Sons 1990), 382.

Nair, P. K. R., ed. *Agroforestry Systems in the Tropics.* (Dordrecht, The Netherlands: Kluwer Academic Publishers, 1989), 672.

Nair, P. K. R. *An Introduction to Agroforestry.* (Dordrecht, The Netherlands: Kluwer Academic Publishers, 1993), 499.

Toman, M. A. and P. M. S. Ashton. Sustainable forest ecosystems and management: a review. *For. Sci.* 42(1996): 366–377.

Smith, D. M., B. C. Larson, M. J. Kelt, and M. S. Ashton. *The Practice of Silviculture: Applied Forest Ecology.* (New York: John Wiley and Sons, 1997), 537.

Wills, C., R. B. Condit, and S. P. Hubbell. Strong density and diversity-related effects help to maintain tree species diversity in a neotropical forest. *Proceedings of the National Academy of Sciences* 94(1997): 1252–1257.

Worwald, T. J. Mixed and Pure Forest Plantations in the Tropics and Subtropics. *FAO Forestry Paper No. 103,* Rome: FAO, 1992), 152.

# Index

## D



## E

## F



## G

## H



## I



## J



## K



## L

# O



# P

## Q



## R

## S

## T



## U

## V



## W



## Y



## Z